A TEXTBOOK OF AUTOMOBILE ENGINEERING

A TEXTBOOK OF AUTOMOBILE ENGINEERING

KHALIL U SIDDIQUI
Department of Mechanical Engineering
Integral University, Lucknow, UP
(*Life Member of ISTE,* INDIA)

Committed to Educate the Nation

Publishing Globally
IN INDIA

NEW AGE INTERNATIONAL (P) LIMITED, PUBLISHERS
LONDON • NEW DELHI
Bangalore • Chennai • Cochin • Guwahati • Hyderabad • Kolkata • Mumbai
Visit us at **www.newagepublishers.com**

Published by New Age International (P) Ltd., Publishers
First Edition: 2011
Reprint: 2022

GLOBAL OFFICES

- **New Delhi** **NEW AGE INTERNATIONAL (P) LIMITED, PUBLISHERS**
7/30 A, Daryaganj, New Delhi-110002, (INDIA), **Tel.:** (011) 23253472, 23253771
Visit us at www.newagepublishers.com • e-mail: contactus@newagepublishers.com

- **London** **NEW AGE INTERNATIONAL (UK) LTD.**
27 Old Gloucester Street, London, WC1N 3AX, UK
Visit us at www.newacademicscience.co.uk • e-mail: info@newacademicscience.co.uk

BRANCHES

- **Bangalore** 37/10, 8th Cross (Near Hanuman Temple), Azad Nagar, Chamarajpet, Bangalore- 560 018
Tel.: (080) 26756823, **E-mail: bangalore@newagepublishers.com**

- **Chennai** 26, Damodaran Street, T. Nagar, Chennai-600 017, **Tel.:** (044) 24353401
E-mail: chennai@newagepublishers.com

- **Cochin** CC-39/1016, Carrier Station Road, Ernakulam South, Cochin-682 016, **Tel.:** (0484) 4051304
E-mail: cochin@newagepublishers.com

- **Guwahati** Hemsen Complex, Mohd. Shah Road, Paltan Bazar, Near Starline Hotel, Guwahati-781 008,
Tel.: (0361) 2513881, **E-mail: guwahati@newagepublishers.com**

- **Hyderabad** 105, 1st Floor, Madhiray Kaveri Tower, 3-2-19, Azam Jahi Road, Near Kumar Theater, Nimboliadda Kachiguda, Hyderabad-500 027, **Tel.:** (040) 24652456, **E-mail: hyderabad@newagepublishers.com**

- **Kolkata** RDB Chambers (Formerly Lotus Cinema) 106A, 1st Floor, S N Banerjee Road, Kolkata-700 014
Tel.: (033) 22273773, **E-mail: kolkata@newagepublishers.com**

- **Mumbai** 142C, Victor House, Ground Floor, N.M. Joshi Marg, Lower Parel, Mumbai-400 013
Tel.: (022) 24927869, 24915415, **E-mail: mumbai@newagepublishers.com**

- **New Delhi** 22, Golden House, Daryaganj, New Delhi-110 002, **Tel.:** (011) 23262368, 23262370
E-mail: sales@newagepublishers.com

ISBN: 978-81-224-3072-1
C-21-08-12793

Printed in India at Ajit Printing Press, Delhi.
Typeset at Kalyani Printers, Delhi.

NEW AGE INTERNATIONAL (P) LIMITED, PUBLISHERS
7/30 A, Daryaganj, New Delhi - 110002
Visit us at **www.newagepublishers.com**
(CIN: U74899DL1966PTC004618)

Dedicated

To

Prof. Syed Waseem Akhtar

Vice Chancellor, Integral University, Lucknow.

A man of strong will who transformed His Visions into Reality

Preface

Safe, comfortable, and rapid means of transportation have become a symbol of progress and prosperity of any country. Advancement of transportation systems is the barometer of a civilization. Faster industrialization and opening of world markets for multinational companies to do business globally has given a great boost to automobile industry. The subject of automobile engineering has acquired immense importance and created a lot of scope for job seekers in the field. The innovations and evolution of electronics and use of computerized activities in the functioning of automobiles have not only maximized the comfort level of motor driving, but also minimized the air pollution in the atmosphere. It is a matter of pride that India has emerged as a giant producer of modern and best performing eco-friendly vehicles.

The present book deals with the fundamental knowledge of functioning of automobiles in a very simple manner and in easily understandable language. The book covers the syllabi of *seventh semester,* B. Tech. (Mech. Engg.) subject of the automobile engineering of U.P. *Technical University and Integral University, Lucknow and other Universities of India*. It will also be useful to anyone who has some interest in studying automobile engineering. Efforts have been made to describe the latest technologies and developments in the auto industry. I hope the book will be of great use for mechanical engineering students, teachers as well as to persons employed in the automobile industry.

This book is an outcome of the motivation and blessings of Prof. S.M. Iqbal, Pro Vice Chancellor, Integral University, Lucknow, whose encouragement to write a book on the subject gave me an elderly push.

I would also like to thank the following well-wishers:

1. Prof. Dr. M.I. Khan, Dean, Faculty of Engineering, I.U., Lucknow.
2. Er. Manoj Kumar Singh, A.I.E.T., Lucknow.
3. Mr. Mohd. Rashid Siddiqui, Faculty of Pharmacy, I.U., Lucknow.

4. Mr. Tanveer Ahmad, Deptt. of Mechanical Engg. I.U., Lucknow.

5. Lastly, Mr. Mohd. Hakim, Motor Mechanic, who explained to me working of various automobile devices.

It will be lacking on my part if I do not thank M/s New Age International (P) Limited, Publishers and the editorial team to bring out the book within a short period with nice get up.

I shall pay my personal gratitude to the teachers and students of the subject, who would care to point out the mistakes, misprints and forward their valuable suggestions for inclusion of some new topics in the present book.

—KHALIL U SIDDIQUI

Syllabus

FINAL YEAR B.TECH. (MECH. ENGG.), 7th SEM.
INTEGRAL UNIVERSITY AND U.P. TECHNICAL UNIVERSITY
LUCKNOW

SUBJECT : AUTOMOBILE ENGINEERING

LTP
310

UNIT–I : Power Unit and Gear Box

Principles of Design of main components. Valve mechanism. Power and Torque characteristics. Rolling, air and gradient resistance. Tractive effort. Gear box. Gear ratio determination. Design of Gear box. **7**

UNIT–II : Transmission System

Requirements. Clutches. Torque converters. over Drive and free wheel. Universal joint. Differential Gear Mechanism of Rear Axle. Automatic transmission. Steering and Front Axle. Castor Angle, Wheel camber and Toe in Toe out etc. Steering geometry. Ackerman Mechanism. Understeer and Oversteer. **8**

UNIT–III : Braking System

General requirements, Road, tyre adhesion. Weight transfer. Braking ratio. Mechanical brakes, Hydraulic brakes. Vacuum and air brakes, Thermal aspects: **5**

Chasis and Suspension System:

Loads on the frame. Strength and stiffness. Various suspension systems. **3**

UNIT–IV : Electrical System

Types of starting motors, generator and regulators, lighting system. Ignition system, Horn, Battery etc. **5**

Fuel Supply System

Diesel and Petrol vehicle system such as Fuel Injection Pump, Injector and Fuel Pump, Carburettor etc. MPFI. **4**

UNIT–V : Automobile Air Conditioning

Requirements, Cooling and heating systems **2**

Cooling and Lubrication System:

Different type of Cooling and lubrication system. **2**

Maintenance system:

Preventive Maintenance, break down maintenance, and over hauling system. **2**

Contents

CHAPTER 1

Automobile Engines

BRIEF INTRODUCTION

In 1769, the first self-propelled vehicle was introduced in *France* by *Nicholas—Joseph Cugnot.* In 1885, *Daimler* and *Benz* used a gas engine working on *Otto cycle* to drive a *four wheeler* vehicle. The *Peugeot family* then introduced the first Gasoline or petrol engine car in France during 1891. It was in 1895 that *Henry Ford* run his car by a 2-stroke gasoline engine. By the end of 1904 there were more than 100 four wheeler manufacturers in America alone.

Now the automobile industry is growing with such a tremendous speed (more than 30,000 lacs on roads) that any country's progress is being measured in terms of the auto vehicles running on its roads.

In India at the time of independence, only two names of car manufacturers were famous Hindustan Motors, manufacturing Ambassador Cars and Premier Automobiles Ltd. selling Fiat Cars. Now quite a good number of auto vehicle-makers are pushing their sales of cars in Indian market—Ford India Ltd., General Motors India Ltd., Hindustan Motors Ltd., Honda Siel Cars, Hyundai Motors India Ltd., Mahindra & Mahindra Ltd., Mahindra Renault Ltd., Maruti Udyog Ltd., Mitsubishi Ltd., Nissan, Skoda Auto India Ltd., Toyota Kirloskar Motors Ltd., B.M.W. India Pvt. Ltd., Daimler Chrysler India Ltd. and Fiat India Automobiles etc. are few to name; almost all of them are fielding their car models from smaller cars to costlier luxury cars.

Maruti Udyog Ltd. is still the biggest car manufacturer in India with their MARUTI 800, ZEN models, ESTEEM, WAGON R, ALTO, SWIFT, etc. and also serves an export hub for Suzuki.

India's Tata Motors are well established in car market with their INDICA, INDIGO, SUMO and SAFARI vehicles. In November 2007, Mr. Ratan Tata put forth for exhibition his dream car—'TATA NANO', the so called Lakhtakia small car, which is struggling to hit Indian roads. In March 2008, Mr. Ratan Tata achieved another *first* by acquiring the most charming world famous cars—JAGUAR and LAND ROVER of United Kingdom.

Two wheeler market in India is also very fast flourishing. Bajaj Auto Ltd., Hero Honda Motors Ltd., Honda Motorcycle, Scooter India Pvt. Ltd., TVS Motorcycles, Yamaha Motor India Ltd., are the major two wheeler manufacturers. Quite a good number of 2-wheelers are running on batteries claiming to be eco-friendly.

The automobiles of the modern age provide their owners a safe driving, more reliability and lot of comfort along with devices of I.T. and entertainment. There are two kinds of vehicles manufactured in India—One for defence purposes and other for civilians. The vehicles for defence are generally produced at Jabalpur Plant and Passenger Cars manufacturing units are located in Mumbai, Kolkata, Chennai, Jamshedpur, Pune and Gurgaon.

Automobile Engines

Internal combustion engines are extensively used for driving motor vehicles. All the engines used in propelling automobile vehicles are called automobile engines or simply auto engines. An *engine* is a device which converts heat energy of fuel into mechanical energy. In an internal combustion engine or I.C. engine, the heat energy is produced by burning the fuel (Petrol, Gas, Diesel oil, or any other fuel) inside the engine cylinder itself. Examples of I.C. engines are Petrol engines, Diesel engines, and Gas engines. Some engines are classified as external combustion engines. In these types, the combustion of fuel takes place outside the engine cylinder such as steam engines and steam turbines. In these engines the combustion takes place inside the boiler furnace and the heat of the steam such produced is utilized for running the engines.

Advantages of I.C. Engines over External Combustion Engines

- Greater mechanical simplicity.
- Lighter in weight.
- High overall efficiency.
- Initial cost is low.
- Compact in size and acquire less space.
- Easy start from cold.

CLASSIFICATION OF I.C. ENGINES

The internal combustion engines are classified on the basis of the following factors:

1. On the basis of working cycle

- Otto cycle engines.
- Diesel cycle engines.
- Mixed cycle engines.

2. On the basis of fuel used

- Petrol engines.
- Diesel engines or oil engines.
- Gas engines (C.N.G., Coal gas, Producer gas, Hydrogen and Furnace gases may also be used).
- Hybrid engines—using more than one fuel such as electricity and any other fuel.

3. On the basis of ignition method

- Spark ignition engine (S.I. Engine).
- Compression ignition engine (C.I. Engine).

4. According to number of strokes per cycle

- Two stroke engines.
- Four stroke engines.

5. According to the method of cooling

- Air cooled engines.
- Water cooled engines.

6. According to number of cylinders

- Single cylinder engines.
- Multi cylinder engines.

7. According to the arrangement of cylinders

- Horizontal engines.
- Vertical engines.
- V–type engines.
- Radial engines.

8. According to their uses

- Stationary engines.
- Portable engines.
- Auto engines.
- Marine engines.
- Aero engines.
- Space craft engines.

9. According to method of fuel supply

- Carburettor type.
- MPFI type.
- Solid injection type.
- Air injection type.

10. According to their speeds

- Low speed engines (up to 100 rpm).
- Medium speed engines (100–250 rpm).
- High speed engines (above 250 rpm).

11. According to the movement of piston/rotor

- Reciprocating engines (S.I. and C.I. engines).
- Rotary piston engines—Wankel engine.
- Gas turbines (Rotary engines).

12. According to method of governing

- Hit and Miss governed engines.
- Quality governed engines.
- Quantity governed engines.

13. According to valve arrangement

- Overhead valve engines.
- L–head type engines.
- T–head type engines.
- F–head type engines.

14. According to place of combustion of fuel

- Internal combustion engines.
- External combustion engines.

USES OF I.C. ENGINES

The I.C. engines are used for driving vehicles in ground, water and air.

- Road vehicles (*e.g.*, cars, scooters, motorcycles, buses, etc.)
- Air craft.
- Locomotives (Railway engines).
- Construction equipment in civil engineering such as bulldozer, scraper, power shovels, road rollers, cranes, etc.
- Pumping sets.
- Cinemas—Cooling and lighting systems.
- Hospitals—Stand-by units of power supply.
- Several industrial prime-movers, pumping sets, etc.
- Electricity Generation—used as prime-movers.
- Agriculture—Tractors, water pumping sets.
- Marines, ships, boats, etc.

WORKING OF A FOUR-STROKE I.C. ENGINE (PETROL ENGINE)

The construction of internal combustion engine is shown in the Fig. 1.1. The cylinder which is closed at upper end is filled with a mixture of fuel and air. As the crankshaft turns it pushes up the connecting rod and the piston is forced up and compresses the mixture in the space at the inside top of the cylinder. The mixture is burnt by a spark produced by spark plug and, as it burns, the gases expand and create a high pressure on the piston, forcing it downward in the cylinder. The piston pushes on the connecting rod which pushes on the U-shape crank. The crank is thus given a rotary (turning) motion.

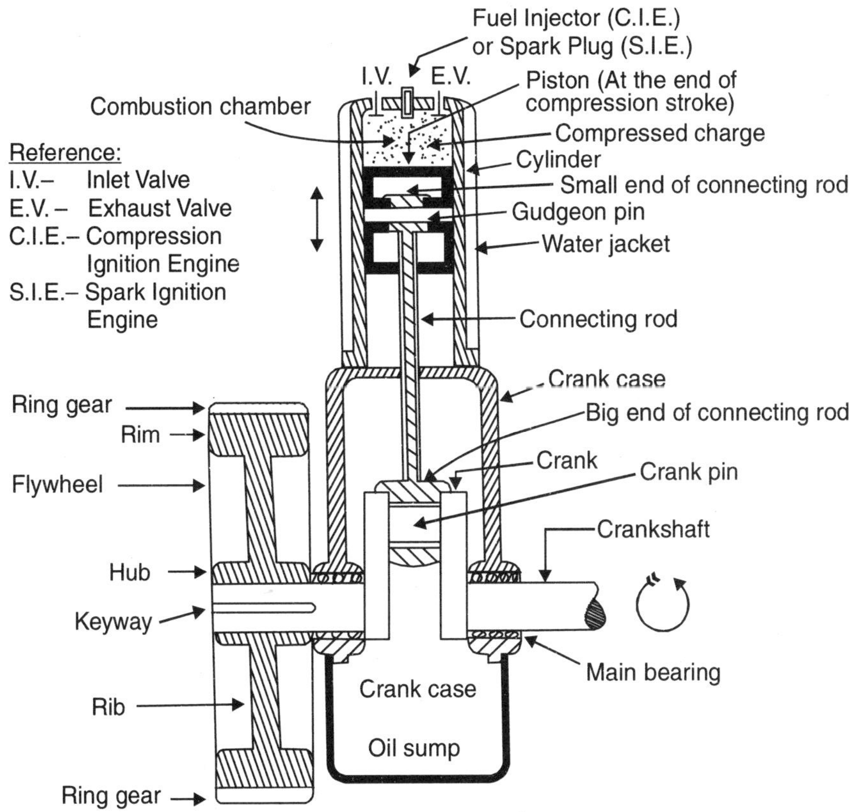

Fig. 1.1 Structure of I.C. engine (single cylinder)

A flywheel, fitted on one end of the crankshaft, stores energy and keeps the crank turning steadily. The upper closed end is called *'Cylinder head'*, carries inlet and exhaust valves and also fitted with fuel injector in case of a diesel engine or a spark plug in case of a petrol engine. The electric spark produced by spark plug, ignites the compressed air fuel mixture. The cylinder is surrounded by a cooling water jacket in case of water-cooled engines or by metallic projections, called *fins*, in case of air-cooled engines as shown in the Fig. 1.2.

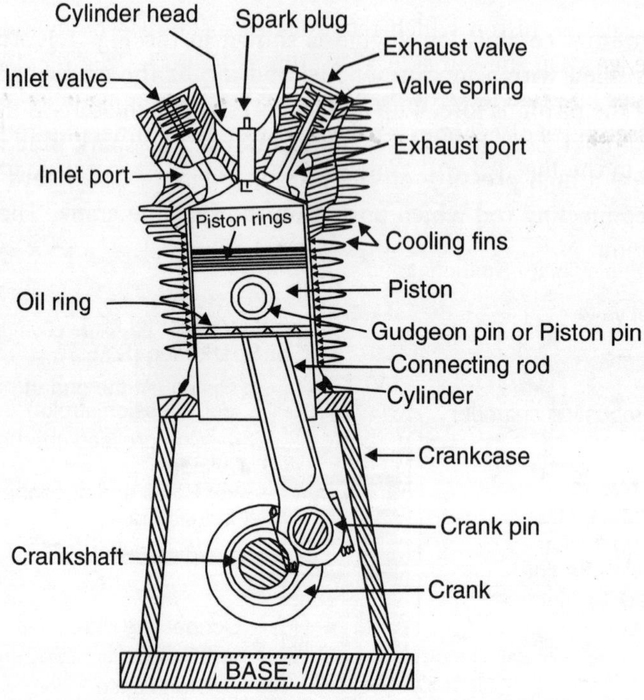

Fig. 1.2 An air-cooled 4-stroke petrol engine

Main Parts of an I.C. Engine

An I.C. engine has the following main parts:

- Cylinder—1, 2, 3, 4, 6, 8 or 16 cylinders.
- Cylinder head.
- Piston.
- Connecting rod.
- Crank.
- Crankshaft.
- Flywheel.
- Crankcase.
- Valves and valves operating mechanism.
- Water jackets or Fins.
- Spark plug (Petrol engine) or Fuel injector (Diesel engine).
- Carburettor or MPFI system (Petrol engine) or Fuel Injection Pump (Diesel engine).

Description of Main Parts

Cylinder: It houses the piston which reciprocates inside it. The cylinder provides working volume for the charge/gas. On the top side, it is attached with a cylinder head which houses the spark plug in S.I. engines or a fuel injector in the case of the diesel engine and the inlet and exhaust valves. The various assemblies of cylinder—piston, piston and connecting rod and connecting rod with crank are shown in the Fig. 1.3 (b). Water-cooled engines have wet-liners or dry liners fitted

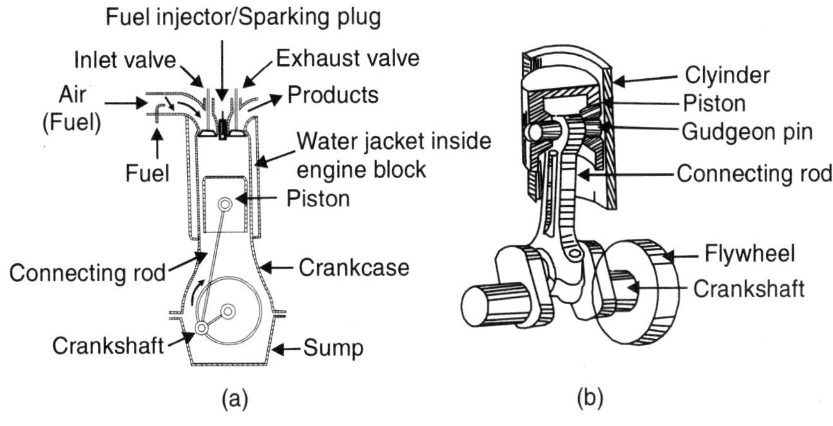

Fig. 1.3 Various engine parts assemblies

inside engine block with interference fit. They are replaced where worn or damaged. They are generally made of Grey cast iron. Liners save the engine block from damage. Both types of liners are shown in Fig. 1.3(c). Engine blocks are made of cast iron or aluminium alleys.

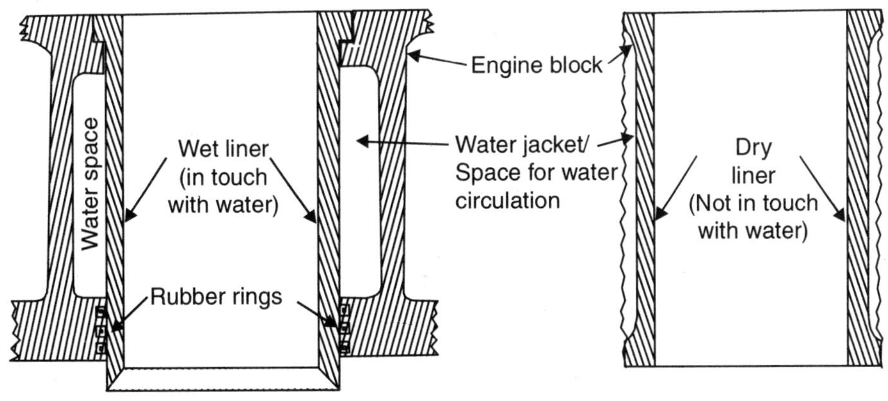

Fig. 1.3 (c) Water-cooled engine cylinder with wet liner/dry liner

Piston Assembly: Reciprocating motion of the piston inside the cylinder executes the working process. Piston transfers the work to the crank shaft through the connecting rod. The piston construction and nomenclature is shown in the Fig. 1.4 (a). Piston rings, fitted into the slots around the piston, provide a tight seal between the piston and the cylinder wall thus preventing leakage of combustion gases. Gudgeon pin forms the link between the small end of the connecting rod and the piston as shown in the Fig. 1.3 (a and b). Modern engine pistons are made of aluminium alloy having small amount of Silicon which gives good mechanical strength and low coefficient of expansion.

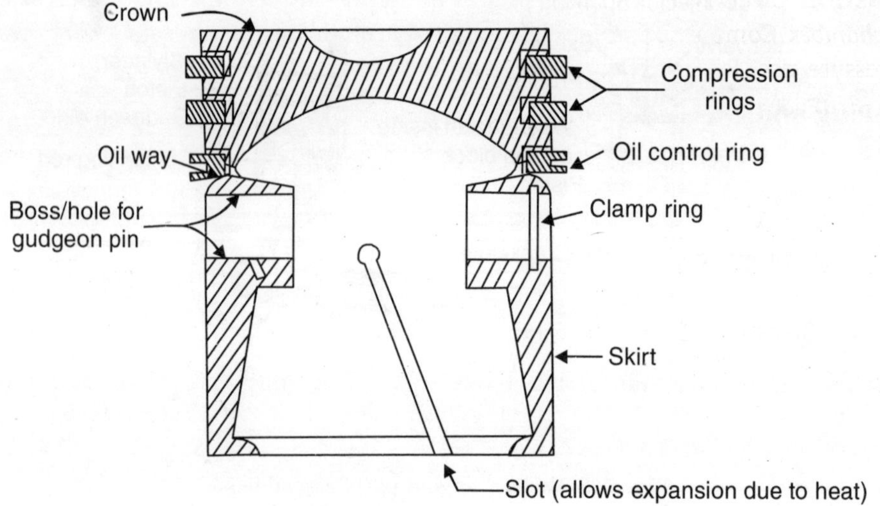

Fig. 1.4 (a) A split skirt piston

Piston Rings: The piston inside the engine cylinder has a clearance fit to allow free movement of piston. To provide a complete sealing between the piston and cylinder walls to hold the compression pressure, the piston is provided with piston rings. Rings are made of cast iron of fine grains or alloy spring steel. The two ends are kept split/apart so that they can be expanded and slipped over the upper end (top) of the piston to fit into the grooves cut in the outer periphery of the piston. The two upper piston rings give a good sealing as the split ends of the both rings are 180° opposite direction in their grooves. Higher the pressure in the combustion chamber, the better is the seal.

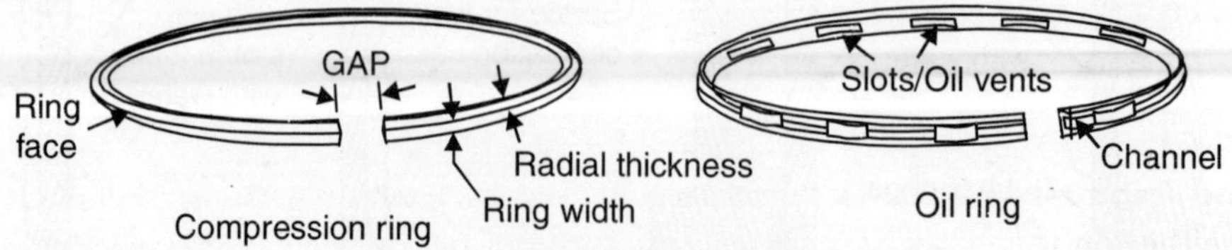

Fig. 1.4 (b) Piston rings

The piston rings are of two types:

- The upper two piston rings are called *'Compression Rings'*. They hold the compression pressure successfully (Fig. 1.4 (b)).

- The lower extra third ring is called *'Oil Ring'.* It is employed only in the four stroke cycle engines (Fig. 1.4 (b)). The oil rings scrap out the extra/excess lubricating oil from the cylinder walls, thus preventing it to enter into the combustion chamber where it would burn and create many troubles and black smoke in exhaust gases.

Gudgeon Pin or Piston Pin: The small end of the connecting rod is fitted over the piston pin whose two ends rest in piston bosses. The piston pin allows the free swiveling motion of the small end of the connecting rod. It is made up of hardened steel. It is simply a round bar spindle type pin. See Fig. 1.3 (b).

Combustion Chamber: The space above the piston inside the cylinder at TDC called the *combustion chamber.* Combustion of fuel inside this space releases the thermal energy resulting in a sudden pressure rise, leading to work done on the piston. (See Fig. 1.1).

Spark Plug and Fuel Injection Nozzle: In spark ignition engine (S.I. engine), the spark plug produces a strong spark at the end of compression of the charge to initiate combustion process as shown in the Fig. 1.3 (a). In the case of diesel engine, spark plug is replaced by a fuel injection nozzle which injects diesel oil into the hot compressed air to initiate combustion process.

Inlet Manifold: The pipe which connects the intake systems (Air cleaner, Carburettor/Fuel Injector) to the Inlet valve of the engine and through which air or air-fuel mixture is drawn into the cylinder is called *Inlet Manifold.* It is a simple steel pipe.

Exhaust Manifold: The pipe which connects the exhaust systems to the exhaust valve of the engine and through which the products of combustion escape into the atmosphere is called the *exhaust manifold.* See Fig. 1.5 (a).

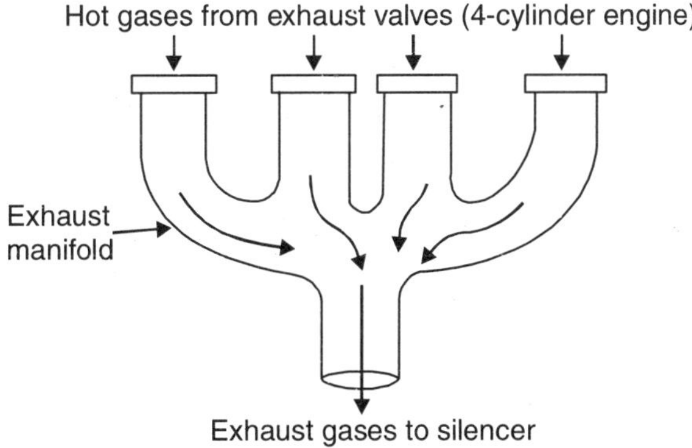

Fig. 1.5 (a) Exhaust pipe or exhaust manifold

Inlet and Exhaust Valves: Valves commonly used are mushroom shaped poppet type. They are provided either on the cylinder head or on the side of the cylinder for regulating the charge coming into the cylinder (inlet valve) and for discharging the products of combustion (exhaust valve) from the cylinder as shown in the Figs. 1.5 (b) and 1.5 (c). Their locations are shown in Figs. 1.1, 1.2 and 1.3. Exhaust valve is made more thick and strong to withstand high temperatures. Temperature at inlet valve is just more or less atmospheric.

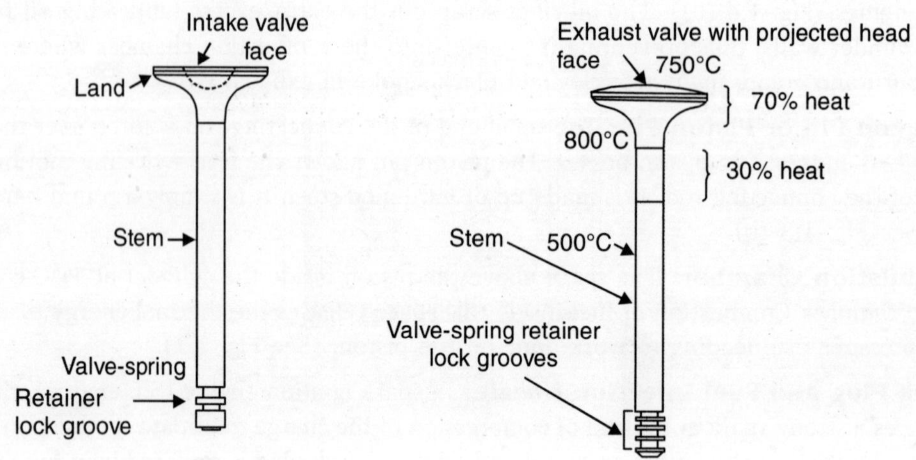

Fig. 1.5 (b) Inlet valve

Fig. 1.5 (c) Exhaust valve showing temperature zones

Flywheel: The net torque imparted to the crankshaft during one complete cycle of operation of the engine fluctuates, causing a change in the angular velocity of the shaft. In order to achieve a uniform torque, an inertia mass in the form of a flywheel is attached to the output shaft as shown in the Fig. 1.1 and this wheel is called the *flywheel.* The variation of net torque decreases with increase in the number of cylinders in the engine and thereby the size of the flywheel also becomes smaller. This means that a single cylinder engine will have a larger flywheel whereas a multi-cylinder engine will have a smaller flywheel. It also facilitates the starting of the engine. It has a ring gear over its periphery to engage with starting pinion of the starting system. Figure 1.6 shows the details of construction of flywheel.

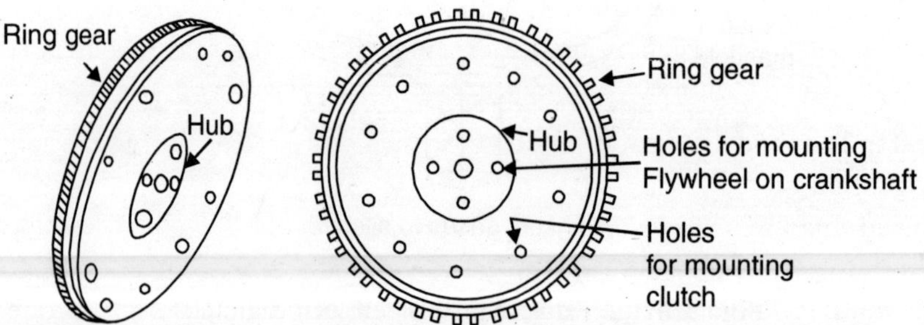

Fig. 1.6 Flywheel

Connecting Rod: The function of the connecting rod is to convert the reciprocating motion of the piston into the rotary motion of the crankshaft. The connecting rod connects the piston to the crankshaft through piston pin and crankpin. The expanding gas force on the piston is thus transmitted to the crank pin through this rod.

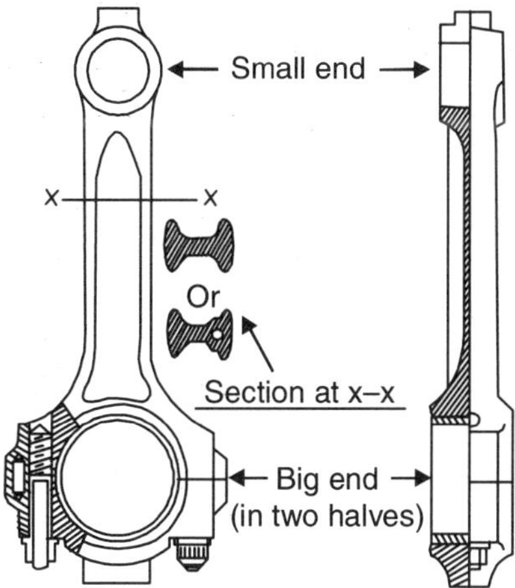

Fig. 1.7 Connecting-rod—half in section

The piston pin fits into the small end of the connecting rod. The big end, which is split into two halves as shown in the Fig. 1.7, accommodates the crankpin. The connecting rod shank is usually of I-section with breadth to depth ratio of 0.6 to 0.7 (approximately). The shank of the connecting rod is subjected to the force of inertia of reciprocating parts and the force due to gas pressure inside the cylinder. The stress produced is cyclic. The connecting rod shank is also subjected to bending stress due to axial shifting of the crankshaft.

Connecting rods for S.I. engines are generally made of medium carbon steel (0.35 to 0.45 per cent carbon). For heavy-duty diesel and supercharged diesel engines, chrome-nickel or chrome molybdenum steel is used. In all cases the connecting rods are forged in dies and heat-treated after machining the ends. Malleable or spheroidal-graphite iron castings or sintered forgings are used in small to medium sized petrol engines. A popular material used for both rod forgings and their clamping bolts or studs are manganese molybdenum steel, its composition being carbon (0.35%), manganese (1.5%), molybdenum (0.3%) and the remainder being iron.

Crankshaft: The connecting rod and crank convert the reciprocating motion of the piston into the rotary motion of the crankshaft. The crankshaft is subjected to the external forces and moments produced due to gas force and inertia of moving parts. The actual complicated shape of the crankshaft is shown in the Fig. 1.8 (a). The gas force produces concentration of stresses at many local points. In multi cylinder engine, crankshaft is subjected to different torque at the same moment due to shift in firing time of the cylinders, which produces unequal torsional deformation at different sections of the crankshaft. This creates inertia moment of the crankshaft elements between the masses and some torsional vibrations are also set up in the crankshaft, making it to vibrate at some frequency during rotations. The crankshaft has to be balanced statically and dynamically to minimize vibrations for smooth running. Crankshaft of a single cylinder has been shown in the Fig. 1.1 in the beginning.

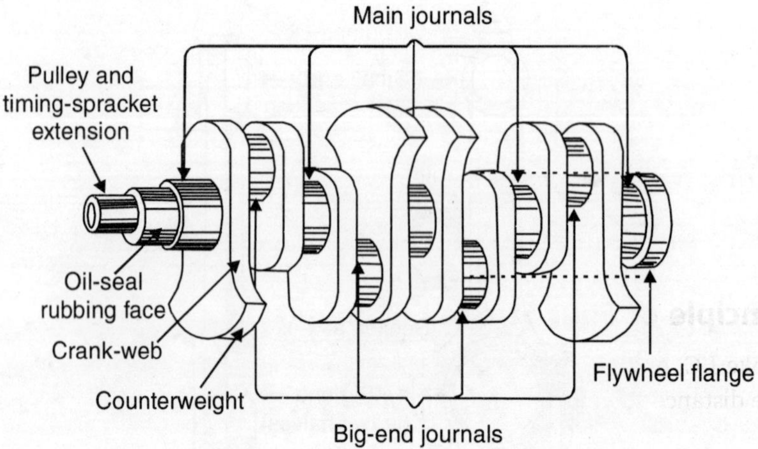

Fig. 1.8 (a) Crankshaft for four cylinder engine

Main journals of the crankshaft are supported by strong bearings fitted in the crankcase walls and these form the axis for the rotation of the crankshaft. Oil holes are drilled from main journals to the crankpins through crank webs to provide lubrication for big end bearings. The centrifugal forces acting at each crank pin due to rotation of both the crankshaft as well as the big end of the connecting rod tend to bend the crankshaft. To counter this tendency, counterweights are either formed as integral part of the crank web or attached separately, on the sides opposite to the crank pin. The flywheel keeps the crankshaft rotating at a uniform speed. The flywheel has teeth in its outer periphery, which mesh with the pinion of the starting motor.

The crank transforms the reciprocating motion of the piston and connecting rod into the rotary motion for rotating the crankshaft and flywheel. The crankshaft is composed of one or more cranks between the ends. The crankshaft is subjected to all sorts of forces developed within the engine, and it is therefore necessary to make it very strong, by forgings from some extremely strong steel alloy or chromium nickel molybdenum steel. A simple diagram of a crankshaft for four cylinder engine is given in the Fig. 1.8 (b).

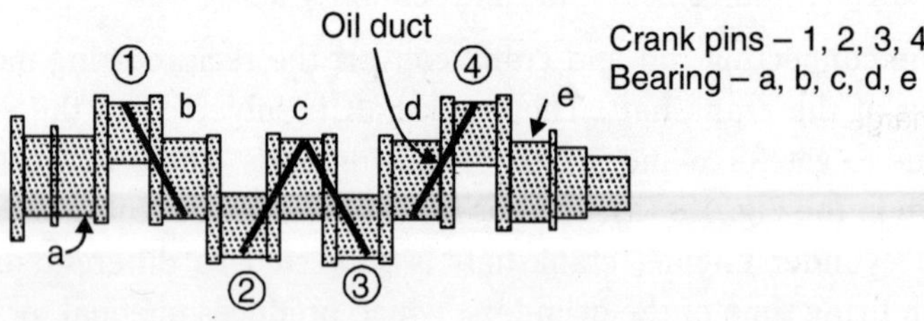

Fig. 1.8 (b) Simple line diagram of a crankshaft for four cylinder engine

Cam Shaft: A simple cam shaft for four cylinder engine has been shown in the Fig. 1.8 (c). It is driven either by a chain or by the timing gears mounted between the crankshaft and the cam shaft. It rotates at half the speed of the crankshaft in 4-stroke engines (inlet or exhaust valve operates once in two revolutions of the crankshaft). Its function is to operate the valve lifting mechanism and to drive F. I. pump (in diesel engines).

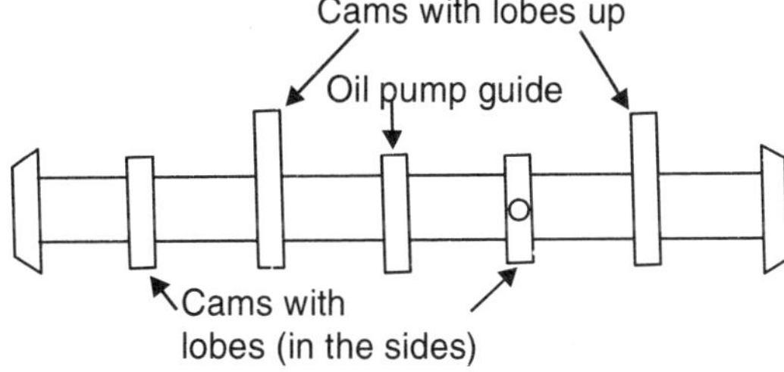

Fig. 1.8 (c) Cam shaft for four stroke two cylinder engine

Working Principle of Four Stroke I.C. Engine

In four stroke I.C. engine, one cycle of operations is completed in four strokes of the piston. One stroke is the distance travelled by the piston from one end of the cylinder to another end during half revolution of the crankshaft. The topmost and bottommost positions up to which the piston can travel inside the cylinder are known as *top dead centre* (TDC) and the *bottom dead centre* (BDC) respectively.

The distance between TDC and BDC is known as *stroke length.* One cycle of operation in a four-stroke engine consists of four strokes. They are:

- Suction stroke,
- Compression stroke,
- Power stroke, and
- Exhaust stroke.

1. Suction Stroke: During this stroke, the piston moves from TDC to BDC creating a vacuum inside the cylinder. The inlet valve is opened and the exhaust valve is kept closed. The vacuum created inside the cylinder draws the charge (air and petrol vapour for S.I. engine and air alone in C.I. engine) into the cylinder through the inlet manifold and the open inlet valve as shown in the Fig. 1.9. This process is known as *suction* and is performed till the piston reaches BDC. The suction of charge carried out during one stroke length of piston movement is called *suction stroke* and its volume is called *swept volume.*

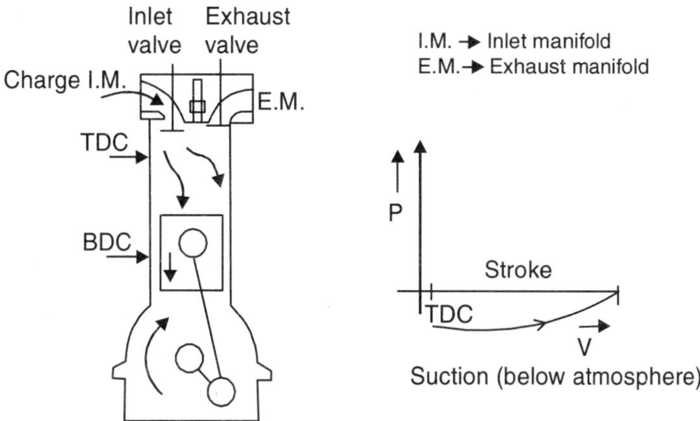

Fig. 1.9 Suction stroke—inlet open and exhaust valve closed

2. Compression Stroke: After the completion of the suction stroke, the piston moves from BDC to TDC. Both inlet and exhaust valves are kept closed during this stroke. The volume reduction of charge during this stroke increases the pressure. This process is known as *compression* and this stroke is called the *compression stroke.* See Fig. 1.10.

If V_1 = Volume of air or charge sucked in during suction stroke (c.c.) = Stroke volume

V_2 = Volume of air or charge after compression stroke is over (c.c.) = Clearance volume

then Compression ratio, $r_c = \dfrac{V_1}{V_2} = \dfrac{\text{Stroke volume}}{\text{Clearance volume}}$

For S.I. engines, r_c = 6 to 8

For C.I. engines, r_c = 16 to 18

In multi-fuel engines, r_c may be up to 21.

Clearance volume: The space left in the cylinder (between cylinder head and piston crown) after completion of the compression stroke when piston is at TDC.

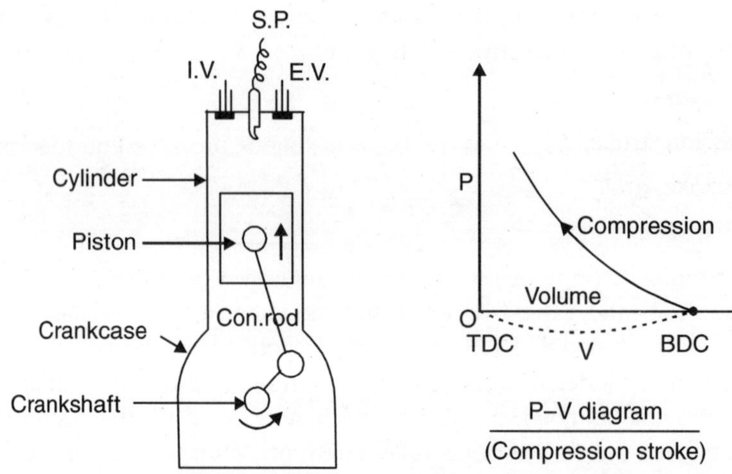

Fig. 1.10 Compression stroke—both valves are closed

At the end of compression stroke, a spark is introduced by the spark plug in S.I. engine to initiate the combustion in which the flame produced by the charge consumes the charge to increase the pressure instantaneously. In the case of C.I. engine, diesel is injected into the compressed air at the end of the compression stroke. The compressed air which is at high pressure and temperature, transfers heat to the diesel vapour. Diesel vapour on receiving heat reaches itself at its ignition temperature quickly and the combustion process begins.

3. Power Stroke: Refer to Fig. 1.11. Combustion process increases the gas pressure suddenly and creates an impact on the piston due to which the piston moves from TDC to BDC, generating work. This stroke is known as *power stroke.* During this process also both inlet and exhaust valves are kept closed. It is also called *'Expansion stroke'.*

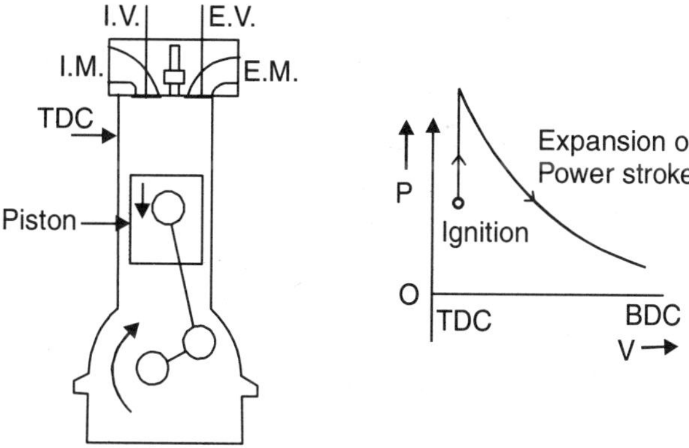

Fig. 1.11 Power stroke—both valves are closed

Only during this stroke the engine develops power and for doing the compression and suction strokes engine gets power from its crankshaft and flywheel. Power required for compression and suction strokes is only a fraction of the power developed by the engine and the balance power is transferred through the crankshaft to the drive line of the vehicle.

4. Exhaust Stroke: After the power stroke, the burnt or exhaust gas is pushed out of the cylinder to atmosphere through the exhaust valve which is kept opened during this stroke. The pressure during exhaust stroke is little above the atmospheric pressure. But theoratically, both inlet and exhaust pressures are taken equal to atmospheric pressure.

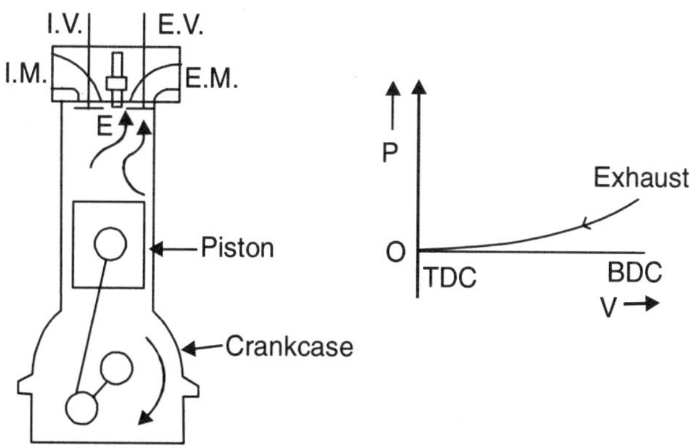

Fig. 1.12 Exhaust stroke—inlet valve closed and exhaust valve opened

Piston moves from BDC to TDC pushing the exhaust gas to the exhaust manifold. Inlet valve is kept closed during this stroke to avoid the entry of exhaust gas into the inlet manifold. Complete removal of the exhaust gas is very important for the efficient combustion during the next cycle. The exhaust stroke is shown in the Fig. 1.12. After exhaust stroke, the cycle is complete and the same strokes are repeated again and again during running of the engine.

Application of Four Stroke I.C. Engine

- Four stroke petrol engines are used in automobiles (two wheelers bikes and cars) and motor boats, domestic generator sets etc.
- Four stroke diesel engines are used in high capacity automobile (cars, buses, lorries, trucks, ships, and small aircrafts).
- Four stroke diesel engines are also used for the electric power generation (diesel generator sets).
- Four stroke diesel engines are also used in operating pumps and machine tools.

Theoretical and Actual Indicator Diagrams for the Four-Stroke I.C. Engines

(i) Otto cycle engine

The variation of pressure with respect to the volume inside the cylinder of four stroke S.I. engine is shown with the help of theoretical processes in Fig. 1.13.

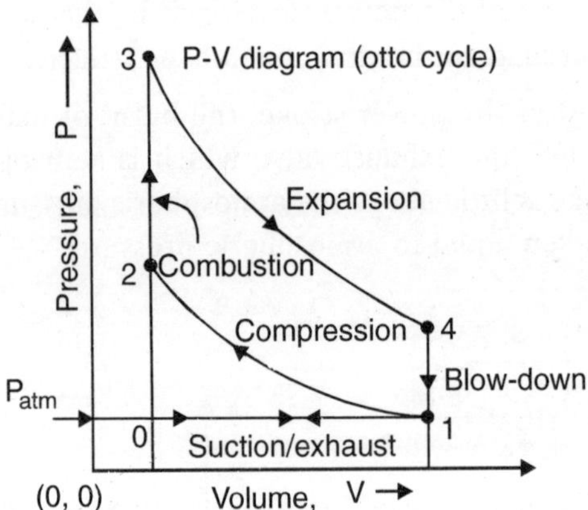

Fig. 1.13 Theoretical indicator diagram for four stroke petrol engine (otto cycle engine)

The minimum working volume inside the cylinder of the engine is known as *clearance volume.* This is the volume above the piston when it reaches the TDC. When the piston moves during the suction stroke from TDC to BDC it creates another volume V_S known as *stroke volume* or *swept volume* during which the pressure almost remains as the atmospheric pressure. During the compression stroke, the pressure increases to 15 to 50 kg/cm^2 with volume reduction based on any one of the compression laws, that is isentropic ($PV^\gamma = C$), polytropic ($PV^n = C$) or isothermal ($PV = C$). At the end of compression, a spark is initiated which creates heat at constant volume and raises the gas pressure instantaneously. The gases expand and power stroke takes place. The expansion of gases is usually adiabatic or isentropic. At the end of the power stroke, the exhaust valve suddenly opens and the pressure drops instantaneously to atmospheric pressure. It then remains almost at the same level during the exhaust stroke. This cycle is based on the theoretical *Otto cycle.*

In actual cycle, there are some deviations. Pressure during the suction stroke is less than atmospheric pressure and the pressure during the exhaust stroke is slightly more than the atmospheric pressure. Spark ignition is just before the end of compression and the heat addition is not exactly at constant volume. The actual indicator diagram is shown in the Fig. 1.14.

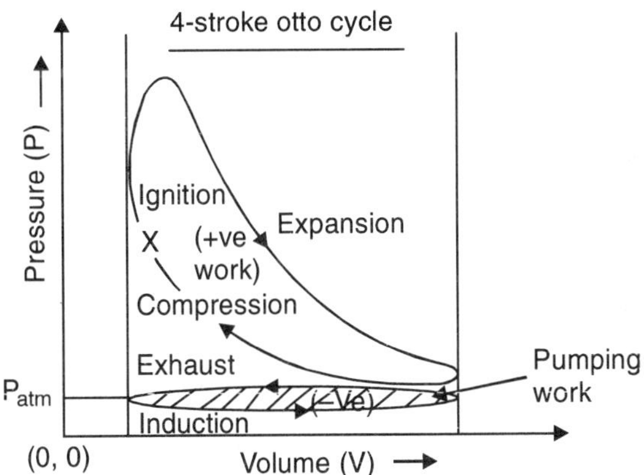

Fig. 1.14 Actual indicator diagram for four-stroke petrol engine

The process of combustion is not an explosion and does not occur at once. It is an orderly burning of the charge. The spark ignites the charge in the immediate vicinity of the spark plug and the flame front proceeds spherically outwards from there at a more or less uniform speed. Ahead of the flame front is unburned mixture and behind it are the products of combustion. The flame front takes certain time to travel from the spark plug to the farthest reach of the combustion chamber. It is thus necessary to start the ignition considerably before TDC in order to complete the combustion a little bit after TDC. The highest efficiency is obtained when the point of ignition and the point at which combustion is complete are symmetric with respect to TDC.

As the piston approaches BDC at the end of the power stroke, the exhaust valve is opened just before BDC and the pressure falls due to rushing out of exhaust gas from the cylinder. This difference in pressure drop between the actual and ideal cycle at this point in the cycle, represents unavailable work and is termed as *exhaust blow down loss.*

The pressure drops below atmospheric pressure as the piston moves down during the suction stroke, drawing air into the cylinder but it is above atmospheric pressure during the exhaust stroke in order to drive the gases from the cylinder. This loop, called *negative loop,* represents work that must be done on the exhaust gases and hence it is also called as *pumping work.* After deducting pumping work from the work obtained during power stroke, we get network for conversion into mechanical energy for driving the vehicle.

(ii) Diesel cycle

This cycle is similar to Otto cycle or petrol engine cycle except that the pure air is sucked during suction stroke and the fuel injection and combustion take place at constant pressure after completion of the compression stroke as shown in Fig. 1.15. Following are the various events which take place in the diesel cycle:

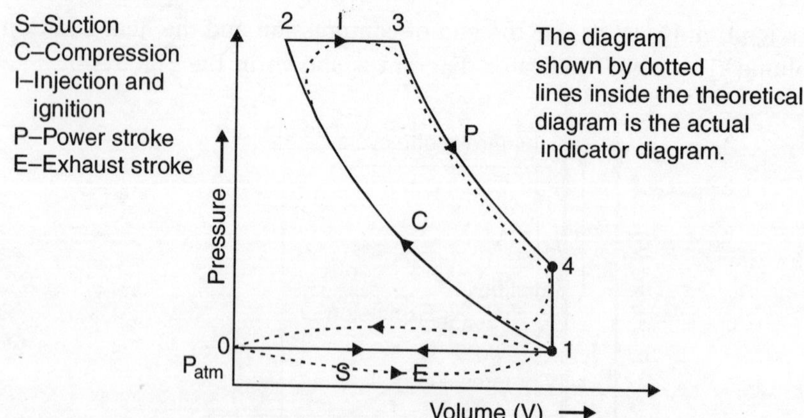

Fig. 1.15 Theoretical diesel cycle indicator diagram

0–1 ⟶ Suction of pure air during suction stroke.

1–2 ⟶ Compression stroke, pure air is compressed adiabatically/polytropically. Pressure ranges from 50 to 75 kg/cm^2.

2–3 ⟶ Injection of fuel (Diesel oil) and its ignition at constant pressure, piston slowly moves back.

3–4 ⟶ Expansion/power stroke, gases expand adiabatically/polytropically.

4–1 ⟶ Release: Sudden drop of pressure due to opening of exhaust valve, also called Blow-Down.

1–0 ⟶ Exhaust stroke, piston drives out burnt gases from the cylinder at near atmospheric pressure.

In processes 1–2, 2–3, 3–4 ⟶ both valves are kept closed.

In process 0–1 ⟶ Suction or Inlet valve remains open.

In process 1–0 ⟶ Exhaust valve remains open.

(iii) Dual cycle or Semi-diesel cycle or Mixed cycle

Modern high speed diesel engines work on dual cycle. In this cycle, ignition takes partly at constant volume (like Otto cycle) and partly at constant pressure (like diesel cycle) as shown in the Fig. 1.16 (a). The efficiency of this cycle is better than diesel cycle due to better atomization and combustion of fuel. Referring to the mixed cycle P–V diagram, following processes take place:

0–1 ⟶ Suction of pure air (Suction stroke).

1–2 ⟶ Polytropic compression of air (Compression stroke)/($PV^n = C$).

2–3 ⟶ Injection and combustion of fuel at constant volume.

3–4 ⟶ Injection and combustion continued at constant pressure.

4–5 ⟶ Polytropic expansion of gases (power stroke)/($PV^n = C$).

5–1 ⟶ Pressure drop at constant volume due to opening of exhaust valve (release).

1–0 ⟶ Burnt gases are driven out of cylinder by piston (Exhaust stroke).

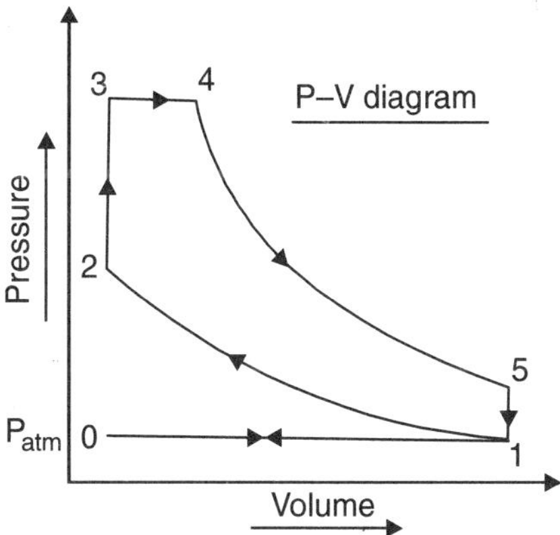

Fig. 1.16 (a) Mixed or dual cycle
(for four stroke diesel engine)

Comparison between Petrol and Diesel Engines

S. No.	Petrol Engine	Diesel Engine
1.	During suction stroke, a mixture of air and petrol is sucked in the engine cylinder.	Pure air is sucked during suction stroke.
2.	To ignite the compressed mixture, a spark plug is employed.	The fuel is sprayed by fuel injector and the mixture gets self-ignited by heat of compression.
3.	Thermal efficiency is about 25%.	Thermal efficiency is higher up to 40%.
4.	Cost of fuel (petrol) is more.	Diesel oil is cheaper.
5.	They are light in weight.	They are heavy engine.
6.	Compression ratio is up to 8.	Compression ratio is up to 21.
7.	Produces less pollution and noise.	Produces more pollution and noise.
8.	Carburettor or multi-point fuel injectors are used for making air-fuel mixture at almost atmospheric pressure at the beginning of the suction stroke.	Fuel injection pump and injectors are employed for injecting fuel under high pressure into the cylinder at the end of compression stroke.

Valve Timing Diagram of a Four-Stroke I.C. Engines

A valve timing diagram (VTD) shows the various angular positions of the crankshaft at which opening and closing of inlet and exhaust valves takes place during one cycle. It helps in adjusting the valves timings and ignition for efficient working of the engine.

A valve timing diagram for a 4-stroke I.C. engines is given in Fig. 1.16(b).

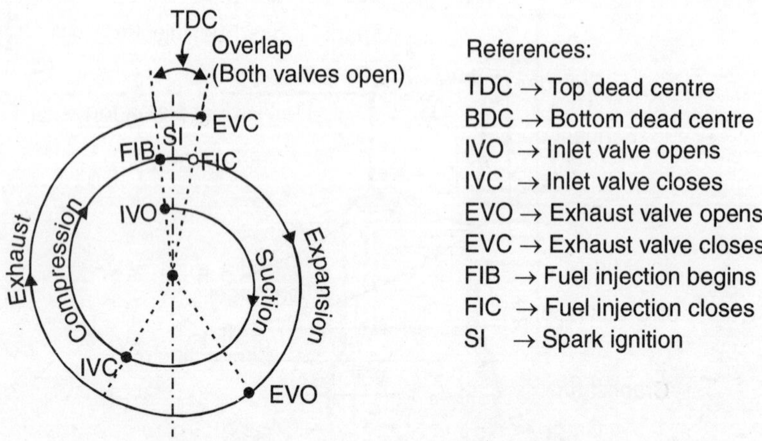

Fig. 1.16 (b) Actual valve timing diagram

For Petrol Engine	For Diesel Engine
IVO ⟶ 10° to 20° before TDC	10° to 20° before TDC
IVC ⟶ 30° to 40° after BDC	25° to 40° after BDC
EVO ⟶ 30° to 50° before BDC	40° to 50° before BDC
EVC ⟶ 10° to 15° after TDC	10° to 15° after TDC

In petrol engines Ignition starts 20° to 30° before. TDC and continues upto 15° after TDC.

In diesel engine fuel is injected 10° to 15° before TDC in the last moments of the compression. The injection of fuel continues 15° to 20° after TDC to complete the combustion.

Inlet valve opening (IVO) is advanced (Lead) before TDC and the inlet valve closing (IVC) is delayed after BDC by few degrees to maximise the suction volume of charge so as to obtain more power. Fuel injection beginning (FIB) and simultaneous fuel combustion in Diesel engines or spark ignition in petrol engines, is advanced to have complete combustion, just after TDC. Exhaust valve opening (EVO) is advanced before BDC at the cost of power or expansion stroke and exhaust valve closing (EVC) is delayed after TDC by few degrees to increase exhaust stroke process in order to achieve maximum removal of burnt gases from the engine cylinder.

WORKING OF TWO-STROKE ENGINE

The two-stroke I.C. engine is similar in construction to the four-stroke I.C. engine except that the valves are replaced by the ports as shown in the Fig. 1.17. One cycle of operation is completed in the two strokes, that is, in one revolution of the crankshaft.

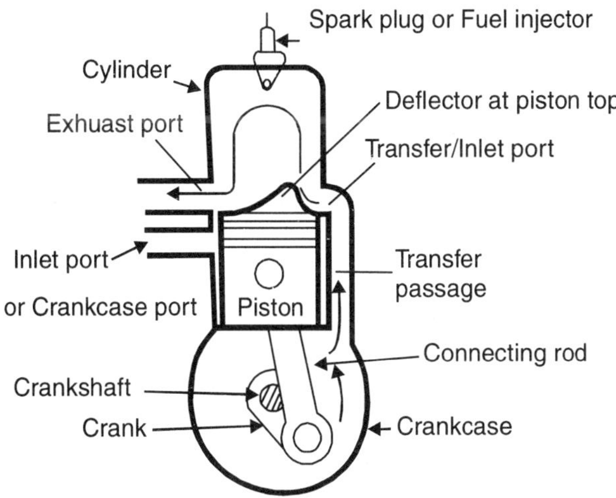

Fig. 1.17 Two-stroke I.C. engine

Working Principle of Two-Stroke Engine

Combustion of fuel in two-stroke engine takes place at TDC similar to a four-stroke engine. Piston moves down executing power stroke after the combustion process till the piston opens the exhaust port. The remaining movement of the piston in downward direction keeps the exhaust port open and allows exhaust gas to move out of the cylinder. During this movement of the piston, transfer port also opens in the opposite side of the cylinder allowing the slightly compressed charge in the crankcase to move into the cylinder through transfer port as shown in the Fig. 1.18. This fresh and slightly compressed charge/air helps to drive cut the exhaust gases through exhaust port. This process is known as *scavenging*.

When piston moves from TDC to BDC during power stroke, the lower side of the piston compresses the charge or pure air sucked inside the crankcase during the piston's upward stroke. This slightly compressed charge is easily transferred to engine cylinder when piston uncovers the transfer port—which connects crankcase to cylinder.

The exhaust and charging takes place simultaneously till the piston covers the transfer and exhaust port in its upward movement. Compression of charge begins after the piston completely closes the exhaust port. The remaining part of the piston movement in the upward direction compresses the charge. At the end of compression, a spark is given (S.I. engine) or diesel is injected (C.I. engine) to produce combustion. When the piston moves up during compression, a slightly low pressure is created in the crankcase which allows the fresh charge to enter the crankcase through the crankcase port via carburettor. Thus, one cycle of operation is completed in two strokes of the piston. In case of diesel engine, fresh air enters in the crankcase during compression stroke of piston. In 2-stroke engines, the suction and exhaust strokes are avoided by providing a system to remove exhaust gases by fresh charge itself.

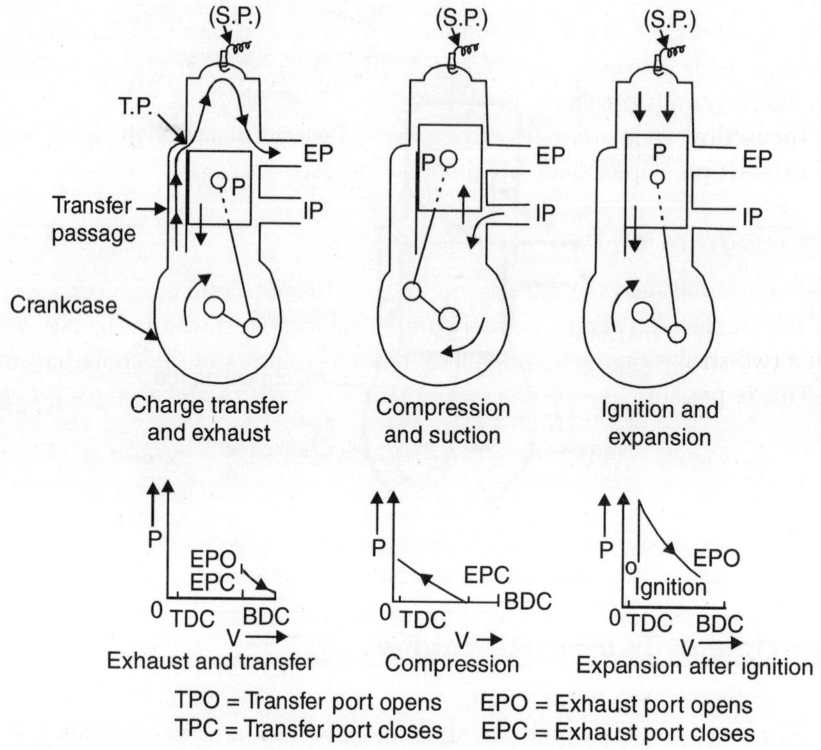

Fig. 1.18 Working of two-stroke cycle engine

Indicator or P–V Diagram for Two-Stroke Engine

The variation of pressure inside the cylinder (above the piston) with respect to volume for two-stroke petrol engine can be represented as shown in the Fig. 1.19.

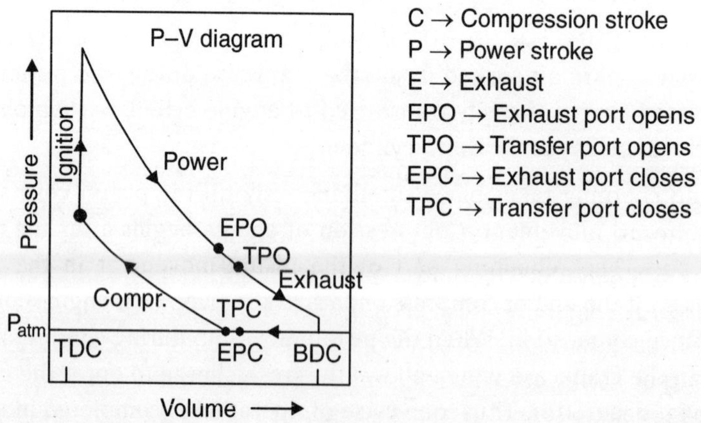

Fig. 1.19 Theoretical indicator diagram for two-stroke petrol engine

The pressure drops closer to atmospheric level from the opening of the exhaust port and thereafter remains at the same level till the piston comes back in its return stroke to close the exhaust port. The compression and expansion follow any one of isentropic, polytropic or isothermal laws. Heat addition is at constant volume in S.I. engine. For two stroke diesel engine the only change in the P–V diagram is the constant pressure heat addition process. The deviations of actual cycle from the theoretical cycle are that (*i*) the pressure is slightly above the atmospheric pressure even after the exhaust port opens and (*ii*) the heat addition processes.

Port Timing Diagram of Two-Stroke Engine

The representation of opening and closing of ports in one cycle of operation with respect to the rotation of the crankshaft is known as *port timing diagram* as shown in the Fig. 1.20. One cycle of operation in a two-stroke engine is completed in one revolution of the crankshaft or two strokes of the piston. This is possible due to the overlapping of charging and exhaust processes.

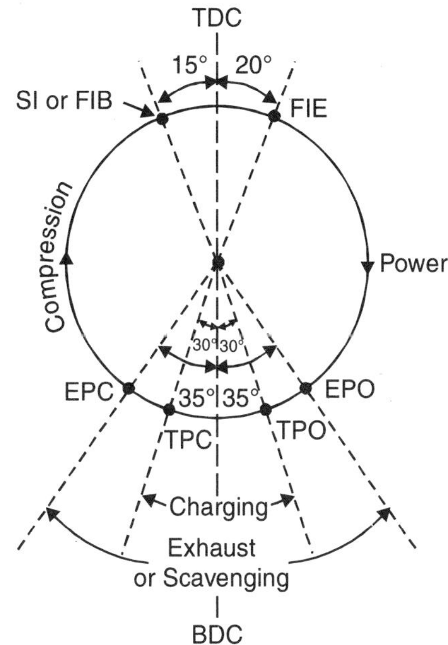

Fig. 1.20 Port timing diagram for two-stroke engine

During this overlapping period, a portion of the fresh charge which helps to push the exhaust gas out also escapes through the exhaust port. The fresh charge that goes out without combustion reduces the thermal efficiency of the two-stroke engine significantly compared to four-stroke engine in which overlapping is minimum. The duration of power stroke represented in the port timing is less than that of four-stroke engine. Hence, even though the two-stroke engine produces two cycles in two revolutions of crankshaft compared to one cycle in two revolutions of crankshaft of four-stroke engine, the output power of two stroke engine is less than twice (≈1.75) that of four-stroke engine. The compression process is also less than that of four-stroke engine making it suitable for low compression petrol engine than for diesel engine which needs high compression for the self ignition of diesel fuel. In case of petrol engine, spark ignition takes place few degrees before TDC such that maximum pressure due to combustion is created when the piston is at TDC. In the case of diesel engine, fuel injection begins (FIB) few degrees before TDC such that the self ignition of diesel takes place when piston reaches TDC and there after the flame produced ignites the injected diesel spontaneously to maintain constant pressure till the fuel injection ends (FIE). The volume expansion inside the crankcase below the bottom side of the piston creates low pressure and it sucks the fresh charge into the crankcase, which is transferred to engine cylinder when piston uncovers the transfer port. This fresh charge being at slightly higher pressure, drives out exhaust gases from the engine cylinder, this process of driving out gases is called **'Scavenging'**.

Gasket

In the assembly of the engine, proper gaskets (made of leather, rubber, cork, metal covered rubber and asbestos) are always placed between two stationary metal surfaces to provide sealing action. Between cylinder and cylinder head, one piece embossed steel rubber gaskets are widely used.

Valve Lifting Mechanism

The valve operating gear opens and closes each valve at the right point in the cycle. The drive comes from the camshaft. In the four-stroke engines, the camshaft is driven by the crankshaft, at half the engine speed. The crankshaft drives the camshaft through a chain or set of gears called **'Timing Gears'**. As the Camshaft rotates, each cam lifts a tappet or rocker arm through push rod. ln the case of a side valve engine, the tapppet lifts the valve stem against the spring force and opens the valve. In the case of an overhead valve engine, the push rod placed above the tappet operates a rocker arm to pivot and push the valve down against the spring force. The valve is closed by the valve spring. This closing happens when further rotation of the cam allows the tappet to descend. Some engines have two springs one inside the other for the each valve. The valve operating mechanisms can be seen in the Figs. 1.21 (a) and 1.21 (b).

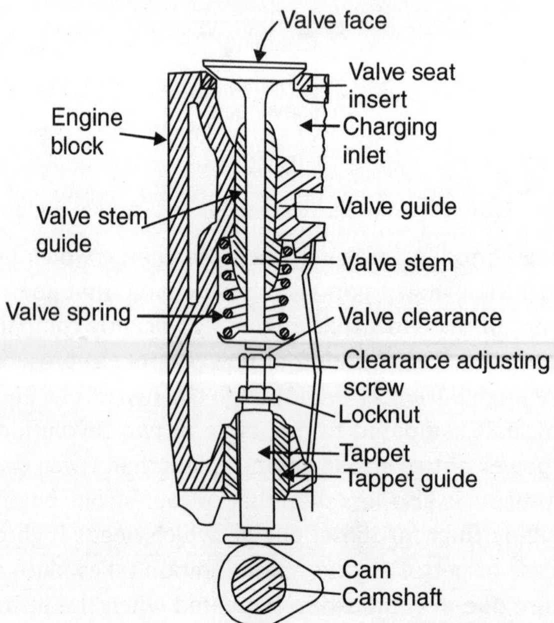

Fig. 1.21 (a) Side valve lifting mechanism

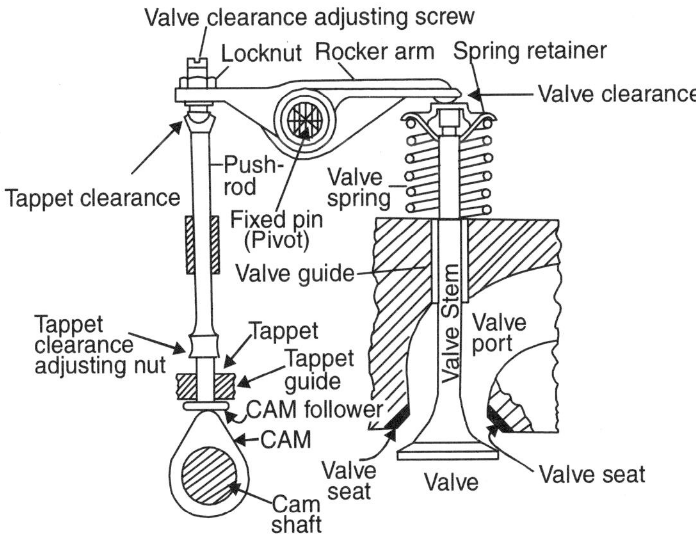

Fig. 1.21 (b) Overhead valve lifting mechanism

Tappet Clearance

For proper operation, the valve must return fully to its seat. To ensure this, a gap is provided between the end of the valve stem or push rod (when the valve is in the closed position) and the end of the tappet or rocker arm which operates on it. This gap is called *tappet clearance.* This gap allows the valve to expand when it becomes hot.

The tappet clearance of each valve should be checked periodically and adjusted to the precise value as recommended by the manufacturer. Greater clearance results in noise. Insufficient clearance may lead to failure of valve closure and hence the failure of engine operation.

Valve Heating

The heating of valve and its seat during engine operation may cause distortion of the valve. Remember they may be round and true when the engine is cold, but may not be so when the engine attains operating temperature. This distortion will be aggravated due to hot spots in the cylinder head or engine block near the valves or unequal tightening of the cylinder head bolts. The temperature of the exhaust valve varies from 500° to 800°C in S.I. engines. For cooling purpose, the exhaust valve is closely seated with water jacket. Sodium cooled valve are also used. Sodium is filled in a special hollow stem, which transfers heat quickly. The valve temperature may rise due to one or more of the following defects:

- Valve seat too narrow
- Loose valve seat insert
- Worn valve guides and/or valve stem
- Sticking valve stems
- Valve guides loose in the block

- Weak valve springs
- Incorrect tappet clearance.

MOUNTING OF ENGINE ON CHASSIS FRAME

The mounting of engine at the chassis has its own importance. The engine produces complex vibrations which are followed by many other stresses and movements of the vehicle such as—bounce and yaw vertically, moving to and fro horizontally, rolling about vertical axis and some lateral movements.

The engine mountings support the dead weight of engine itself and also isolate the chassis from engine vibrations and bear the torque reactions. Rubber pads are commonly used as spring material between frame and engine. Metal bond rubber pads used as packing between the mounting surfaces help in damping vibrations and noise.

Engine may be mounted at three places at the frame:

1. *Engine in Front*. In majority of auto vehicles, the engine is mounted on the front side of the chassis and the drive is given to either front wheels or rear wheels. The engine is placed longitudinally or transversely on the frame.
2. *Engine at Centre*. Many vehicles, particularly air-conditioned buses (Royal Tiger World Master Buses) are fitted with engine at the centre of frame. Due to mounting of engine at center, a lesser road clearance is provided. These vehicles are suitable for running on plain pacca roads only. Quite good number of scooters (Kinetic Honda) and bikes have their engines fitted in the mid of chassis frame. Racing cars are fitted their engines at centre.
3. *Engine at Rear*. Some vehicles are having their engines fitted on rear portion of frame in India and abroad (Germany and U.S.A.). This rear side mounting eliminates the use of a long propeller shaft, but gear shifting and accelerating long linkages pose problems. Majority of scooters are mounted their engines at rear side. Some city buses and taxis are also fitted engines at rear portion of chassis.

FEW TERMS RELATED WITH PERFORMANCE OF I.C. ENGINES

1. *Cetane Number*. It is the percentage by volume of cetane in a mixture of cetane ($C_{16}H_{34}$) and α-methyl-naphthalene that has the same performance in a standard test engine as that of the actual fuel used. It shows the readiness of a diesel oil to ignite. High speed diesel engine should have cetane number greater than 50.
2. (*i*) *Octane Number*. The anti-knock or anti-detonation characteristics of Petrol/Gas engine fuel are measured in terms octane number. The octane number of a fuel (Petrol) is the percentage by volume of iso-octane in a mixture of iso-octane (C_8H_{18}) and *n*-heptane (C_7H_{16}), which will exhibit the same anti-knock characteristic of the fuel under test when tested in a standard engine under a set of standard test conditions. Its best values are 65 to 80.

(ii) *Highest Useful Compression Ratio (H.U.C.R.).* Petrol is also rated by H.U.C.R. It is the highest compression ratio at which fuel can be ignited under compression without detonation in a test engine under test conditions. But it is only applicable to a particular set of engines operating under similar conditions. It varies from 4 to 10.

3. *Firing Order.* Firing order refers to the sequence in which the charge in various cylinders of a multi-cylinder engine is ignited. The order of firing is so chosen to ignite the charge in various cylinders at the alternate ends of the crankshaft to balance the turning moments on it. This enables the crankshaft to be stressed almost uniformly along its length during a cycle of operation in the multi-cylinder engines.

Firing order for some engines:

For 3 cylinders → 1-3-2

For 4 cylinders → 1-3-4-2 or 1-2-4-3

For 6 cylinders → 1-5-3-6-2-4 or 1-4-2-6-3-5

In case of Vee engines the firing order is so selected that there is no concentration of subsequent explosions near one location of the crankshaft.

One of the Vee engines with 8 cylinders has following firing order →

1A-4B-2B-2A-3B-3A-4A-1B

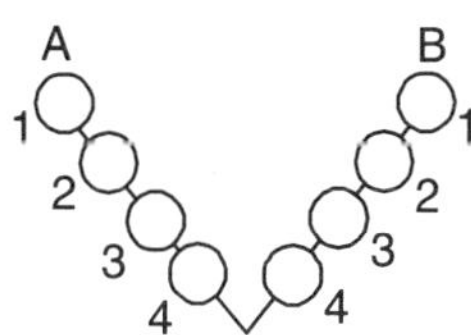

8 Cylinder Vee engine with 4 cylinders on each side—A and B

4. *Pre-ignition.* The ignition of air-fuel mixture should not take place before the spark occurs. If the ignition starts earlier due to any other reason, when piston is still completing compression stroke, the phenomenon is called *Pre-ignition.* It results in loss of power, violent thumping and mechanical damage to the engine.

5. *Detonation.* Just before the flame, started from spark plug, has spread across the combustion chamber, the remaining unburnt charge fires spontaneously without external source, resulting in rapid increase of local pressure which sets up pressure waves that hit the cylinder walls producing high notes of sound. It is this PING that manifests Detonation in S.I. engines. Few chemicals known as DOPES are mixed with petrol to control Detonation. Tetra-ethyl lead (one of the dopes) is mixed with petrol in ratio of 1 : 900 (one part of dope to 900 parts of petrol).

6. *Diesel knock.* In C.I. engines, the auto ignition and uncontrolled combustion of initially injected fuel, that is, the fuel injected during delay period causes knock. It is associated with high rise of pressure and metallic sound. Delay period is the time interval, in milliseconds, between the commencement of fuel injection and the beginning of ignition and combustion. In practice, it is as low as 0.006 seconds. Diesel knock is an undesirable activity.

7. *Drawbar Horse Power.* Some part of the brake horse power is consumed in overcoming various resistances acting on a moving vehicle. The actual power which is available for propelling the vehicle on the road, is called Drawbar Horse Power (DHP).

$$\text{DHP} = \text{BHP} - (\text{Power Lost in Resistances})$$

8. **Taxable Horse Power (THP).** In America in 1908, an agreed upon empirical formula for taxation purposes was promulgated:

(*i*) Taxable h.p. or R.A.C. rating $= \dfrac{D^2 \times n}{2.5}$

$= (\text{Dia. of engine cylinder in inches})^2 \times (\text{No. of cylinders})/2.5$

All auto vehicles owners were asked to pay road-tax to the government on the basis of calculation of road tax from the above formula.

The following assumptions were made in delivering the above formula for THP:

Speed of piston = 1000 feet per minutes.

Mean effective pressure = 90 pounds per square inch (1 pound = 454 gms)

Mechanical efficiency = 75%

[1 inch = 2.54 cms., 1 foot = 12 inches = 30.5 cms.]

The formula was later on accepted by almost all countries of the world. If we put D(diameter) in mm., then Thp = $D^2 \times n/1612.9$, n = number of cylinders in an engine.

The actual horse power of the engine is almost four times more than the taxable horse power. The indicated power formula in f.p.s. (foot, pound, second) system:

$$\text{I.H.P.} = \frac{P_{mep} \times LAN \times n}{3300}$$ was used to calculate the taxable power.

Where

P_{mep} = Mean effective pressure

L = Length of stroke

A = Area of piston = $\dfrac{\pi D^2}{4}$

N = R.P.M. of engine, n = Number of cylinder

The taxable power is also termed as R.A.C. *rating.*

(*ii*) Society of Automobiles Engineers (S.A.E.) made some changes in the THP. For comparision between auto vehicles on the basis of no. of cylinders and their diameter, S.A.E. derived the same formula by running the engine without generator, air cleaner and cooling fan. It was named as **S.A.E.H.P. rating.**

(*iii*) **DIN rating.** In Germany, DIN rating is used for the calculation of road tax on vehicles. The engine is run in actual conditions with generator, air cleaner and cooling fan and the horse power is determined. The value of power such obtained is converted to DIN rating by substituting standard temperature and pressure. It is a factual type of rating.

(*iv*) In India, the road tax is based on cubic centimeter capacity of the engine and cost of vehicle.

GOVERNING OF I.C. ENGINES

To regulate speed according to load on the vehicles is called *Governing.* Auto vehicles do not require a separate speed governing system. In automobiles the speed and load are matched by shifting the gears and regulating the speed by accelerator.

A governor is always fitted with fuel injector pump on a diesel engine used for other purposes than transportation. The governor automatically controls the fuel delivery to meet the operating conditions. It is the governor that directly controls the amount of fuel injected into the engine cylinder, according to the variation of load so as to keep the speed more or less constant. The adjustment lever movement changes only the setting of the governor to achieve the speed control within the desired range. If a governor is not used on stationary engines, the diesel engine may stall at low speed or it will run so fast to destroy itself on no loads.

Commonly two types of governors are used on I.C. engines:

- *Mechanical (Centrifugal type) Governor.*
- *Pneumatic Governor.*

Mechanical Governor: A mechanical governor has fly-weights that spin with the camshaft which also drives the fuel injection pump (F.I. Pump). The to and fro or up and down movement of the governor sleeve is transmitted through the control rack to the plunger of F.I. pump to adjust the quantity of fuel to the injector. The governor is incorporated within the assembly of the fuel injector pump. A Hartnell Mechanical Governor is shown in the Fig. 1.22. The balls or fly-weights fly upwards when speed increases and come downwards when speed decreases, the sleeve slides by this movement of balls.

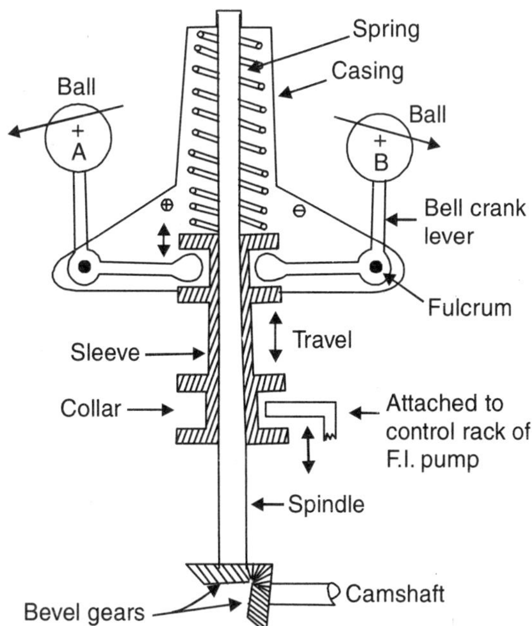

Fig. 1.22 Hartnell mechanical governor or centrifugal governor

The sleeve is also attached to the F.I. pump control rack and gear which partly rotates the F.I. pump's plunger in order to decrease or increase the fuel supply to injector.

Pneumatic Governor: Figure 1.23 shows the layout of a pneumatic governor fitted on the F.I. pump body. The throttle valve provides a vacuum signal to the governor. A hose from a venturi tube in the intake manifold connects to a vacuum chamber in the diaphragm. The control rack of F.I. pump is attached to the vacuum chamber diaphragm. Any change in vacuum influences the governor. As the throttle opens and closes, the varying vacuum causes the diaphragm and control rack to move. This movement rotates the plungers of F.I. pumps to vary the amount of fuel delivered. In a diesel engine, the air-fuel ratio varies from 100 : 1 at idle speed to 20 : 1 under full load. The governor keeps the air-fuel ratio within these limits. Richer air-fuel ratio causes unwanted smokes in the exhaust.

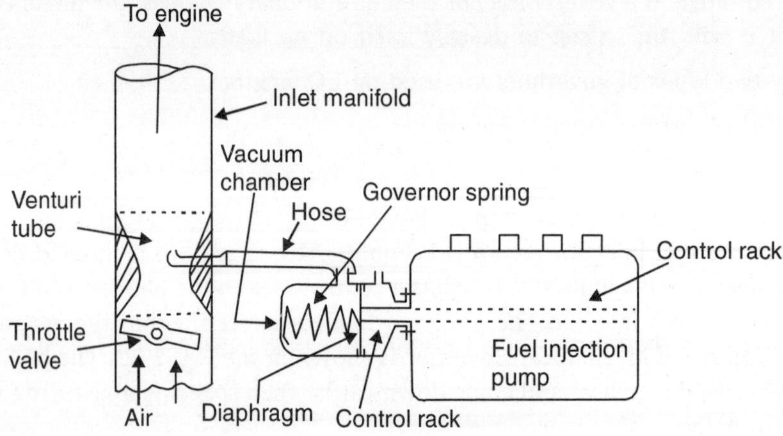

Fig. 1.23 Pneumatic governor

A governor regulates output of the engine by controlling the supply of the fuel or charge and maintains a constant speed at all loads. A fly wheel stores the available mechanical energy when it is in excess of the load requirements and supplies the same stored energy when it demanded by the engine. In short, flywheel dampens the fluctuations of speed during the thermodynamic cyclic operations in the engine cylinder.

HYBRID AUTO ENGINES

These are the engines running on more than one fuel.

GREEN is the new buzzword in the automobile industry today, and hybrids are the latest weapons touted to fight emissions and more importantly global warming. It is known fact that any vehicle, which combines two or more sources of power that can directly or indirectly, provide propulsion power, it is hybrid (Refer to Fig. 1.24). However, less known is that hybrid implementations are as varied as different kinds of engine layouts available today, with each manufacturer having a signature approaches to a hybrid. A hybrid passenger vehicle typically has an on-board Rechargeable energy storage system (RESS) combined with a fuelled power source such as an internal combustion engine or a fuel cell for propelling the vehicle.

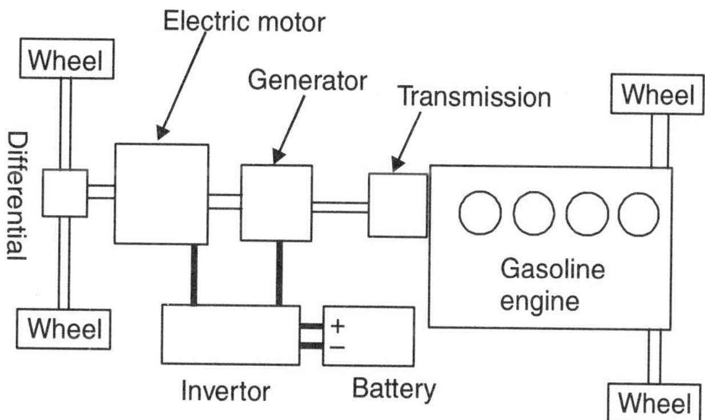

Fig. 1.24 Hybrid system (Gasoline + Electricity)

Simply put, a hybrid vehicle tries to combine the power and range of conventionally powered motor vehicles with the environment friendliness and economy of an electric vehicle. Or in other words, it attempts to significantly increase the mileage and reduce emissions of a conventional automobile while overcoming shortcomings of an electric car.

1. Electric and I.C. Engine Hybrid Vehicle

These are perhaps the most popular kind of hybrids, usually sporting a petrol engine, although diesels are slowly making their foray. However many so-called hybrids are available globally, where massive engines are mated to an electric motor, throwing all hybrid efficiency out of the window. Case in point here are makers of large SUVs and pick-ups, who seem to have gone on a spree hybridizing their entire line-up of V8s and V6s to clear their eco conscience.

The engine on an ideal hybrid should be sized closer to the average power requirement rather than the peak power what most conventional engines are sized for. The electric motor and the electronic controllers are probably the most sophisticated bits on a hybrid car. For instance, it can draw energy from the batteries to accelerate the car, if needed acting as a generator. While braking though, the electric motor that drives the hybrid can also slow the car, instead of relying solely on the brakes, conserving energy, which is otherwise wasted in the form of dissipated heat while braking. This way, the electric motor acts as a generator and charges the batteries while the car is slowing down, a process known as *'regenerative braking'*.

2. Pneumatic Hybrid (Air engines)

Pure CNG run vehicles are now common these days. Although hard to fathom, this novel hybrid uses compressed air to power a hybrid car. The flywheel is equipped with an electric moto-alternator, which in fact is a multi-functional device. This act as an air compressor, starting motor, alternator, an electric moderator/brake and even as a temporary power supply. Moteur Development International (MDI) in France is currently producing cars running on compressed air, called the *OneCAT.* These engines can be equipped with and run on the dual energies—fossil fuels and compressed air. While the car is running in fossil fuel, the compressor refills the compressed air tanks. This approach is similar to that of a hybrid-electric vehicle where braking energy is harnessed and stored to assist the engine as needed during acceleration.

Recently Tata Motors entered into an agreement with MDI France for Indian application and licensing of MDI's pneumatic technology. The agreement is being rumoured to enable development of a pneumatic engine for Tata's Nano, which means that Nano will have virtually no running costs, in addition to its extremely affordable acquiring costs, adding another feather in Tata's cap. In a different spec, the engine may find its way onto the new Indica as well. Maruti is selling its cars running on Compressed Natural Gas (CNG) and Petrol in Indian market.

3. Fuel Cell Hybrid

Fuel cell hybrids imply a hydrogen fuel cell generator teamed up with familiar lithium-ion batteries. However many significant technical hurdles need to be overcome before a fuel cell hybrid become a production viable option. Fuel cell vehicles remain expensive due to the high cost of lithium-ion batteries.

4. Hydraulic and Nitrogen Gas Hybrid

A hydraulic hybrid vehicle uses hydraulic and mechanical systems. A variable displacement pump is used in place of motor and a highly compressed Liquid Nitrogen Gas accumulator is used to store energy. The liquid Nitrogen Gas accumulator is cheaper and long lasting than conventional batteries.

DESIGN CONSIDERATIONS FOR I.C. ENGINES

Design of I.C. engines plays a very important role on their performance. I.C. engines are manufactured in a wide range of sizes—big ship engines to small industrial engines. Speed ranges from 50 rpm for very large engines to 5000 rpm for very small engines. Engine size for a typical four stroke S.I. engine is given below:

Bore (mm)	80
Stroke (mm)	80
Displacement (litre/cylinder)	0.402
Number of cylinder	4
Output/cylinder (kW)	10
Rated speed (rpm)	4800
BMEP (atm)	7.5

Diesel engines are favoured for their high efficiency over petrol engines. The efficiency of diesel engines is higher due to relatively high compression ratios (12 to 22) and lean mixtures. The power-to-volume ratio of the engine may be greatly increased by the use of turbo charging. By putting more air in the cylinder and fuel, *high brake mean effective pressure* (BMEP) can be achieved in the range 2.0 to 2.4 MPa. Diesel engines are heavy, noisy, and expensive and have serious emission problems.

The design of combustion systems for diesel engines depends on very complex processes that take place in diesel combustion. Some amount of swirl may be introduced by the intake port shape and as the piston moves upward, the cylinder air is pushed into the bowl of the piston producing a squish flow. For a naturally aspirated engine without a turbo-charger, a compression ratio of 16 will raise the cylinder gas temperature and pressure to about 560° C and 2.25 MPa at 20° crank degrees before TDC, which is a typical injection timing for best fuel economy at 2000 rpm.

Balancing and Vibration Control

The reciprocating motion of piston is converted into rotary motion of crankshaft through connecting rod. An I.C. engine consists of a piston which reciprocates, a crankshaft that rotates and a connecting rod that does both. Counterweights are placed on the crankshaft to balance the rotating forces while it is not possible for the reciprocating masses. The connecting rod mass can be equated to two masses, one at the gudgeon pin end and one at the crank-pin end. The sum of these masses must equal the mass of the connecting rod and the masses must be apportioned so that their mutual centre of gravity is the same as that of the connecting rod. Then the mass at the gudgeon pin end may be added to the mass of the piston to make a new reciprocating mass and the mass at the crankpin end can be added to the mass of the crankshaft for a new rotating mass. As a rule of thumb, the split of the two masses is approximately 1/3 and 2/3; 1/3 being at the gudgeon pin end. The counterweights on the crankshaft have to balance both the crankpin and the fraction of the mass of connecting rod assigned to the crankpin, that is 2/3 of the connecting rod mass.

The crankshaft can be balanced by two shafts, symmetrically located relative to the plane containing the cylinder axes, rotating in opposite direction at twice the crankshaft speed, each carrying an unbalanced weight symmetrically located from front to back of the engine.

Thermal Stresses Produced During Running

Components subjected to high rates of heat flow and thermal stresses are the cylinder head, the cylinder liner and the piston. Pistons will have some heat transfer problem in all engines. Air cooled cylinder heads and barrels always present a challenge and for a vehicle engine rating, meticulous design and high quality foundry work are required.

The designer should use stress analysis techniques to approximate sizing of simple components based on the temperature distribution in a cylinder head or the stresses in cylinder blocks. The engine should be designed using classical analysis, built and then be rigorously tested. Components which fail should be modified and retested. The whole process thus spans from evolution to the final product.

WANKEL ROTARY ENGINE

No significant developments have taken place in the internal combustion engine except for some modifications here and there since the inception of the cycles of operations introduced by Otto in 1887. The Wankel engine and its operation was first presented by Felix Wankel in 1954. The engine is purely rotary, having no reciprocating parts or pistons. It makes use of a rotor of triangular shape, but with curved sides. The chamber in which this rotor turns is of epi-trochoid form.

The rotor is mounted on 3 lobes A, B, and C and sealed tightly against the side of the oval chamber. The engine is water cooled.

Working of Wankel Engine

The four events, viz. suction, compression, power and exhaust happen simultaneously for every revolution of the rotor.

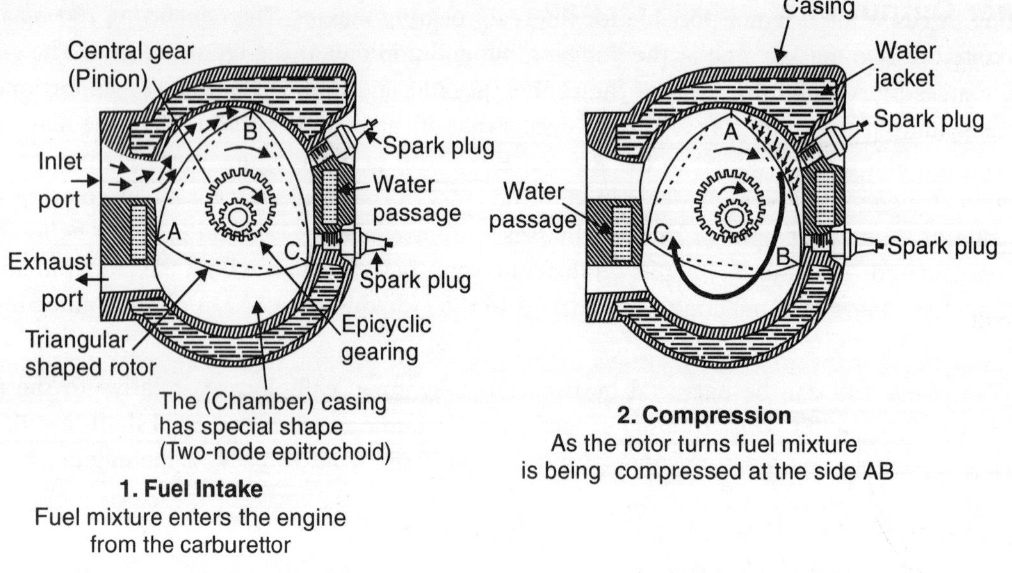

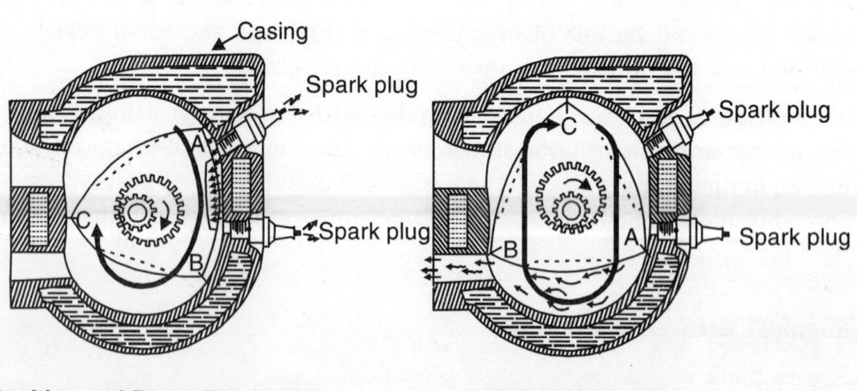

Fig. 1.25 Wankel rotary engine

Figure 1.25 (1) represents the fuel intake, in which the rotor side AB creates suction and fuel mixture from the carburettor enters the suction chamber. As the rotor turns clockwise, the fuel mixture is compressed between the rotor and the chamber followed by ignition and expansion as represented in Fig. 1.25 (2, 3, and 4). During the firing 'stroke' Fig. 1.25 (3), the mixture is burnt and as it expands, the leading lobe takes the thrust and turns around the central gearing, producing enough power for transmission. The cycle goes on in all the three chambers simultaneously.

It is now seen that there are three power cycles for every revolution of the rotor. The rotor revolves about its own axis while it rotates about the geometrical casing at $2/3^{rd}$, the speed of the main shaft. The eccentric motion of the rotor develops vibrations which are eliminated by the use of two symmetrically mounted flywheels.

Power Output of a Wankel Engine

Wankel engine's power output is higher but speed, being too high, creates problems. The power output for a Wankel engine per kilograms of its weight is considerably higher and running cost is comparatively lower than reciprocating engines.

The engine has separate working chambers (corresponding to the cylinders) for every revolution of the rotor, but its output shaft is driven at a ratio 3 : 1 making the rpm of the shaft three times that of the rotor. There are three power impulses for each revolution of the rotor. Eccentric or output shaft rotates at three times the speed of the rotor. The regulation of this speed is practically very difficult.

Wankel engines run at very high speeds. Excepting for this high speed, the other characteristics of ordinary I.C. engines are more or less the same for Wankel engines.

Advantages of Wankel Engine

- The Wankel engine is smaller in size in comparison with other reciprocating I.C. engines.
- Its volumetric efficiency is very high, often exceeding 100%.
- Many of the working parts like the connecting rod, crankshaft, valve mechanism etc. are absent in the Wankel engines. It can be made cheaper and constructional details are simple for mass production.
- As this type of engine does not have any reciprocating parts, the problems of balancing etc. do not arise. Road tests have proved that this engine is practically free of vibration or torque fluctuations.
- It is easier to balance revolving weight by the addition of counter weights.
- The power output per kg of engine weight is considerably higher.
- Running costs are lesser.
- Power is not wasted in overcoming friction.
- As very high speeds are obtainable, no overdrive is necessary.
- Correct timings for opening and closing of parts are achieved.

Disadvantages of Wankel Engine

- The very high engine speed is a disadvantage for automobile engines, and therefore, reduction in speed is necessary in transmission.
- Higher oil consumption per B.P. per hour.
- Lower torque at low road speeds.
- Braking effects of the engine is far less.
- Higher fuel consumption at lower speed ranges.
- Cylinder may distort due to close proximity of inlet and exhaust ports.
- Very high temperature of exhaust gas (about 600°C) poses lot of problems in the exhaust manifold and silencer.
- The designing of engine parts is a tedious job.

These are the reasons which have become a major obstacle in the production and use of Wankel in automobiles.

QUESTIONNAIRE

1. (*i*) What is an I.C. engine? (*ii*) Classify I.C. engines.
2. Draw a neat diagram of an I.C. engine and explain its working.
3. (*i*) Describe the construction of a split skirt piston of an I.C. engine.

   (*ii*) Show how the cooling of a wet liner is done.
4. Draw neat sketches and explain the uses of the following:

   (*i*) Piston rings, (*ii*) Flywheel,

   (*iii*) Crankshaft, and (*iv*) Cam shaft.
5. Draw a neat sketch of a connecting rod of an I.C. engine and describe its working.
6. Describe with sketches the working of a petrol engine operating on the 4-stroke cycle.
7. Explain the working of 4-stroke diesel engine with suitable line diagrams.
8. Explain the working of a 2-stroke petrol engine. How it differs from a 4-stroke engine?
9. (*i*) Give a comparison between a petrol engine and a diesel engine.

   (*ii*) What is Dual cycle?
10. What is the use of valve timing diagram? Explain any valve timing diagram.
11. What is pumping work? How it affects the net power obtained in any cycle?
12. With the help of neat sketches, explain the working of a valve lifting mechanism of a 4-stroke I.C. engine.

13. Explain the following terms:

    (*i*) Firing order, (*ii*) Pre-ignition,

    (*iii*) Detonation, and (*iv*) Diesel knock.

14. (*i*) What is Taxable horse power?

    (*ii*) What are R.A.C. rating and DIN rating?

15. Explain the working of a Hartnell Mechanical Governor.

16. Describe a Pneumatic Governor with a neat sketch.

17. Describe various design consideration for I.C. engine.

18. Explain the working of a Wankel Rotary engine with suitable sketches.

19. Describe any Hybrid engine working on more than one source of energy.

20. Explain the following terms:

    (*i*) Octane number, (*ii*) Cetane number,

    (*iii*) Compression ratio, and (*iv*) Port timing diagram.

21. (*i*) What is Drawbar Horse power?

    (*ii*) At how many places an I.C. engine can be mounted on the vehicle frame?

22. Write short notes on various aspects of design of I.C. engines.

Chapter 2

Fuel Supply and Ignition Systems

FUELS FOR AUTO ENGINES

Petrol

Gasoline or petrol is the most common fuel used in automobile engines. Sometimes *Gasohol*, a mixture of *alcohol* (10%) and *gasoline* is also used. Petrol engine vehicles may also be run on LPG (Liquified Petroleum Gas) with special feed kit. *Gasoline* is a hydrocarbon, made up largely of hydrogen and carbon compounds. It is made from petroleum crude oil by a refining process that also produces lubricating oils for engines, diesel oil and so many other products. During the refining process, several additives or *Dopes* (Tetra Ethyl Lead 0.001%) are also mixed with gasoline to improve its burning properties. A gasoline which is required for automotive engines should have the following characteristics:

- Proper volatility for quick vaporisation but should not be too volatile to cause vapour–lock in the line.
- Resistance to Detonation (high octane gasoline), Low-octane gasoline causes detonation.
- Oxidation inhibitors, which prevent formation of sludge.
- Anti-freezer, prevents carburettor icing and fuel line freezing.
- Detergents to keep carburettor and fuel system clean.
- It should not have 'Lead' when used in vehicles with catalytic converters.
- Some dye for identification from other oils.

Diesel Oil

Diesel oil is the fuel used in Diesel engines. It is produced by refining the crude petroleum oil. Diesel fuel is a light oil with proper volatility, viscosity and cetane number suitable for running diesel engines. *Cetane number* of diesel fuel may be defined as the ease with which the fuel ignites.

High cetane number fuel is preferred to avoid knocking. The diesel fuel must be very clean, even invisible dirt particles may clog the holes of injection nozzles and cause poor engine performance.

Now more refined diesel and petrol namely Power, Speed, or Xtra mileage fuels are available in the petrol depots. These fuels are having better octane and cetane ratios, thereby rendering better combustion and more mileage. Cars running on LPG give a saving of about 33% and those running on CNG give a saving of about 70% as compared to consumption of petrol and diesel oil.

Installing a diesel engine on the chassis of a petrol car is not allowed by R.T.O. The petrol car chassis is lighter than a diesel car chassis.

FUTURE FUELS

Sun Diesel

Sun diesel is a biogenic diesel fuel, produced from wood chips made up of shredded waste wood and shavings from trees rather than using food crops such as Soya-bean and rape seed. Using a multi stage process, the wood chips are first carbonised to form a biocoke and a low temperature carbonisation gas. These are used to produce the synthetic gas which is then converted into liquid sun diesel fuel by mean of a classic synthesis process called the *'Fischer-Tropsch process'*. Sun diesel generate very less emissions of harmful exhaust gases(very low CO_2) and particulate compared to crude diesel oil. It has a high Cetane number and therefore gives much better ignition performance. It has no aromatics or to say that sulphur and other pollutants in exhaust emissions are very low. It can be used without any change in the existing structure of conventional diesel engines.

Major environmental effects of diesel oil combustion

1. Green house effect which increases temperature.
2. Formation of photo-oxidants. The compounds that form the so called summer smog.
3. Eutrophication (over fertilisation) of bodies of water.
4. Acidification of land and water.

Sun diesel perform better in all the above environmental categories. It reduces 87% CO_2 emission, 90% reduction in photo-oxidants, 13% in over fertilisation and 27% in acidification.

Bio-Diesel Oil

A plant named *Jatropha* is being cultivated in India for extraction of bio-diesel oil which is being used after mixing with mineral diesel oil in varying ratios in locomotive diesel engines by the Indian Railways.

Hydrogen Gas

It will be ideal future fuel producing no pollution. But it is still a difficult and costly process to obtain hydrogen from water. Its storage also posses a great problem because it occupies very large space compared to petrol or diesel.

Solar Energy

Vehicles with solar panels are running under test conditions. With the exhaustion of petroleum products, the future equipments will be bound to run on solar energy.

Electrical Energy

Many bikes and scooters can be seen on roads running on battery fed electrical motors. Experiments to run four wheelers fitted with batteries and motors are going on. Batteries, which are quite expensive, have to be replaced after about two years of use.

FUEL SUPPLY SYSTEMS IN AUTOMOBILES

It is a must to have a proper supply system of fuel to the engine. The fuel is burnt in the engine cylinder and the heat of combustion of fuel is converted into mechanical energy by the engine. Mechanical energy is used as the driving force for the vehicles. There are two popular systems of fuel supply in auto engines:

Gravity Feed System

This system is used on scooters, mopeds and motor cycles only. The fuel tank is placed above the level of carburettor and the fuel flows to intake manifold by gravity as shown in the Fig. 2.1.

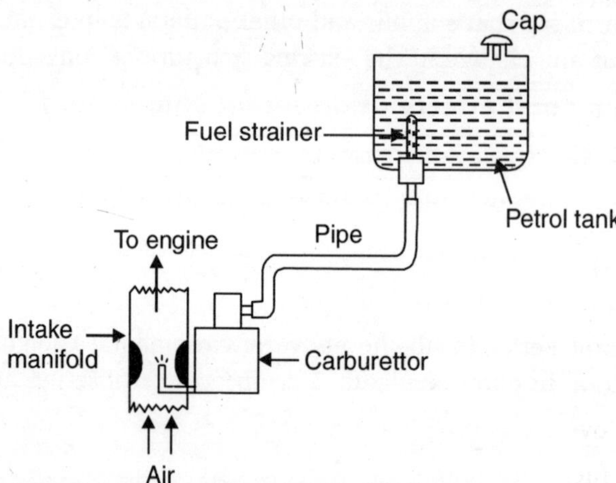

Fig. 2.1 Gravity feed system

Forced Feed System

This system is widely used on all four wheelers. In petrol run vehicles, the petrol is first filled into a fuel tank made up of steel sheet with anti rust property. The tank is generally fitted below the level of carburettor or MPFI system. From fuel tank, it passes through a small diameter pipe to a fuel filter. The filter removes the dirt in the petrol. The cleaned petrol now is sucked by a fuel pump (refer to Fig. 2.2) and then delivered to carburettor. The fuel pump may be operated mechanically as shown in the Fig. 2.5. Electro-magnetic and electrical fuel pumps are also being used on most of vehicles.

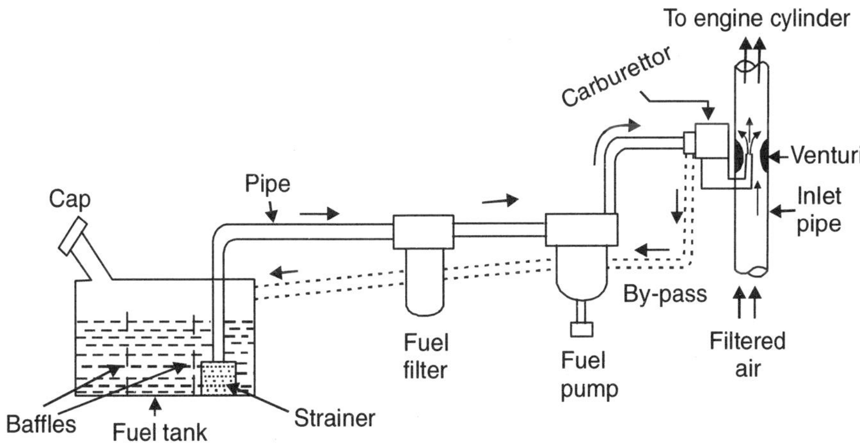

Fig. 2.2 Petrol feed system (forced feed type)

FUEL DELIVERY SYSTEMS IN S.I. ENGINES

Fuel delivery system for S.I. engines can be classified into two groups, carburetion and fuel injection, comprising four basic configurations: Conventional carburettor, Feed back carburettor, Single point or throttle body fuel injection and Multipoint or individual port fuel injection.

The feed back carburettor functions similar to a conventional carburettor, however, it is equipped with an actuator. The actuator allows the carburettor to modulate the air fuel ratio by controlling the fuel flow through its internal metering channels in response to signals from an onboard microprocessor.

A fuel injection system works in unison with a microprocessor to precisely control air fuel ratio by modulating injector pulse time. A single point fuel injection system uses a single injector in each bore of a throttle body mechanism. A multipoint fuel injection system uses a separate intake manifold runner mounted with injector for each cylinder.

The fuel injection system used in the present day modern car engines, offers the following functional advantages over carburetion:

1. Reduces air fuel ratio availability for improved emission control and performance.
2. Improves matching to specific operating requirements, including variations in altitude and temperature.
3. Reduces evaporative losses because unlike the conventional carburettor, there is no fuel bowl in MPFI system.
4. Prevents engine stalling due to fuel starvation, stopping and cornering.
5. Eliminates engine run on since fuel flow stops immediately when the engine is shut off.

DESCRIPTION OF MAIN COMPONENTS OF FUEL SUPPLY SYSTEM OF A PETROL ENGINE

1. Petrol Filter

Particles of dirt, dust and some other unwanted particles may be present in the petrol. The petrol should be clean and free from all the unwanted particles for the smooth working of the carburettor. The dirt particles are responsible for choking the carburettor jets. A petrol filter consists of a sediment bowl of steel sheet and a very fine strainer made up of Nylon, Copper, or, Bronze mesh which arrests the particles and allows only the clean petrol to pass. A simple petrol filter has been shown in the Fig. 2.3.

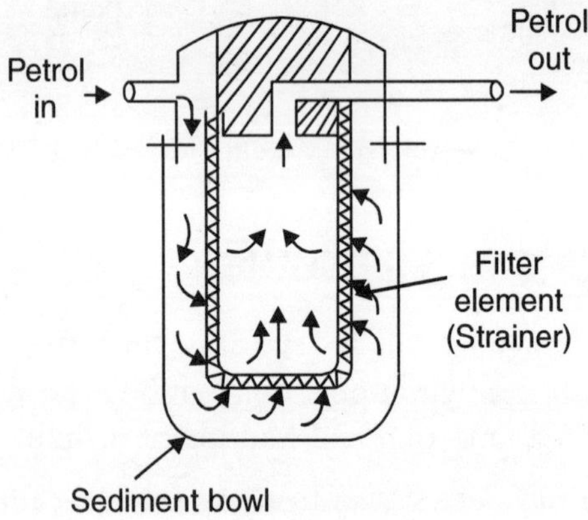

Fig. 2.3 Petrol filter

2. Diaphragm Type Fuel Transfer Pump or Mechanical Fuel Pump

The fuel system of a car uses a fuel pump to lift fuel from the tank and to supply the same to the carburettor or fuel injection system. Mechanically operated diaphragm type fuel pumps are commonly employed. These are operated by an eccentric mounted on the camshaft of the engine.

The fuel pump can be seen in Fig. 2.4. It consists of a spring loaded flexible diaphragm actuated by a rocker arm. The rocker arm is actuated by the eccentric. Non return valves are there in the inlet and outlet of the pump. These valves ensure flow of fuel in the proper direction. The internal parts of the pump are made of suitable materials to withstand exposure to fuel, oil, low and high temperatures also wear and tear, vibration etc.

As the rocker arm is moved by the eccentric, the diaphragm is pulled down against the spring tension. This movement causes a partial vacuum in the pump chamber. Because of low pressure and the construction of valves, the outlet valve remains closed while the suction or inlet valve opens. This admits fuel into the pump chamber. At the maximum position of the eccentric, the diaphragm reaches the end of its stroke. After this, further rotation of the eccentric will release the rocker arm. Now the rocker arm will simply follow the eccentric by the action of the return spring. The diaphragm spring will now push the diaphragm upwards and force the fuel to flow out, opening the outlet valve into the delivery tube. Now the suction/inlet valve remains closed. This action is repeated as the eccentric revolves.

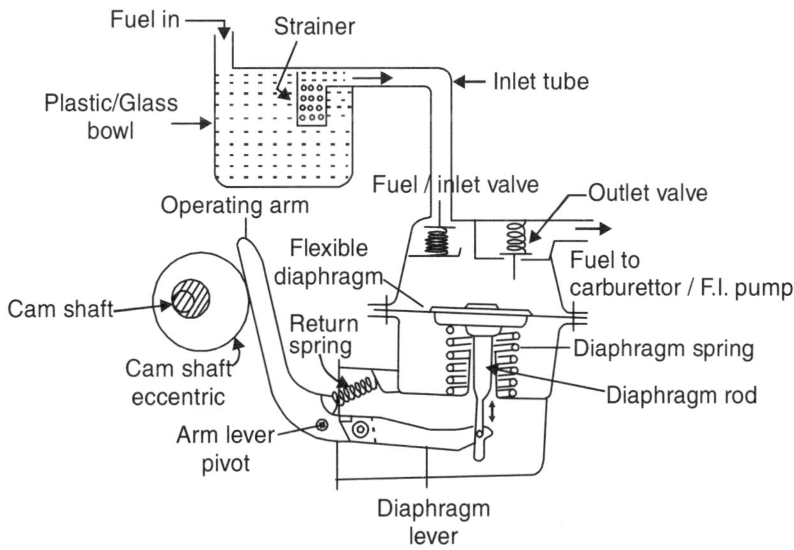

Fig. 2.4 Mechanical fuel pump

In these pumps, the downward movement of the diaphragm is caused by the rocker arm, while the delivery stroke is achieved by the tension of the diaphragm spring. The diaphragm spring is so designed that the fuel pressure is suitably balanced by the boyancy of the float system of the carburettor. As such, when the needle valve closes with carburettor bowl becoming full, the fuel pump cannot deliver fuel to the carburettor. In this case, the rocker arm simply continues to rock while the diaphragm remains at or near its end of travel. However, as the carburettor uses the fuel, the needle valve opens to admit fuel. Now the diaphragm moves down by the rocker arm action and sucks the fuel to deliver back the same when required. This self regulating feature helps the pump to deliver the correct quantity of fuel at all operating condition.

Sometimes lot of petrol vapour is formed in the pipe line and the vapours do not allow the flow of petrol in the fuel feed system. This phenomenon is called **'vapour lock'**. To remove vapours, the pipe line is first removed to release the vapours and then refitted. Electromagnetic and electrical fuel pumps are also being employed in many vehicles.

3. Carburettor

A carburettor is said to be the *'HEART'* of a petrol engine. To understand the functioning of carburettor, we shall first study a simple carburettor. A simple carburettor as such is not used in vehicles.

Working of a Simple Carburettor

The function of a carburettor is to atomize and meter the liquid fuel and mix it with the air as it enters the induction system of the engine, maintaining under all conditions of operation, fuel-air ratio suitable to those conditions.

All carburettor are designed according to the Bernoulli's Theorem,

$$V^2 = 2\,gh$$

where, V = velocity in m/sec

g = acceleration due to gravity in m/sec^2

and h = head causing the flow in meters of height of a column of a fluid.

The equation of mass rate of flow is given by,

$$\boxed{m = 1/\rho A\sqrt{(2gh)}}$$

where, ρ = density of fluid,

and A = cross-sectional area of fluid stream in m^2

In simple carburettor as shown in the Fig. 2.5, F is the float chamber for fuel. Inside the float chamber, a float is fitted with a needle to open or close the petrol port. A float is a hollow, lightweight and air-tight small box made of thin metal sheet or plastic. It is attached to float chamber through an arm hinged at chamber wall. The fuel is supplied under gravity or by a fuel pump. The fuel enters the float chamber through a strainer S. When fuel reaches a particular level, the needle valve closes the fuel supply. When the fuel level drops in the chamber, the float comes down, consequently the needle valve opens the fuel supply and the chamber is again filled with fuel. Thus the float and the needle valve maintain a constant fuel level in the float chamber. M is the main jet from which the fuel is spread into the air stream as it enters the inlet manifold P. When the inlet air passes through the venturi R, its velocity is increased due to narrow passage, which creates a low pressure at the main jet and causes the fuel to come up in the form of a fine spray. The fuel particles in the spray gets mix up with air and the mixture enters the engine cylinder. The fuel level is always maintained slightly below the outlet of the main jet when the engine is not operative.

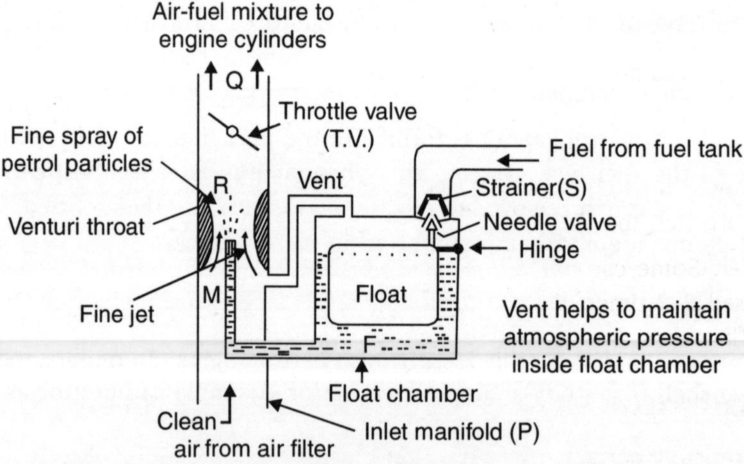

Fig. 2.5 Simple carburettor

As the piston moves down in the engine cylinder, suction is produced in the cylinder as well as in the induction manifold Q as a result of which air flows through P and the carburettor jet. The velocity of air increases as it passes through the neck at the venturi R and pressure decreases due to conversion of a portion of pressure head into kinetic energy. Due to decreased pressure at the venturi and hence by virtue of difference in pressure (between the float chamber and the venturi) the jet issues fuel oil into air stream. Since the jet has a very fine bore, the oil issuing from the jet is in the form of fine spray; it vaporises quickly and mixes with the air. This air fuel mixture enters the engine cylinder; its quantity being controlled by the movement of the throttle valve TV.

Limitations of a Simple Carburettor

- Although theoretically the air-fuel ratio supplied by a simple (single jet) carburettor should remain constant as the throttle goes on opening, actually it provides increasingly richer mixture as the throttle is opened. This is because of the reason that the density of air tends to decrease as the rate of flow increases.
- During idling, however, the nearly closed throttle causes a reduction in the mass of air flowing through the venture. At such low rates of air flow, the pressure difference between the float chamber and the fuel discharge nozzle becomes very small. It is not sufficient to cause fuel to flow through the jet.
- Carburettor does not have any arrangement for providing rich mixture during starting and warm up.

In order to correct these faults:

- Number of compensating devices are used for controlling speed and loads.
- An idling jet is used which helps in running the engine during idling.
- Choke arrangement is used for starting.

All these features are incorporated in *SOLEX* and *S.U. Carburettors.*

Devices for Ease of Starting

Some extra devices are provided on carburettor such as:

- *Choke*: Choke is provided on scooters and motorcycles for quick and easy start. It is a butterfly valve or choke valve. When operated, it chokes the air passage and a slight vacuum is created at the jet, due to which extra fuel is issued out and makes the starting mixture rich for easy ignition.
- *Tickler:* Some carburettors a spring controlled pin is provided on float chamber. When pressed 2–3 times, it strikes the float which then sends extra quantity of fuel to the jet which makes the mixture rich at the time of starting the vehicle.

Air-Fuel Mixtures

The theoretically correct mixture of air and petrol is 15 : 1 (by weight). Thus, the uniform supply of such mixture would result in burning without leaving excess of air and fuel. But it is difficult to get such a mixture in actual practice. When there is insufficient air, some of the fuel goes unburnt or simply changed to carbon which deposits in the combustion chamber. When there is too much air in mixture, it burins slowly and erratically and there is a loss of power.

There is however a *range of mixtures* between which combustion will take place.

- The *'lower limit'* is approximately 7 : 1 to 10 : 1. This mixture is barely explosive. It is called as *'Rich Mixture'*.
- The *'upper limit'* is approximately 20 : 1. The mixture burns irregularly. It is called as *'Poor Mixture'*.

The above limits will also vary with the characteristics of the fuel, the shape of the combustion space and the temperature and pressure in the combustion space.

Mixture requirements of automotive engines

- For *"average crusing speeds"* the air-fuel ratio is approximately from 15 : 1 to 7 : 1.
- In order to obtain 'maximum power' to be able to accelerate the engine quickly, a richer mixture of the ratio of about 12 : 1 is desirable. This is also called *maximum power ratio.* To start the engine from cold even a mixture richer than this but above the lower limit is obtained.
- For 'maximum economy', *i.e.*, less fuel consumption for unit power, the mixture ratio may be approximately 16 : 1 to 17 : 1. This ratio, however, *entails some loss of power.*

Hence, it is essential that the carburettor should be such designed that proper proportions of air and fuel are obtained to meet the varying operating conditions of the vehicle. Some of the commercially used carburettors are being described below:

(a) Solex Carburettor

The process of preparing a proper airfuel mixture in S.I. engine outside the engine cylinder is known as *carburetion.* The device used for this purpose is known as carburettor.

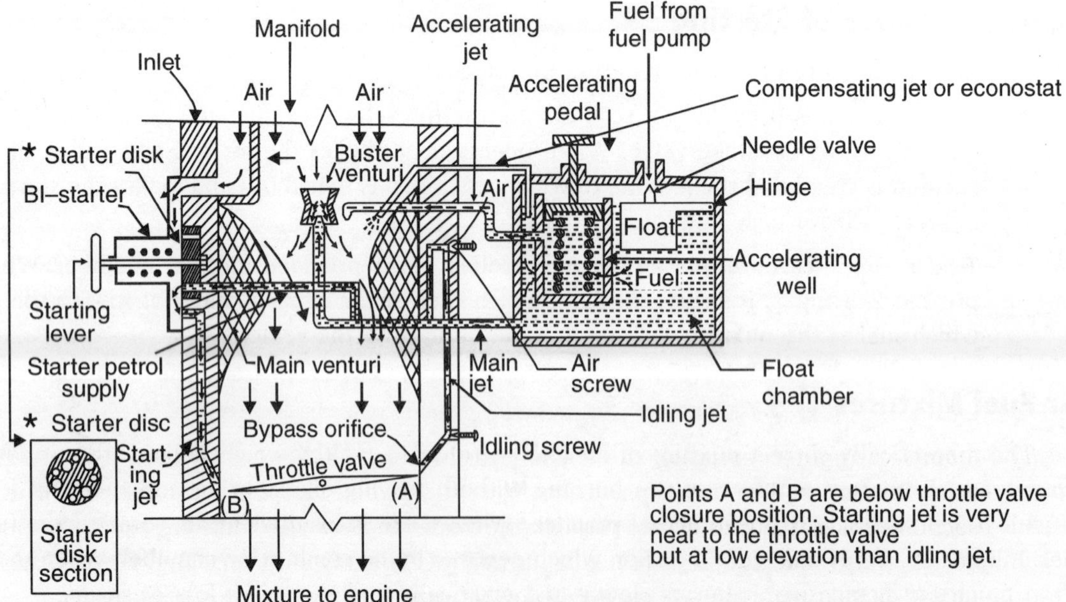

Fig. 2.6 Solex carburettor (down draught type)

Solex carburettor is the most common type of carburettor which is used in the petrol engines especially in Fiat and standard cars. It is famous for easy starting, good performance and reliability. This is also provided with basic jets required for starting and normal running. In addition to this, it is also provided with separate idling jet and accelerating jet. A simple diagram of Solex carburettor is shown in Fig. 2.6.

(i) *Starting Jet*: The unique feature of solex carburettor is the provision of bi-starter or progressive starter. Starter valve is in the form of disk with holes of different sizes. These holes are connected to the petrol jet. The air for starting is also supplied over the starting disk. The mixture passage is connected to the downside of the throttle valve. The quantity of fuel supplied through this disk depends upon either bigger or smaller holes which are opened in the starting passage. The starter lever which rotates the starter disc, when operated by driver, the holes are opened in the starting passage. At the beginning, very rich mixture is used by bringing bigger holes to open for starting, and as the engine gains speed, then smaller holes are brought in contact as the required richness of mixture is reduced. After the engine has gained the rated speed, the starter lever is brought to 'off' position.

(ii) *Compensating Jet*: With the increase in speed, the main jet as well as compensating jet deliver more fuel and mixture becomes richer. The characteristic of the compensating jet of making the mixture rich decreases within a very short period as the throttle opening increases. The characteristics of main jet and compensatory jet are shown in the Fig. 2.7. The discharge from this jet will take place only when depression within the Econostat body is sufficient enough to lift the metered quantity of petrol through Econostat jet from the float chamber upto the injector jet in the buster venturi.

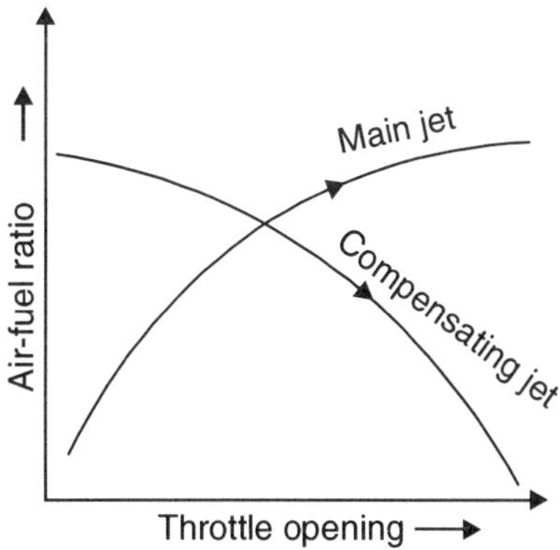

Fig. 2.7 Characteristics of main jet and compensating jet

(iii) *Main Jet*: Another feature of this carburettor is a provision of buster venturi in addition to main venturi. The fuel from the main jet is supplied into the buster venturi. During suction strokes the air is passed through the buster venturi where the air velocity becomes very high, so that the fuel injected from main jet is sprayed in very fine particles and is mixed quickly with the air entering into the engine cylinder through inlet valve.

(*iv*) *Idling Jet*: A rich mixture is required for idling jet. In solex carburettor, a separate idling jet is provided. A fuel connection from the main line leading to main jet is taken and opened to down side of the throttle valve through an air-screw. Fuel is drawn and mixed with air coming in through an air-screw and is supplied to the engine. The volume control screw controls the flow of air and also controls the richness of the mixture supplied with high accuracy.

To ensure the smooth transfer from idle circuit to main jet circuit without occurrence of the flat spot, a by-pass orifice towards venturi side near the idle jet is provided. As the throttle is wide-open, the suction near idle jet decreases, but the suction is applied now at the by-pass which then affects the loss of the suction at the idle port and thus flat spot is avoided. Flat spot is a 'No Charge' gap in between the spray stream.

(*v*) *Accelerating Jet:* For achieving sudden speed rise, a separate acceleration pump is provided. Whenever, acceleration pedal is pressed, an extra fuel is supplied through an accelerating jet. When the pedal is released, it is taken up by the spring provided in the accelerating well. This causes little vacuum in the accelerating well and petrol from the main float chamber admits the petrol into accelerating well and pushes it through a duct into the main venturi. And such the vehicle gets a sudden speed rise.

(*b*) *S.U. Carburettor*

The S.U. carburettor is an example of the constant vacuum carburettor. It employs a single jet of variable type in which a needle operates. The area of the throat of the carburettor is varied by means of a piston which slides up and down. The S.U. carburettor has been shown in the Fig. 2.8.

In this carburettor the area of venturi continuously remains varying by the up and down movement of piston attached with a needle valve. The needle valve moves up and down inside the main jet to regulate the supply of petrol into the entering air. The upper body which houses the piston is connected to atmosphere through a hole.

When throttle valve is partially opened, the piston stays in its down position and narrows the venturi. The incoming air, during stroke of the engine, takes out more petrol from the main jet for starting or idling. When throttle valve is fully opened, as shown in the Fig. 2.8, the inside air of the piston is also drawn out through a air-vent made in the piston which creates a partial vacuum above the piston. The atmospheric air lifts the piston in the suction chamber, thereby increasing the area at venturi and allowing more air to pass. The needle valve regulates the quantity of petrol spray according to load and speed.

When the accelerator is operated, the throttle valve (or needle valve) is opened and more petrol passes through the opening and mixes with the incoming atmospheric air in the venturi throat. The taper form of the needle is carefully made to give the required mixture strength during the different operating conditions. By changing the needle valve and adopting a suitable one, the same carburettor can be used on different engines of varied horse power, to some extent.

The main advantage of this type is the rapid response during acceleration and hence this type is fitted in racing cars and in general, in some of the scooters and motorcycles.

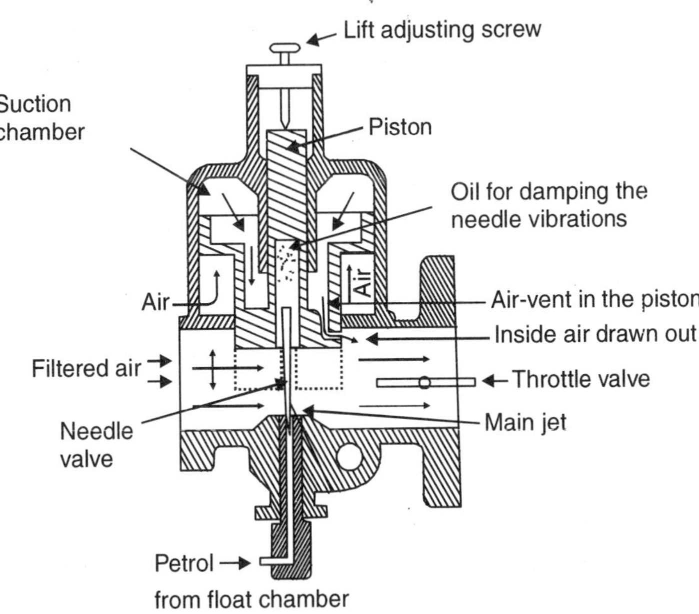

Fig. 2.8 S.U. carburettor (constant vacuum type)

Multi-Carburetion

Most engines with eight or more cylinders are provided with duel carburettors. In these there is only one float chamber, whereas there are two venturi tubes and two throttles. At times, two or three separate carburettors are synchronised to give the engine a balanced fuel supply. There are engines having four barrel carburettors also.

DEVELOPMENT IN CARBURETTOR DESIGN

In conventional carburettors, the air-fuel ratio vary about ± 5 over the operating range in actual running conditions. In multi-cylinder engines, cylinder to cylinder air-fuel ratio also changes.

In recent years quite a good number of improvements in carburettor design have come up. The aim is to achieve (*i*) metering of an exact quantity of fuel and air required for combustion at different speeds and loads and (*ii*) vaporising the fuel and (iii) mixing it with air for delivering to the individual cylinders of the engine.

Two latest carburettors are:

(*i*) *Sonic Carburettor:* It aims at improved fuel atomisation by increasing the velocity of air at the throats. It has following advantages:

(*a*) Excellent atomisation and proper mixing.

(*b*) Very small negative power required/wasted.

(*c*) Improved fuel economy.

(*d*) Very large reduction of air-pollutants.

(ii) *Ultra Sonic Carburettor:* In this carburettor, the fuel atomising function is performed by mechanical agitation of an ultrasonic driver mounted in the throat (venturi) to break up the injected fuel spray. It is more durable and has independence from ambient temperature for cold starting alongwith all advantages of sonic carburettor.

Atomisation of Fuel

When fuel leaves the carburettor's jet, it is more or less in the form of a stream. This stream is turn apart into ligaments. These ligaments break up and contract to form tiny drops of various sizes. This process is called atomisation of the fuel jet. These fuel droplets must be properly vaporised to the maximum extent in the inlet pipe and manifold itself. Atomised fuel spray forms a perfect gas when mixes with air.

It depends upon the following factors:

(i) Density of fuel

(ii) Surface tension of fuel.

(iii) Relative velocity of fuel and air stream.

The degree of atomisation is a function of the relative velocity squared.

Vaporisation of Fuel Spray

It is the process of changing liquid atomised fuel into vapour the vaporised fuel well uniformly mixes with air and burn effectively. To vaporise the fuel spray, following methods of heating inlet manifold are employed:

1. Outer surface of inlet manifold is heated by exhaust gases or by hot cooling water.
2. By contact of hot walls of cylinder head and piston head.
3. Mixing the fuel with some residual gases inside the engine cylinder.
4. By heat of the incoming air.
5. By compression heat of the air/charge. In diesel engines heat of compression atomises the fuel injected.

Down Draught and Updraught Carburettors

Most of the carburettors are of the down draught type in which the fuel mixture is drawn in a downward direction into the induction manifold and for this the air enters through the top of the carburettor. Down draught carburettors are kept above the engine level. Updraught carburettors were used in some of the earlier engines in which the air enters the bottom or side of the carburettor and the air-fuel mixture leaves at the top. It can be installed low at the side of the engine, but accessibility to it will be less for adjustments or repairs. At the same time, this method does not permit petrol entering in the cylinders when starting etc.

As the updraught carburettor is kept at low level near the engine, it is likely to become too hot and will result in excessive vaporisation and fuel loss.

Advantages of Down Draught System

- The fuel flow is unopposed.
- A large in-take manifold can be adopted. If a manifold of larger cross-section is used in the updraught type, the suction effect will be considerably less.
- The location of the carburettor above the engine, spares it from engine heat and fumes.
- The air which enters the carburettor is cooler.
- The carburettor is more accessible for inspection and adjustments.

Trouble Shooting

When engine performance is not satisfactory, trouble can be located by careful investigation. It should be remembered that automobile troubles of any kind can be relegated to one of the following categories.

Some mechanical troubles are easily found as they can be either seen or heard. Electrical troubles often result in complete stoppage of the vehicle without any advance warning, while the troubles in the carburettors are often preceded by spitting and misfiring before the engine finally stops. The electrical troubles can be immediately checked by removing the spark plug and testing it for sparks. If good sparks are obtained at the plug, it is probable that the system is in order and electrical trouble may be ruled out.

Systematic Checking

When the trouble has been located in the carburettor, the first step is to make sure that the fuel reaches the carburettor; if not, check the fuel supply lines, fuel pump and the petrol tank, if necessary. The defects in the carburettor are mainly due to the dirt entering the system with the fuel or air or it may be due to incorrect adjustment. A carburettor should be serviced every 5000 miles or 8000 km on the following points:

- The petrol pipe should be checked carefully.
- The filter should be removed and washed in petrol.
- All passages in the carburettor be cleaned by petrol and compressed air.

When it is found necessary to overhaul the complete carburettor, it should be done only by an experienced mechanic with the necessary tools. Disconnect the carburettor from the engine and fuel lines and dismantle it as far as possible completely and wash it in petrol until all the parts are clean. By blowing out with compressed air, any possible grit that may be present can be removed from small holes, jets etc. When the parts are re-assembled, use new gaskets and replace worn parts, if necessary.

Flooding

Flooding in carburettors occurs due to the flow of petrol from main jet. The causes are as follows:

- Too much petrol pressure in the float chamber on account of the fuel pump, or relief valve sticking.

- Improper fitting of the carburettor. When the jet level is below that of the float chamber, natural flow of fuel takes place.
- Damaged float.

When the float is punctured it may sink and the level of petrol rises in the float chamber which will cause flooding. A new float may be used, or for temporary adjustment, the hole may be soldered.

Engine Hunting

This is a phenomenon in which the engine runs in a series of surges and the speed of the engine fluctuates intermittently. It can be rectified by adjusting the slow speed air-screw.

Popping Back

Popping back through the carburettor occurs due to explosions in the carburettor. It may happen due to a weak mixture or a burnt or sticky valve. As the inlet valve opens, if burning gas enters inside the intake manifold, it is likely to explode there, causing a spitting or coughing sound in the carburettor. It is more likely that old engines will have greater tendency to do so due to incomplete combustion. It can also be caused by retarded ignition, improper seating of inlet valves, weak or broken valve springs; a crack between a pair of combustion chambers in the head, or a crack between cylinder bores.

Back Firing

A faulty exhaust valve or spark plug may leave partially burnt or unburnt gas in the exhaust manifold, which when it comes into contact with atmospheric air, bursts in the silencer. The causes may be too weak or slow running mixture or a faulty exhaust valve.

Trouble Shooting Chart

S. No.	Possible Causes	Remedies
A.	**Difficult Starting**	
1.	Incorrect throttle opening.	Check choke and throttle connections.
2.	Starting control not correct.	Check and adjust.
3.	Poor idling mixture adjustment.	Set for correct idle mixture.
4.	S.U. carburettor piston sticking.	Dismantle and service.
5.	Fuel leak.	Check and replace gaskets, if necessary.
6.	Fuel does not reach carburettor.	Check pipelines for vapour lock.
7.	Vaporisation of fuel in the float chamber.	Use choke while starting.
8.	Over rich mixture caused by pumping of the accelerator pedal.	Adjust.
9.	Flooding.	Adjust the main jet, float level, needle valve etc.

S. No.	Possible Causes	Remedies
B.	**Excessive Fuel Consumption**	
1.	Fuel leakage.	Check and adjust as required.
2.	Flooding (float chamber level too high).	Check and adjust as required.
3.	Accelerating pump too strong.	Check and adjust as required.
4.	Incorrect jets.	Check and adjust as required.

FUEL INJECTION SYSTEM IN DIESEL ENGINES OR C.I. ENGINES

The following are the functions of a fuel injection system in diesel engines:

- To filter the fuel.
- To meter or measure the correct quantity of fuel to be injected.
- Time the fuel injection correctly.
- To control the rate of fuel injection.
- To atomise or break up the fuel to fine particles for complete combustion.
- Properly distribute the fuel in the combustion chamber for smooth ignition.

All the injection systems are manufactured with great accuracy, especially the parts that actually meter and inject the fuel. Some of the tolerances between the moving parts are very small of the order of 1 micron. Such closely fitting parts require special attention during manufacture and hence the injection systems are more costly than other parts.

In compression ignition engines (diesel and semi-diesel) two methods of fuel injection are used.

- *Air injection system*
- *Solid or airless injection system.*

Air Injection System

In this method of fuel injection, air is compressed in the compressor to a very high pressure (much higher than developed in the engine cylinder at the end of the compression stroke) and then injected through the fuel nozzle into the engine cylinder. The rate of fuel injection can be controlled by varying the pressure of the injection air. Storage air bottles which are kept charged by an air compressor (driven by the engine) supply the high pressure air. This method is absolete these days and not in use any more because of being complicated and expensive.

Solid or Airless Injection System

The solid or airless injection is also termed as *mechanical injection.* In this system, a fuel pump is used which supplies a measured quantity of fuel to the atomizer or injector which injects it at a high pressure and a very high velocity into the engine cylinder in the form of sprays. The injection pressure varies from 100 to 145 bar (or even more in some cases). This pressure is produced by the fuel pump.

Fuel injection system consists of the following components:

- *Fuel Injection Pump (F.I. Pump)*
- *Injector or fuel atomizer.*

A schematic diagram of solid fuel supply system has been shown in the Fig. 2.9.

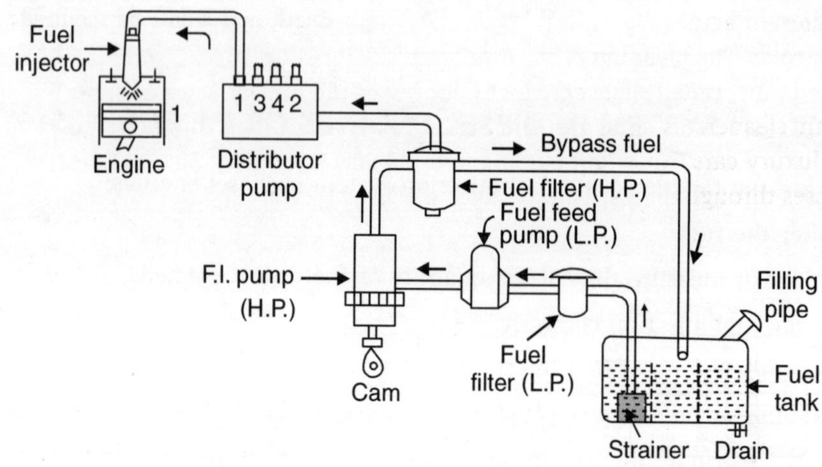

Fig. 2.9 Diesel fuel supply system (solid injection)

The diesel oil is filled into the fuel tank. The fuel feed pump (similar to mechanical fuel pump for petrol engine or a piston type pump) fitted at the chassis draws the oil through strainer and through a low pressure fuel filter, the oil passes on to the fuel injector pump, generally JERK—type, which delivers the fuel at a very high pressure (10 MPa to 20 MPa) to the fuel injector fitted in the combustion chamber of the engine. The fuel injector sprays about 150 cubic millimeter oil per opening at about 20 times per seconds. In some heavy oil engines a high pressure secondary fuel filter is also fitted after the fuel injection pump so that no dirt or foreign particles may pass to the injector nozzles. In multi-cylinder engine, a rotary distributor pump is used to supply the high pressure fuel oil to various fuel injectors fitted in the engine cylinders in the sequence of the firing order of the engine. Fuel feed pump and F.I. Pump with governor and distributor are fitted on a single assembly box. The rotary distributor pump sends fuel to the fuel injectors nozzles in the engine cylinders as the rotor inside the pump rotates and opens the passages according to the firing order.

Diesel Fuel Filter

Diesel injection pump and injectors may develop abrasive wear due to fine particles up to 2 to 4 m. The fuel filter is required to arrest particles of size above 10 m of rust, dust and other materials. Corrosion of injectors due to water in diesel oil is also very common. Fuel filters get choked largely due to formation of organic asphaltene wax-like substances. Water traps are provided on the filters at high pressure side and in vicinity of engine block to take advantage of engine's heat for evaporation of water particles in the fuel. It is similar to a petrol filter.

Air Filter/Cleaner

The air from the atmosphere is sucked into the intake manifold during the suction stroke of the piston. The air in the atmosphere contains lot of dust and dirt particles whose removal is very essential. In the petrol engines the cleaned air is passed on to the venturi around carburettor jets where it mixes with fine spray of petrol and form almost a perfect fuel gas, which is fed to the engine. In the diesel engines cleaned air is directly sucked, through Inlet Valve, into the cylinder during suction stroke. The cleaning of air is done in the *air filter* or *air cleaner* and passes through specially designed pipe, called silencer, which suppresses the hissing noise of entering air. There are three types of air cleaners as shown in the Fig. 2.10. Oil bath type filters are commonly used on heavy vehicles/luxury cars. Air when passing over oil surface, leaves dirt and dust particles in oil. The air then passes through the wire mesh which catches oil drops and the remaining dust particles.

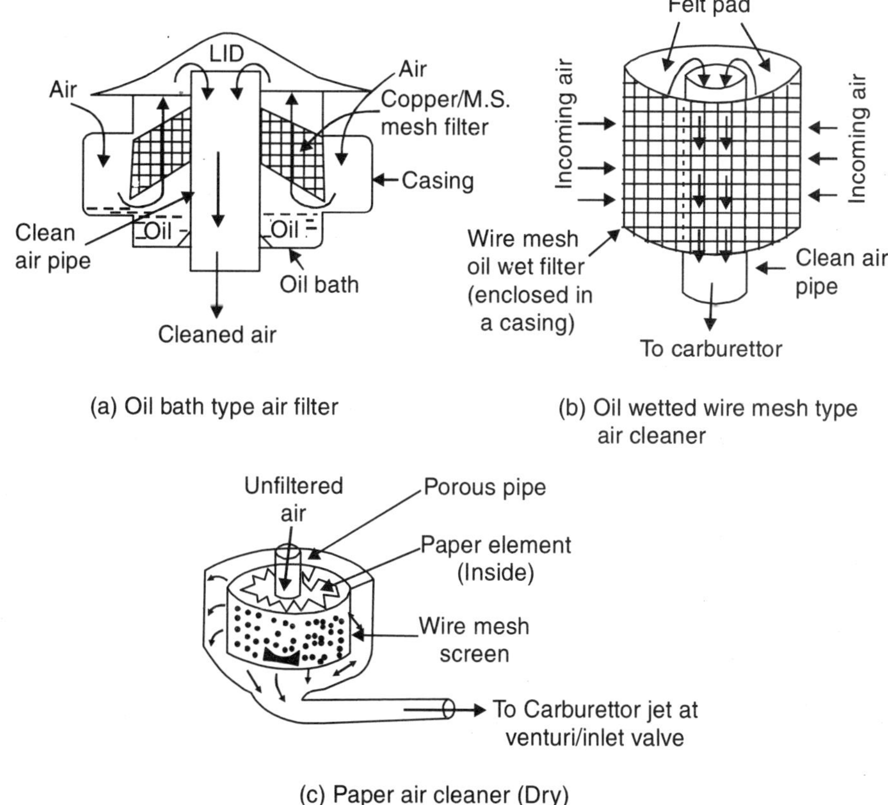

(a) Oil bath type air filter

(b) Oil wetted wire mesh type air cleaner

(c) Paper air cleaner (Dry)

Fig. 2.10 Different types of air filters/air cleaners

Pre-Air Cleaner

Vehicles running in high dust areas/deserts are fitted with a *glass bowl shaped pre-air cleaner.* In addition to usual air cleaner. The air first enters in the glass bowl of the pre-air cleaner and is given a cyclonic movement. The air leaves the heavy dust particles in the glass bowl at the filter mesh fitted at the outlet of the bowl. When glass bowl gets filled up with dust particles, it is unscrewed, cleaned and refitted. The pre-air cleaner saves the actual air cleaner from frequent cleanings.

Fuel Feed or Fuel Supply Pump for Diesel Engine

It is also called *Fuel Transfer Pump.* It delivers the fuel, through filters, to the fuel Injection Pump. Figure 2.11 shows a fuel supply pump of PISTON TYPE. It consists of a casing with a spring loaded U-shape piston. The piston rod is lifted by an eccentric mounted on cam shaft. When eccentric moves to downward position, the piston moves down and the fuel is sucked in through a suction valve from the fuel tank. When eccentric lobe lifts the piston rod the fuel is pushed out by opening discharge valve and is delivered to the F.I. pump. The by-pass pipe leads the excess fuel back to suction chamber. On the upper part of the fuel pump, a hand priming fuel pump is also installed for the bleeding of air bubbles which sometimes are trapped inside the fuel supply system.

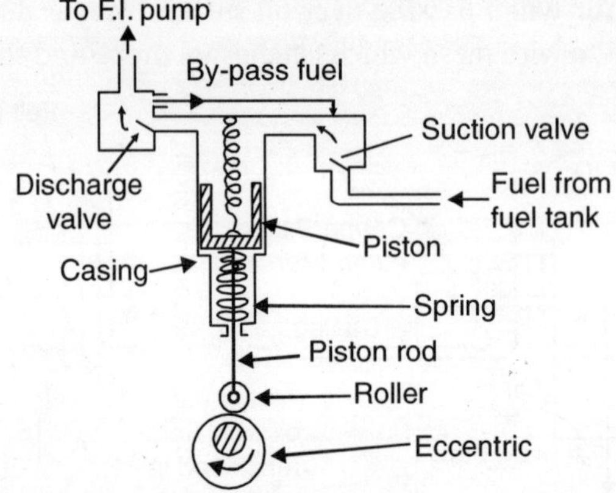

Fig. 2.11 Piston type fuel supply pump (diesel)

Fuel Injection Pump for Diesel Engine

A Jerk type F.I. Pump is shown in Fig. 2.12.

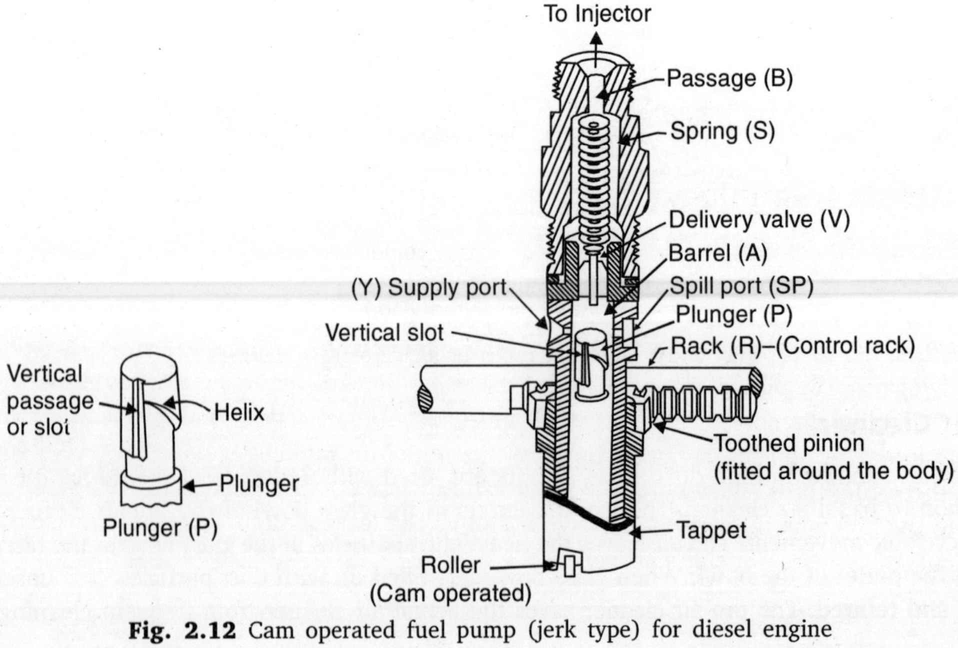

Fig. 2.12 Cam operated fuel pump (jerk type) for diesel engine

P is the plunger which is driven by a cam and tappet mechanism at the bottom. A is the barrel in which the plunger reciprocates. There is a rectangular vertical groove in the plunger which extend from top to another helical groove. V is the delivery valve which lifts off its seat under the liquid fuel pressure and against the spring force (S). The fuel pump is connected to fuel atomiser through the passage B, SP, and Y are the spill and supply ports respectively. When the plunger is at its bottom stroke the ports SP and Y are uncovered and oil from the low pressure pump is forced into the barrel. When the plunger moves up due to cam and tappet mechanism, a stage reaches when both the ports SP and Y are closed and with further upward movement of the plunger the fuel gets compressed. The high pressure thus developed lift the delivery valve off its seat and the fuel flows to atomiser through the passage B. With further rise of the plunger, at a certain moment, the port SP is connected to the fuel in the upper part of the plunger through the rectangular vertical groove by the helical groove; as a result of which a sudden drop in pressure occurs and the delivery valve falls back and occupies its seat against spring force. The plunger is rotated by the rack R which is moved in or out by the governor or by Accelerator pedal. *By changing the angular position of the helical groove (by rotating the plunger) of the plunger relative to the supply port, the length of stroke during which the oil is delivered can be varied and thereby quantity of the fuel delivered to the engine is also varied.* This activity is shown in the Fig. 2.13 (a, b and c):

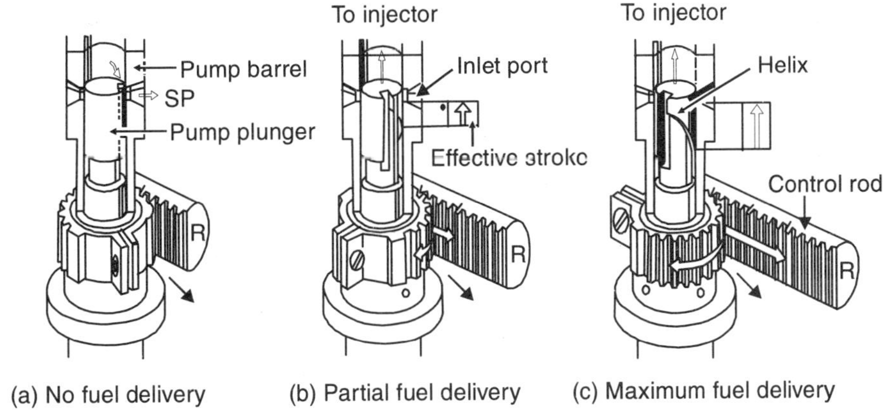

Fig. 2.13 Variation of fuel quantity—The movement of the toothed control rack turns the plunger to vary the amount of the fuel injected in the cylinder

Fuel Atomizer or Injector for Diesel Engine

The injector is shown in the Fig. 2.14. It consists of a nozzle valve V fitted inside the nozzle body (B). The nozzle valve is held on its seat by a spring 'S' which exerts pressure through the spindle E. 'AS' is the adjusting screw by which the nozzle valve-lift can be adjusted. Usually the nozzle valve is set to lift at 135 to 170 bar pressure. FP is the feeling pin which indicates whether valve is working properly or not. The oil under pressure from the fuel pump enters the injector through the passages P and C and lifts the nozzle valve. The fuel travels down the nozzle N and injected into the engine cylinder in the form of fine sprays. When the pressure of the oil falls, the nozzle valve occupies seat under the spring force and fuel supply is cut off. Any leakage of fuel accumulated above the valve is led to the fuel tank through the passage A and B. The leakage occurs when the nozzle valve gets worn out, this causes difficult starting.

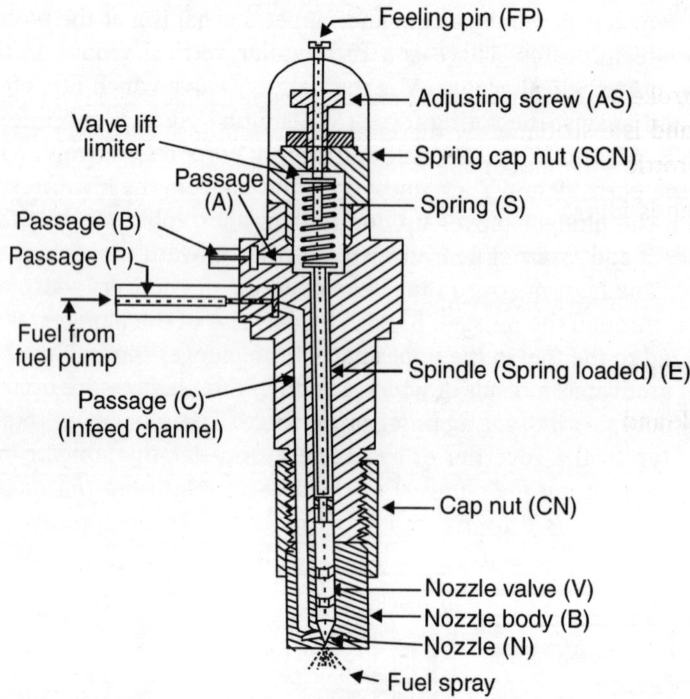

Fig. 2.14 Cross-sectional view of a fuel injector for diesel engines

MULTI POINT FUEL INJECTION SYSTEM (MPFI SYSTEM) IN S.I. ENGINES

Modern cars are invariably employing a more efficient fuel injection system called *MPFI system*. The *MPFI system* used in place of carburettor has been shown in the Fig. 2.15. The main components of the system are as follows:

Engine Control Module (E.C.M.)

It contains of at least two non-serviceable microprocessors with computer software programs to control power output, fuel system, air intake, exhaust gas recirculation, radiator fan control, air condition On/Off system, evaporative emission control system etc.

Air Intake System

It consists of the following parts:

- Air cleaner.
- Air intake pipe.
- Throttle body-controls the amount of air intake.
- Intake air temperature sensor (IAT).
- Idle air control valve (IAC): supply/by-pass air depending on the engine condition/ requirements.
- Fast idle air valve: when engine is cold it supplies the air directly to engine by passing the throttle body.

- Intake manifold.
- Manifold absolute pressure sensor (MAP).

During suction stroke of the piston of the multi-cylinder engine, the air passes from air cleaner to the throttle body and is distributed by the intake manifold into each combustion chamber. Air flow by-passes the throttle valve under the following conditions:

- When engine is idling.
- When engine is cold/started from cold state.
- When idle air control valve is open.

Fuel Delivery System

It has a fuel tank and electronic fuel pump, fuel filter, fuel pressure regulator, delivery pipe and fuel injectors.

Fuel in the tank under pressure is fed by the fuel pump duly filtered by fuel filters through the delivery pipe.

Fuel pressure within the fuel line is about three times higher than the air pressure in the manifold. Fuel is injected into the intake port when ECM signals the injector to open at the correct time.

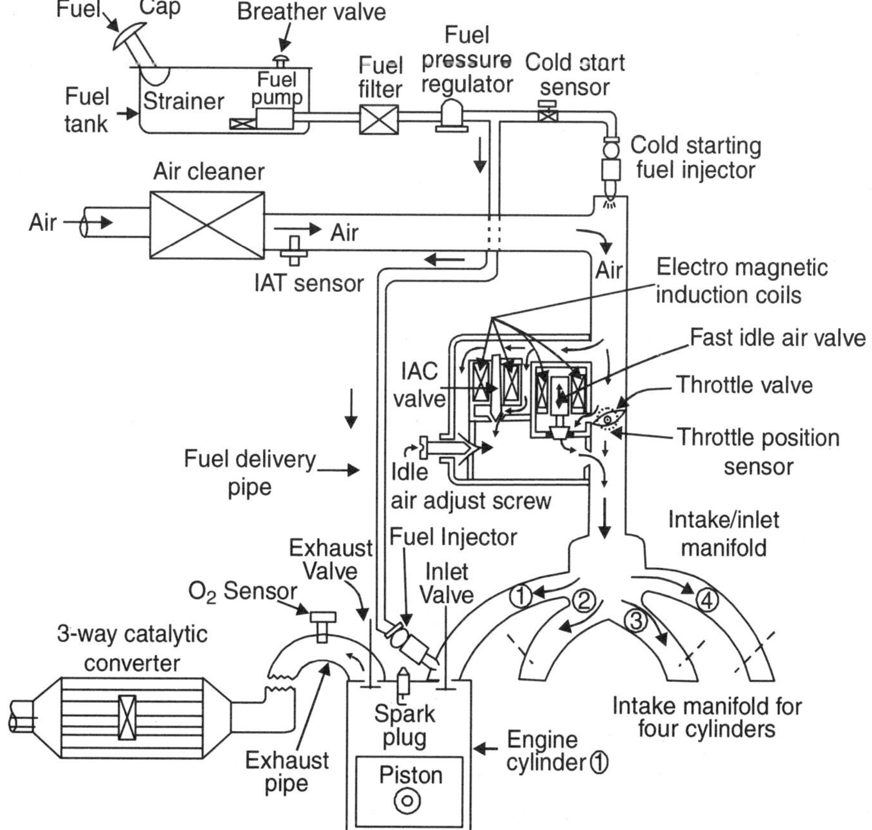

Fig. 2.15 MPFI system for 4-cylinder engine simplified line diagram

Excess fuel returns to the fuel tank via the fuel return line. (Not shown in the Fig. 2.15). There is one injector for each cylinder of the engine as shown.

The engine is operated by an electromagnetic system which when signaled by ECM, lifts the injector plug and allows the fuel to get injected through needle valve into the intake port. The fuel spray immediately mixes with air and vaporised and then enters the engine cylinder during suction stroke.

Electronic Control System

It consists of Engine Control Module (ECM), engine sensors, fuel pump control system, and throttle position sensor. It controls the quantity of petrol injected as per requirements of the vehicle particularly load and speed.

Catalytic Converter

The exhaust gases are passed through a catalytic converter (EURO–II and EURO–III) which absorbs harmful gases such as Nitrogen oxides, Carbon monoxide, Sulphur dioxide, etc. The least harmful exhaust gases are then released to the atmosphere. A schematic diagram of the catalytic converter is shown in the Fig. 2.16.

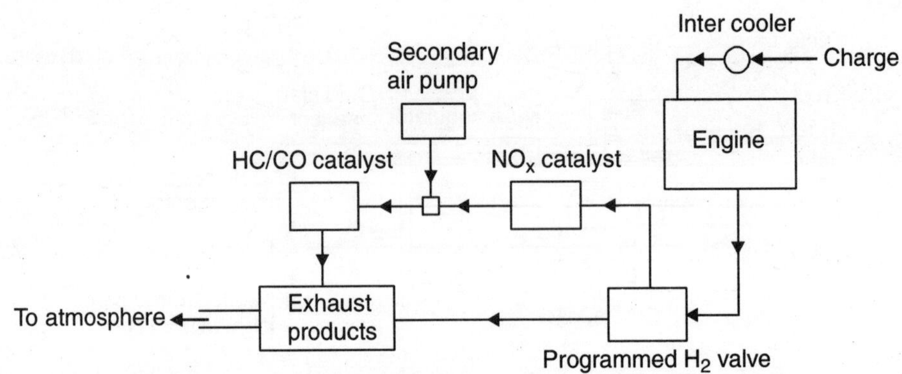

Fig. 2.16 Schematic diagram of a catalytic converter

Advantages

- EFI (Electronic Fuel Injection System) provides better atomization of the fuel mixture through the use of the fuel injectors.
- In carburetion system fuel is mixed by a single carburettor. EFI system introduces fuel at intake valve of each cylinder of the multi-cylinder engine. This allows the EFI system to provide better power and fuel economy.
- EFI provides a better control of ideal fuel air mixture over a wider operating range. This ideal air-fuel ratio called *'stoichiometric ratio'*.
- EFI provides fuel atomization during engine warm up (starting period) to prevent fuel 'padding' on the intake manifold.

- EFI provides better volumetric efficiency due to fewer restrictions on the intake route. High volumetric efficiency increases fuel economy and engine performance and reduces emission level (Level of harmful gases) from the exhaust.
- Cold starting and normal running is a better than carburetion system.
- EFI/MPFI system has better emission regulation because the optimum fuel injection and uniformity of the air-fuel ratio in all the cylinder is easily achieved.

Advantages and Disadvantages of Petrol Injection System

Advantages

- Owing to absence of any restriction (such as venturies and other metering elements in the air passage) there is increased *volumetric efficiency and consequently increased power and torque.*
- Better starting and acceleration.
- Engines having fuel injection system can be used in any tilted position (which will cause surface trouble in the carburettor).
- Higher compression ratios (higher by 1 to 1.5) can be employed (due to lower mixture temperatures in the engine cylinders).
- Blow break and icing are eliminated.
- Lower specific fuel consumption (since mixture to each cylinder is distributed more cffcctivcly and uniformly).

Disadvantages

- It is more noisy due to some moving parts and injection of fuel at intake valve.
- MPFI equipment is more bulky and heavy (than that of a carburettor).
- High initial cost (due to precise and complicated component assemblies).
- Increased service problems and servicing at authorised garages only.

Supporting Devices for Easy and Cold Starting of the Diesel Engines

Pre-heater or Glow Plug

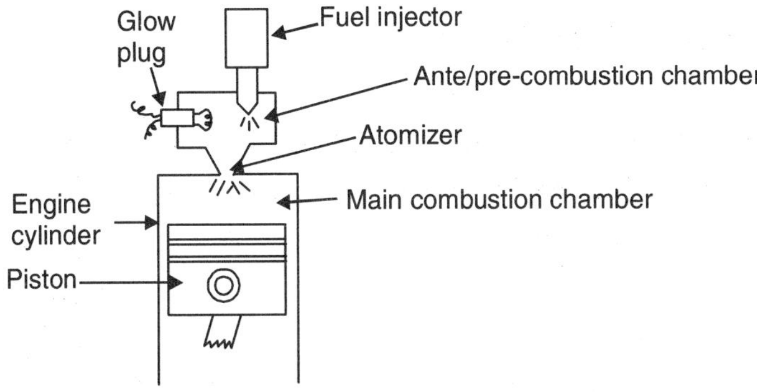

Fig. 2.17 Glow plug or heating plug or pre-heater

These are used on the diesel engine during cold weather. The plugs are fitted in pre-combustion chambers and are electrically heated by battery to enable the engine to take an easy start. Position of glow plug is shown in the Fig. 2.17.

Piezo Electric Fuel Injection System

These are used in luxury vehicles. They act as electronically controlled pump in fuel injection system. They improve cold starting, reduce noise and harmful exhaust emissions.

TURBOCHARGING AND SUPERCHARGING

At high speeds and also at high attitudes, the volumetric efficiency of the engine (during suction stroke) is reduced, which results in power loss. Some engines are provided with a special pump, fitted on the intake manifold, to deliver a 'supercharge' of air-fuel mixture or pure air to the engine. This increases the volume of air-fuel mixture (in S.I. engines) or the volume of the air sucked in the diesel engines which increases power output.

There are two types of supercharging devices for the auto engines. Both devices use a compressor or a rotary air pump. The device which uses a compressor driven by a belt from the engine crank shaft pulley is called a *Supercharger.* If the compressor is driven by the gas turbine run by the exhaust gas of the engine, the device is called as *Turbocharger.* Turbocharger is commonly installed on bigger car engines, especially diesel-engine cars.

In S.I. engines, supercharging also called boosting, is done only on aircraft and racing car engines and vehicles used on high altitudes. Supercharger is placed after carburettor. The increased inlet charge pressure also increases tendency to detonate. Supercharging pressure may vary from 1.1 to 1.5 atmospheres but not more than 2.5 atmos.

Working Principle of a Turbocharger

A turbocharger consists of a *turbine* and a *compressor* linked by a shared axle. The turbine inlet receives exhaust gases from the engine causing the turbine wheel to rotate. This rotation drives the compressor, compressing ambient air and delivering it to the intercooler which cools the air (hot air will cause a loss in power gained) then the air is delivered to the air intake manifold of the engine at colder and higher pressure, resulting in a greater amount of the air entering the cylinder, which results in high volumetric efficiency.

The objective of a turbocharger is the same as a supercharger; to improve upon the size-to-output efficiency of an engine by solving one of its cardinal limitations. A *naturally aspirated* automobile engine uses only the downward stroke of a piston to create an area of low pressure in order to draw air into the cylinder through the intake valves. Because the pressure in the cylinder cannot go below 0 at (vacuum), and because of the relatively constant pressure of the atmosphere (about 1 at), there ultimately will be a limit to the pressure difference across the intake valves and thus the amount of airflow entering the *combustion chamber.* This ability to fill the cylinder with air is called its *volumetric efficiency.* Because the turbocharger increases the pressure at the point where air is entering the cylinder, and the amount of air brought into the cylinder is largely a function of time and pressure difference, more air will be forced in as the inlet manifold pressure increases. The additional air makes it possible to add more fuel (if a turbo is attached without any other engine enhancements than it most likely will cause the engine to run lean, *i.e.*, too much air and not enough fuel), increasing the power and the torque output of the engine, particularly at high engine rotation speeds. A schematic diagram of a turbocharger is shown in the Fig. 2.18.

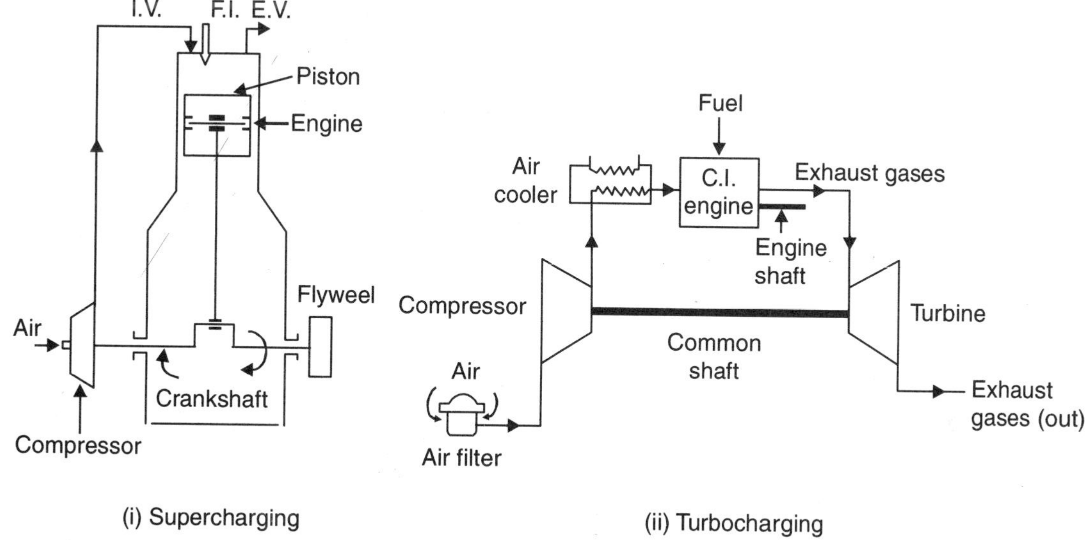

Fig. 2.18 Schematic diagram of a supercharger and a turbocharger

Because the pressure in the cylinder must not go too high to avoid pre-ignition and physical damage, the intake pressure must be controlled and this is done by a *wastegate,* which controls boost pressure by routing some of the exhaust flow, away from the exhaust side turbine. This controls shaft speed and regulates boost pressure in the inlet tract/manifold.

The application of a compressor to increase pressure at the point of cylinder air intake is often referred to *as forced induction.* The *Centrifugal supercharger* operates in the same fashion as a turbo; however, the energy to spin the compressor is taken from the output energy of the engine's rotating crankshaft. Superchargers and turbochargers use output energy from an engine to achieve a net gain, which must be provided from some of the engine's total output. In the case of superchargers, the energy is taken either directly or from a separate smaller engine or electrically driven motor from the main engine's generator current.

Factors Responsible for More Fuel Consumption in the Auto Vehicles

- *Running for short distance only.* Engine takes 10 to 15 minutes running time to become a fuel efficient engine or technically saying-to come out of cold running conditions. Cold running of engine consumes more fuel.
- *Running at high speeds.* Fuel consumption is increased at high engine speeds. It is always advised to drive at moderate speeds 50–60 km./hrs. Driving at low speeds also increases fuel consumption and may also cause knocking which shortens the life of the engine.
- *Too many starts and stops.* If you spend 10 minutes idling in a 30 minutes trip, your fuel consumption may double. Avoid frequent stops and starts.

- *Bad roads.* Bad roads not only shorten life of tyres but also increase fuel consumption due to frequent slowing down of speeds for passing potholes.
- *Weight of the car.* Heavy cars and or over loading also increase fuel consumption.
- *Engine tuning and timing.* Most of the cars of the present period are having MPFI system and Sequential Electronic Fuel Injection (SEFI) systems. These systems need very little tuning unlike old carburettors fitted engines. But the nozzles and injectors do need to be cleaned at prescribed intervals by the authorized service centres. Sometimes adulterated fuel is also responsible for faulty functioning of the injection system of the engine and more fuel consumption.
- *Fuel and oil filters.* Dirty filters also cause lot of troubles in engine running and increase fuel consumption. Get them cleaned/changed as per requirement at intervals prescribed by the manufacturer.
- *Air filters.* If you drive mostly in the dusty atmosphere, air cleaner gets choked frequently and needs cleaning regularly. The clean air increases life of the engine parts, besides helping in reducing fuel consumption.

SAVE ON FUEL

- Do not use clutch too much.
- Switch the vehicle off at red light stops/signals, if stoppage is more than 4 mts.
- Drive in gear corresponding to the speed.
- Get your vehicle serviced regularly.
- A poorly tuned engine can increase fuel consumption by upto 50%.
- Driving at 90–100 km/hrs. may increase the fuel consumption by at least 10%.
- A loaded roof rack will increase fuel consumption by as much as 5% in highway driving.
- Even empty roof rack can increase fuel consumption by 1%. If the carrier is attached to the vehicle, remove it when not in use.
- Avoid "Jack Rabbit" starts, not only do they increase fuel consumption, but are hard on your tyres. Anticipating stops and avoiding abrupt stops will decrease fuel consumption and increase the life of your brakes and tyres.
- Avoid unnecessary steering wheel movement since each sideward movement of the tyres causes fuel-consuming drag.
- Maintain a steady speed.
- Accelerate slowly on gravel or slippery roads.
- Avoid unnecessary braking. Speed control by accelerator is beneficial.
- Reviving the engine just before turning off the ignition costs extra fuel and may cause damage to the engine.
- Take the advantage of rolling resistance rather than heavy braking to help slow your vehicle. This deceleration technique is one of the best for fuel saving.

- Avoid using the air conditioner at lower speeds.
- Cleaning and changing spark plugs regularly can save a lot.
- Fill up fuel in your vehicle only in the morning hours when the ground temperature is cold. When temperature rises, the fuel expands and vaporises and lesser fuel reaches into the fuel tank of your car.
- Do not allow filling of fuel at fast speed. During fast filling some of the fuel gets vaporised.
- Fill the tank before it becomes totally empty. It is advised to fill the tank when it is half empty. In empty tank, air occupies lot of space and vaporises the fuel. The fuel vapours escape out when oil tank cap is opened. During the process of filling an empty tank, the dirt and dust sediments present in the tank get mixed with fuel and choke the fuel filters earlier.

Star Rating

In near future, all the new vehicles will carry '*Star Rating*' labels that will specify how much fuel each vehicle model consumes and how well each fares in comparison to others in the same cost range. Number of stars will show how much fuel efficient is your vehicle. The star rating will conform to a voluntary standard and labeling scheme for the automotive sector and will be first step towards enforcing mandatory fuel efficiency standards for the cars, two wheelers and other four wheelers including commercial vehicles.

IGNITION SYSTEMS IN S.I. ENGINES

The internal combustion engines working on *otto cycle* require a strong spark inside the engine cylinder to ignite the compressed mixture of air and fuel at the end of compression stroke. Therefore, an ignition system in automobile vehicles is must for igniting the compressed mixture of air and fuel. The ignition system produces a spark in each cylinder of the engine in the sequence of firing order without any 'miss'. The requirement of a good ignition system may be summarised as under:

- Spark at the spark plug electrodes must occur timely and continuously without misfire.
- The spark must be strong enough to ignite the compressed mixture of the air and fuel.
- The system should work at all engine speeds and loads.
- The occurrence of spark should not produce any interference with the radio, T.V., or any other electronic device fitted in the car.

The following three common types of the ignition systems are employed in the auto vehicles:

- *Battery and Coil Ignition System.*
- *Magneto Ignition System.*
- *Electronic Ignition System.*

Battery and Coil Ignition System

This system requires a 12 or 24 Volts battery and an ignition coil to boost up the voltage up to 15000 Volts to 20000 Volts for producing a strong spark across the spark plug gap. The battery ignition system consists of the following parts:

1. One 12 or 24 Volts battery.
2. Ignition switch.
3. Ammeter.
4. Ignition coil.
5. Distributor.
6. Vacuum advance mechanism.
7. Spark plugs.

Ignition Coil

It is a type of induction coil which steps up 12 Volts of battery to about 20,000 Volts so that a strong spark may be produced across the spark plug gap. It consists of two types of windings—Primary and Secondary. Both are insulated and wound around a soft iron lamination core. The primary winding is having lesser number of turns but made of a thick wire of 20 gauge diameter. The secondary winding of about 40 gauge diameter of wire, may have 100 to 1000 times more turns than in primary winding. The windings are contained in insulating materials inside a casing with terminals projecting outside. Some coils casings are filled with oil to improve insulation and cooling. The coils are sealed to prevent them from moisture. A simple diagram of an ignition coil is shown in the Fig. 2.19.

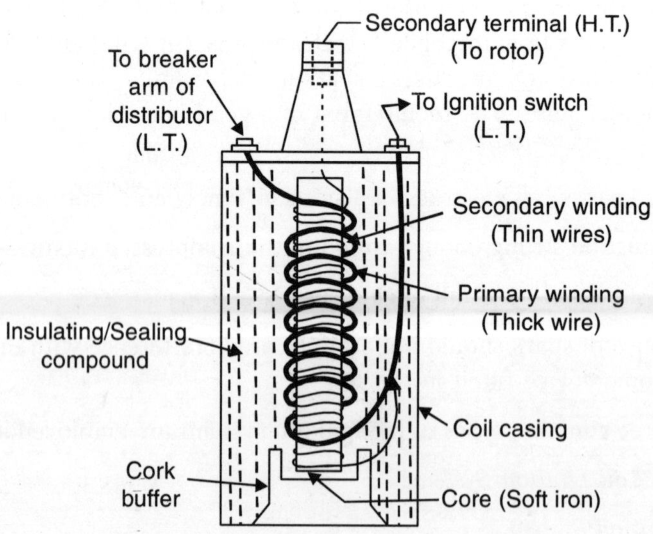

Fig. 2.19 Ignition coil

Distributor

It consists of a housing, a drive shaft with cam for breaker points, a centrifugal governor (Refer Fig. 2.20) for advancing (20° to 30°) or retarding the ignition according to load and speed,

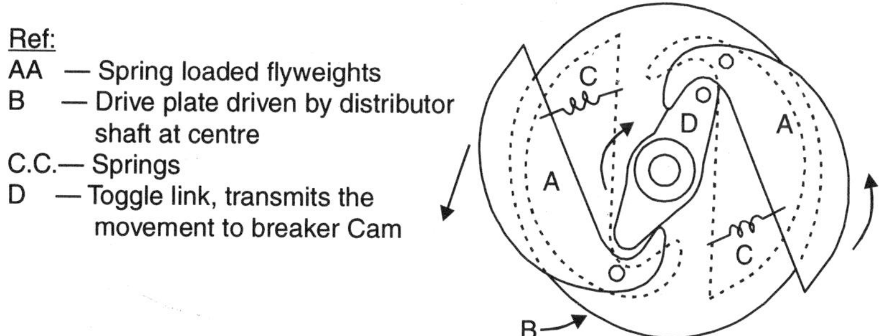

Fig. 2.20 Centrifugal advance mechanism

a breaker fitted with contact points, a condenser which saves contact points from pitting or burning, a rotor arm and a cap.

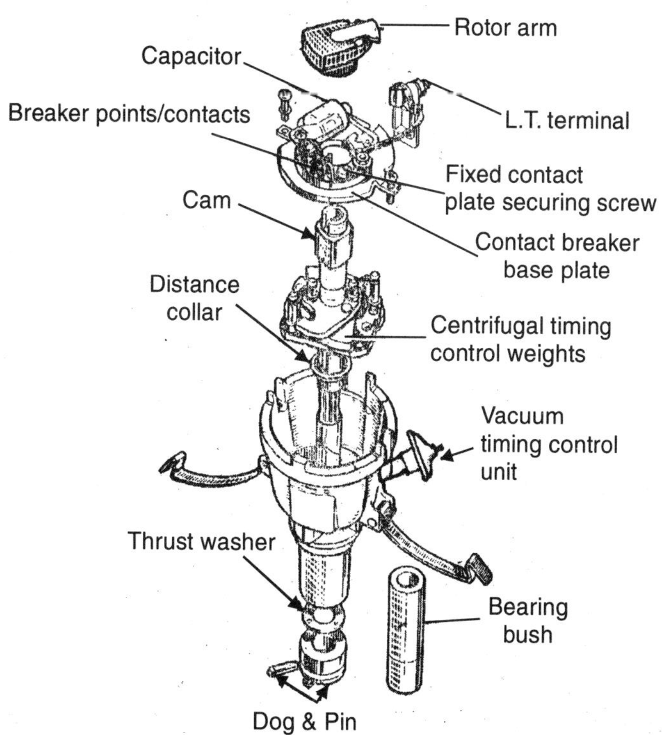

Fig. 2.21 Distributor

The cam shaft is rotated at half the engine speed in four stroke engine. An exploded view of the distributor is shown in the Fig. 2.21.

Working of Battery Ignition System

One terminal (generally negative) of the battery is ground to the frame of the engine. The other terminal of the battery (positive) is connected, through Ammeter and the ignition switch to one terminal of the primary winding of the ignition coil. The other terminal of the primary winding is connected to one of the contact breaker points operated by the cam. The other breaker point is ground to frame. A condenser is fitted parallel to the breaker points and provides a reservoir of charge for the current induced in the primary circuit at the time of contact breaking. It also gives faster break to the circuit when contact breaker points open, which results in a higher voltage in the H.T. circuit (Fig. 2.22).

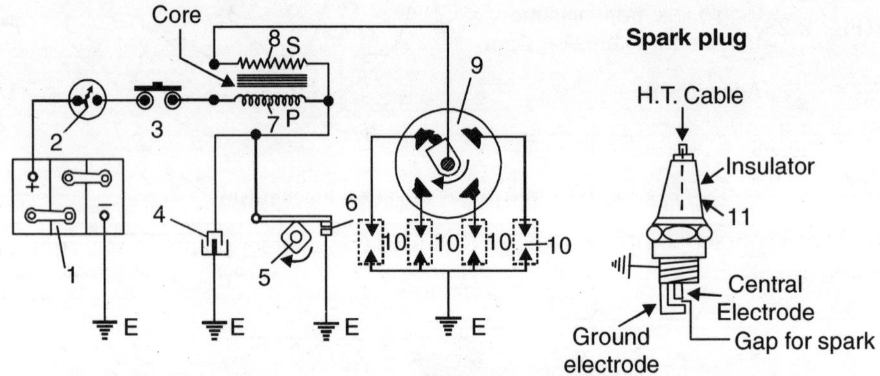

Fig. 2.22 Battery ignition system for a four cylinder S.I. engine
1. Battery; 2. Ammeter; 3. Ignition switch; 4. Condenser; 5. Cam; 6. Contact breaking points (opening 0.35–0.55mm); 7. Primary coil; 8. Secondary coil; 9. Distributor; 10. Spark plug 11. Spark plug (real shape).

The primary circuit of the ignition coil gets completed when contact points of the circuit breaker are closed and ignition switch is 'ON'. The gap between the breaker points is of the order of 0.35 mm to 0.55 mm. The secondary terminal is connected to distributor's rotor arm which according to firing order of the engine, touches distributor points and completes the H.T. circuit by joining with spark plug central electrode. The other electrode of the spark plug is earthed (E) to frame.

As the rotating cam breaks the primary circuit by opening contact points, a change of magnetic field takes place in the primary winding due to which a very high voltage of about 20000 + Volts is induced in the secondary coil. This high voltage, through distributor rotor arm is transferred to central electrode of the spark plug and a strong spark jumps across the gap in the spark plug. The spark such produced ignites the compressed charge inside the combustion chamber.

Dwell Angle: The angle of cam rotation, during which the contact points remain closed is called *dwell angle or cam angle.* It must be enough to allow magnetic saturation in the primary winding. The angle between the cam lobes depends on the number of engine cylinders.

The Battery Ignition System is universally used in auto vehicles due to its maintenance free and low cost characteristics.

Vacuum Advance Mechanism

During idling or partial throttle opening, a small amount of air-fuel mixture is sucked in the cylinder and low compression pressure is obtained, which restricts the quick ignition of fuel. In such condition the spark should be advanced more than that achieved by the centrifugal governor during low speed of the engine. The additional advance (about 10°) is obtained by a vacuum advance mechanism.

It consists of spring loaded flexible diaphragm fitted inside a small chamber. One side of the chamber is in contact with atmosphere through a vent and the other side is connected to inlet manifold at the throttle valve position. A linkage connects the diaphragm with breaker plate of the distributor (Fig. 2.23).

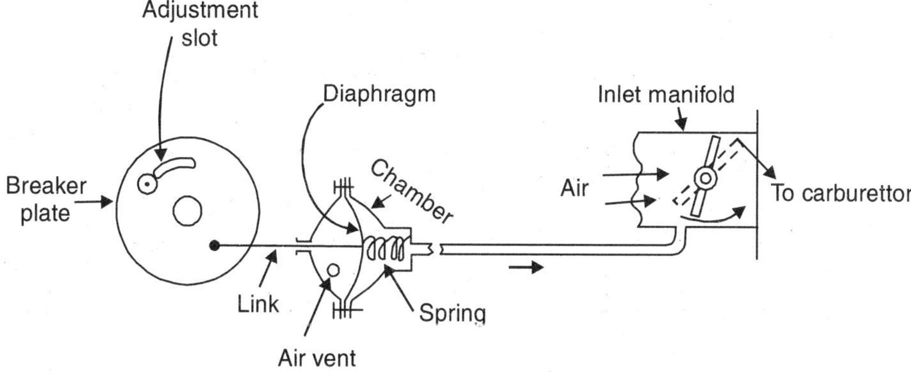

Fig. 2.23 Vacuum advance mechanism

When the engine runs at low speed (idling) the throttle opens partly and the vacuum inside the inlet manifold is quite high. The diaphragm is pushed by atmospheric pressure through vent towards the spring side. This motion of the diaphragm is transferred to base plate through linkage and consequently the cam closes and opens the C.B. points earlier during the compression stroke. The low compression pressure air-fuel mixture when ignited burns slowly, that is why more time is required to burn it up properly. This is achieved by vacuum advance mechanism.

Spark Plug: The spark plug provides a gap between the two electrodes placed in the combustion chamber of the engine. The proper gap between electrodes facilitates the discharge of a high potential electric spark that will ignite the air-fuel mixture at the desired point at the working cycle of the engine (Otto Cycle). The spark plug should have the following properties:

- It must provide a gap across which a strong spark may be produced at a rate of about 40 sparks/seconds.
- It must withstand sudden high pressure at least up to 55 bars.
- It should provide strong insulation between the two electrodes to prevent short-circuiting.
- It should resist erosion, and burning away of the spark points.
- It must be capable of bearing high temperatures up to 2500°C for longer periods.

- The insulating material must posses high electrical resistance, good thermal conductivity and a high mechanical strength to withstand fluctuating temperature and pressure.
- It must be fitted 'gas-tight' and all of its joints must also be 'gas-tight'.
- It must possess a high heat resistance so that electrodes do not become too hot to cause the pre-ignition of the charge.

A spark plug is shown in the Fig. 2.24 below.

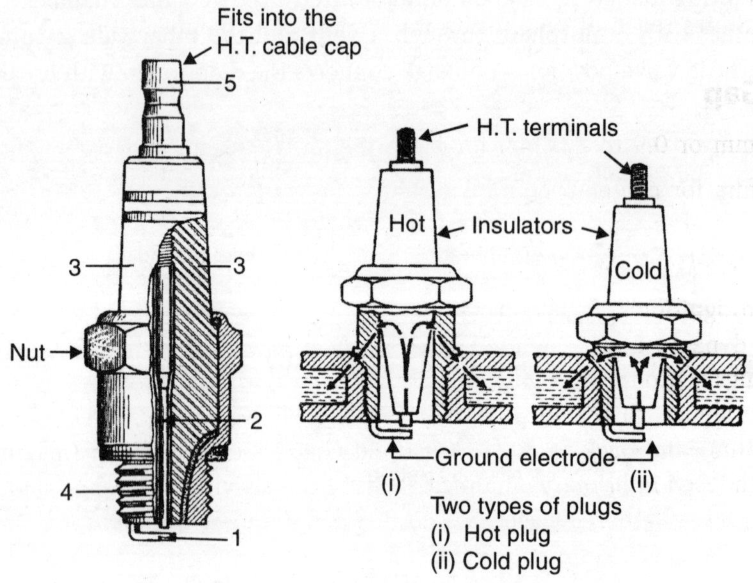

Fig. 2.24 Half section spark plug (1) Ground electrode; (2) Central electrode; (3) Insulator; (4) Plug thread; (5) Terminal (H.T.)

The spark plug consists of a metal shell having two electrodes which are insulated from each other and are having an air-gap of about 0.8 to 1.0 mm. The central electrode is enclosed in a ceramic or porcelain insulator. The earth electrode is welded to the under side of a metallic shell having threads on the outer surface. The threaded portion tightly fits into the combustion chamber wall. Electrodes are generally made of special Nickel alloy mixed with small amount of tungsten or chromium. The spark plugs are also identified by heat range or the relative temperature obtained during operations. They are termed as *hot running* and *cold running* spark plug. Hot running spark plug has a lower rate of heat transfer and its insulator tip is long while a cold running spark plug transfers heat rapidly from its firing end. A cold running spark plug has a short insulator tip and shorter heat rejection path and is used to avoid overheating. It is used in racing cars and heavy vehicles. The spark plug should not run too hot to cause pre-ignition and also should not run too cold to cause rooting and oiling up. It must have a suitable heat range. The manufacturers of vehicles specify the spark plugs to be used in a particular vehicle. The gap of spark plug electrodes is always adjusted by moving up or down the ground electrode. The central electrode is never touched otherwise the insulation will crack which will render the plug useless.

Plug Identification

Manufacturers of a particular model of engine prescribe the type of spark plug. Some spark plugs are marked like: W175Zl. In this,

W – States type of the threads used.

175 – States heat range in degrees celsius.

Z – States type of electrodes used.

I – States reach of plug inside the combustion chamber.

Spark Plug Gap

0.7 to 0.8 mm or 0.9 to 1.0 mm for battery ignition system and,

0.4 to 0.5 mm for magneto ignition system.

Magneto Ignition System

The magneto ignition system is similar to the Battery Ignition System except that it employs a magneto or a dynamo in place of battery. The magnetic field in the core of the primary and secondary windings is produced by a rotating permanent magnet between two poles of the core of the coil. As the magnet turns, the magnetic field is produced from a positive (maximum) to a negative (maximum) and back again. As this field changes from positive (maximum), a voltage and current are induced in primary windings. Rests of the activities are same as in Battery Ignition System. The magnetic system replaces heavy battery and therefore preferred in air craft engines and two wheelers. The magnetic system does not furnish enough voltage at low speeds therefore other starting and idling means became a necessity. A simplified circuit diagram of the magneto-ignition system is shown in the Fig. 2.25. Magnetos are generally 6 Volts or 12 Volts and deliver 50 to 80 Watts power in the two-wheelers.

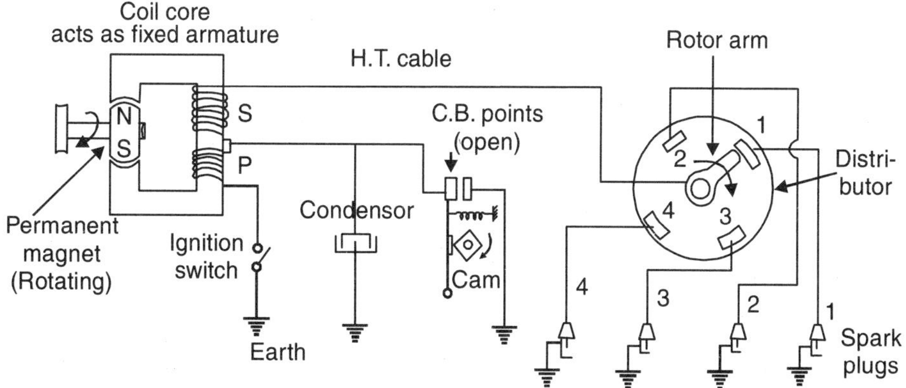

Fig. 2.25 Magneto ignition system

Trouble–Shooting Chart for Coil Ignition System

S. No.	Complaint	Causes	Remedy
1.	Engine misfires on one cylinder.	Detached high tension cable.	Attach high tension cable after cleaning.
		Broken wires in cable.	Replace the wires with new H.T. cables.
		Faculty spark plug.	Replace the spark plug.
		Large gaps in spark plug.	Adjust the gap to the specified value.
2.	Engine misfires irregularly on all cylinders.	Loose L.T. connection.	Tighten L.T. connection.
		Loose contact screw.	Tighten contact screw.
		Dirt or moisture on distributor or ignition coil.	Clean the distributor or ignition coil and remove the dirt or moisture.
		Gaps of spark plug are more.	Adjust the gaps to the specified value.
		Leaking Plugs.	Change the plugs.
		Worn out distributor brush and segments.	Replace them.
3.	Engine does not start. There is also no spark from the coil.	Discharged battery.	Charge/Replace the battery.
		Circuit is open in L.T. wiring.	Close the circuit.
		Dirty contacts.	Clean the contacts.
		Loose contact screw.	Tighten screw.
		Defective condenser.	Replace it.
		Defective ignition coil.	Replace it.
4.	Engine gets unduly hot.	Retarded ignition lever.	Adjust ignition lever to the correct position.
		Automatic governor not moving freely. (Disel engine)	Clean the parts of the governor.
		Faulty plugs/current leakage.	Replace them.
		Incorrect timing.	Adjust for correct timing.

Trouble-Shooting for the Magneto System

S. No.	Trouble	Causes	Remedy
1.	Engine does not fire.	Dirty contacts	Clean the contacts.
		Contact braker out of adjustment.	Adjust the gap of C.B. points.
		Incorrect starting settings.	Correct the starting controls.
		H.T. lead faulty.	Replace the lead.
		Fuel line choked.	Check and clean the line.

2.	Engine misfires.	Dirty or pitted contacts. Dirty spark plug. Contact breaker out of adjustment.	Clean the contacts. Clean the plugs and adjust the gap. Adjust the C.B. gap according to the specifications.

Comparison between Battery Ignition System and Magneto Ignition System

S. No.	Items of Comparison	Battery Ignition System	Magneto Ignition System
1.	Cost of manufacturing.	Low	Higher
2,	Maintenance	Easier	Not so easy
3.	Quality of spark at start.	Better	Poor at low speed
4.	Spark timing adjustment.	Not required	Must
5.	Weight, size, and volume.	More	Light, compact, less
6.	Reliability.	Lesser	More
7.	Repair.	Cheaper	Costly
8.	Suitability to high speed engines.	Less	More readily
9.	Need of a low tension (L.T.) cable.	Needed	No, *i.e.* less complexity
10.	Dependency.	Fully on battery, Stops working at discharged battery.	Independently contained
11.	Used on.	Cars, buses, and, SUV's.	Racing cars, motor cycles, scooters and, few air-crafts.

Few short terms related with engines

- MVs — Multiple Valves.
- DOHC — Double Over Head Cams.
- DFI — Direct Fuel Injections.
- MPFI — Multi-Point (or Port) Fuel Injection.

Transistor Assisted Ignition System

Conventional C.B. type ignition system's output depends upon two factors:

- The amount of current in the primary winding.
- The period during which the contact points remain closed or dwell angle of cam.

As the engine speed increases, the contact period goes on decreasing. The amount of current that can be interrupted by the conventional tungsten contact points is around 5 amperes for a reasonable contact point's life. The transistor assisted ignition system gives easier starting and better running at high speeds. In this system (Fig. 2.26), a special high voltage transistor relieves the contact breaker points from the function of making and breaking the primary circuit. The C.B. points do the light duty of turning the transistor "ON" and "OFF" at proper timings according to firing order. This prevents arcing at C.B. points and the need for a condenser is eliminated. The

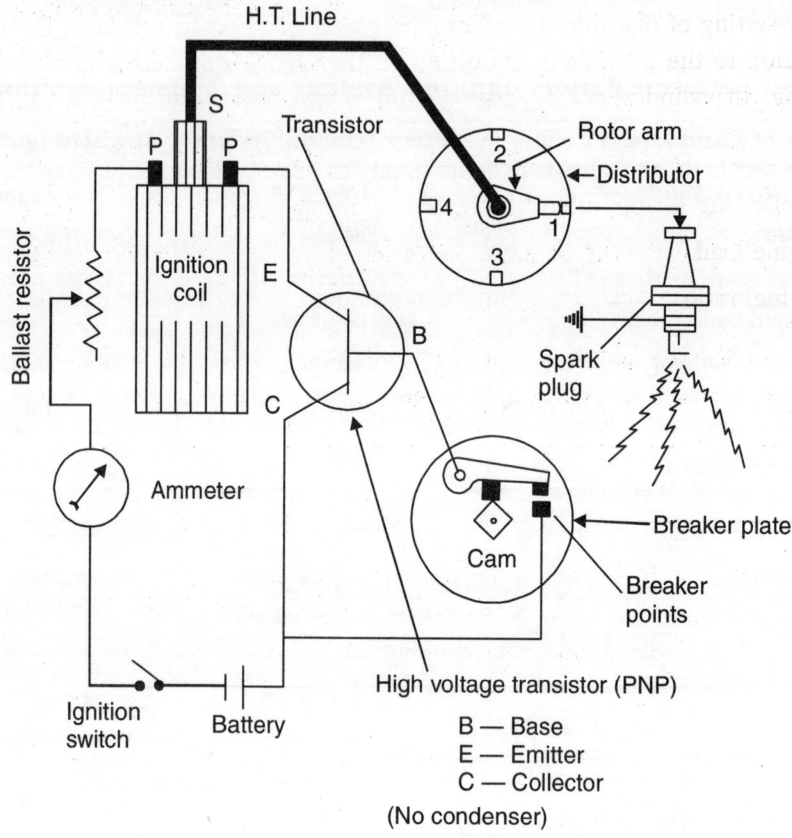

Fig. 2.26 Transistor assisted ignition system

controlled current is thus increased. The reduced amount of primary current at C.B. points during "ON/OFF" function of transistor, helps to increase the life of the contact points. The transistor unit can easily be installed into the Battery Ignition System, to convert it into a transistor assisted ignition system. A ballast resistor is also used in series with circuit to control the over-rise of the primary current, the ballast resistor increases the life of the C.B. points. During starting from cold, it is by-passed to allow more current to flow in primary circuit.

Fully Transistorised or C.B. Less Ignition System

In this system contact breaker points are also eliminated, that is why this system is also known as *'Breakerless'*. In this system the two stages of transistors are triggered by electrical pulses produced by an electro-magnetic pick-up. The field is obtained by mounting pole-pieces on the fly wheel of the engine. It is capable of producing well-timed sparks up to 1000 per seconds. The system is used on racing cars.

Ignition Timings

The charge must be ignited before TDC for complete combustions of fuel and to obtain maximum power.

Correctly setting of distributor with engine cam shaft after ascertaining the position of C.B. points in relation to the position of piston inside the engine cylinder is known as setting of the Ignition timing. In cylinder no. 1, when piston is fast approaching T.D.C. during end of the compression stroke, the timing mark on Flywheel or crank pulley aligns with the pointer, the position of contact breaker points is fixed as "ready to open". This process fixes the initial ignition timing. The following factors affect the ignition timing:

1. Engine load
2. Speed
3. Compression ratio
4. Air-fuel ratio
5. Cylinder bore
6. Temperature and
7. Quality of the fuel

Stroboscopic timing light is also used to correct or check the ignition timing. It gives a very correct ignition timing. It is connected with the circuit of the spark plug as shown in the Fig. 2.27. When there is spark in the cylinder, simultaneously, the light is flashed out at flywheel ignition

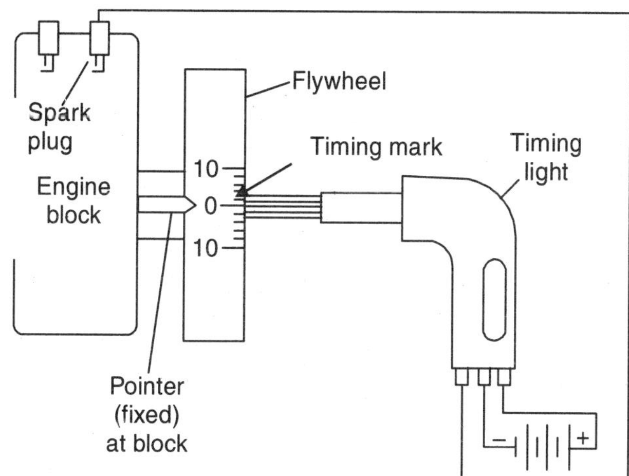

Fig. 2.27 Stroboscopic timing light

mark; the flywheel mark must look stationary at high speeds and correct timings. If correction is required, the distributor is adjusted by the turning it manually. Faulty ignition timing results in loss of power and faulty running of the engine. After overhauling of the engine, checking/setting ignition timing is must for maximum output.

QUESTIONNAIRE

1. Draw a neat sketch showing the lay out of a petrol feed system of a car and explain its working.
2. Explain air-fuel mixture requirement in the petrol engines. When is a weak mixture desired?
3. Describe, with a neat sketch, the working of a simple carburettor.
4. What is carburetion? How it differs from fuel injection? What are the functions of a carburettor?
5. What is vapour lock? How is it removed?
6. Describe a Solex carburettor with a neat sketch.
7. Differentiate between
   (*i*) Air injection and Solid injection.
   (*ii*) F.I. pump and Fuel injector.
   (*iii*) Fuel feed pump and Fuel injection pump.
   (*iv*) Air cleaner and Fuel strainer.
8. Describe with a sketch, the fuel feed system of a diesel car.
9. Draw a neat sketch of a fuel injection pump and explain its working.
10. Explain the construction and working of a fuel injector used in the diesel engines.
11. Differentiate between Supercharging and Turbocharging. How turbocharging is achieved?
12. What are the factors responsible for more fuel consumption in the car engine?
13. Suggest some points in order to save the fuel consumption.
14. Draw a neat sketch of a MPFI system used on today's car engines and briefly describe its function.
15. What are the advantages of MPFI system over the traditional carburettor's air-fuel mixture supply system?
16. What are the basic requirements of an ignition system?
17. Describe a Battery Ignition System with neat sketch.
18. Describe a Magneto Ignition System. What are its advantages?
19. Explain the construction of a spark plug. How are they classified?
20. What are the function of the following:
    (*i*) Condenser, (*ii*) Distributor, (*iii*) Ignition coil.
21. Describe with neat sketch the construction of an ignition coil.
22. What do you mean by ignition timing? How is it adjusted?
23. Describe a transistor assisted ignition system.

24. Describe the following:

    (*i*) Dwell angle, (*ii*) Specification of a spark plug, (*iii*) Use of Ballast resistor.

25. Describe the working of a stroboscopic timing light with sketch.
26. Describe the vacuum advance mechanism in an ignition system.
27. What do you understand by faulty ignition timing? What are its affect on the performance of the engine?
28. What do you understand by 'Firing order'? What is its importance?
29. Explain 'Ignition advance'. How and why is it advanced?
30. Describe the requirements of the good contact breaker points.
31. Differentiate between a transistor-assisted ignition system and a fully transistorized ignition system.
32. Why condenser is not required in a transistor-assisted ignition system?
33. What is the role of a spark plug in an ignition system? Describe its construction with a neat sketch.
34. Describe various methods of obtaining automatic ignition advance. Why is it needed?
35. Discuss the construction and working of primary and secondary windings of an ignition coil.
36. Give a comparison between Battery ignition system and Magneto ignition system.
37. What do you understand by:

    (*i*) Atomisation of fuel,

    (*ii*) Vaporisation of fuel.

CHAPTER 3

Engine Cooling Systems

NEED OF COOLING

Cooling system keeps the engine at its most efficient operating temperature at all speeds and under all running conditions. During combustion of fuel, the temperature inside the engine may go higher up to 2500°C+, which if not controlled, will damage the engine parts. Cylinder-wall temperature should not exceed beyond 260–280°C, otherwise, lubricants film will loose its lubricating properties, consequently the friction between piston and cylinder walls will increase abnormally and may cause piston-seizer, distortion of engine parts, cracking and melting of certain parts, bearings seizer, warping of valves particularly exhaust valve. The fresh charge entering the engine cylinder would get quickly heated leading to detonation, reduction in volumetric efficiency and engine output. Excess carbon will get deposited on the walls of the combustion chamber and on the crown of the piston which will lead to pre-ignition. Therefore it is essential to cool the engine. Cooling systems are designed to remove the excess heat (about 30 to 35%) from the engine cylinder and maintain the temperature around 250–260°C.

During combustion, the temperature of engine cylinder reaches an alarming state from 2000 to 2500°C. At a rough estimate, one-third of heat of combustion is converted into mechanical power, one-third heat goes out through exhaust gases and the remaining one-third energy is absorbed by the cylinder, cylinder head, piston and valves. Overheating of these engine parts results in engine break down, seizure of piston inside the cylinder, distortion of valves and cylinder liner etc.

Following factors affect the flow of heat into the various engine parts:

- Duration of combustion of fuel mixture/charge.
- Speed of the engine.
- Temperature during combustion.
- Design of combustion chamber.
- Turbulence or voilent air-circulation in the cylinder.
- Timing of ignition or fuel injection.
- Size of the engine cylinder.
- Surroundings temperature.

Overheating causes the following troubles in the engine:

- Evaporation of lubricating oil which causes piston seizure due to heat added by the excessive friction between piston and cylinder walls.
- Thermal stresses are set-up in the cylinder head and piston, which may develop cracks in the parts.
- Piston rings get stucked in the piston grooves leading to increased blow-by gases (gases leaking through piston rings into the crankcase) and loss of thermal efficiency.
- Damage to the piston crown and combustion chamber.
- Distortion and warping of exhaust valve.
- Pitting and burning of piston crown (top).
- Volumetric efficiency is reduced due to heating of incoming charge which results in the low power output.
- Overheated cylinder leads to pre-ignition in S.I. engines.

To overcome all the above problems, proper cooling of the engine parts is a necessity. But it is also to be remembered that excessive cooling is very harmful to the engine parts. It will reduce thermal efficiency and mechanical efficiency, there will be corrosion of engine parts and improper vaporisation of fuel which will increase wastage of fuel.

TYPES OF ENGINE COOLING

Air Cooling: Small engines used in the two-wheelers, *i.e.* scooters and motorcycles are cooled by atmospheric air. To have effective cooling, the surface area of the engine cylinder and cylinder head is increased by casting the fins (extended surfaces) over their outer walls/surfaces as shown in the Fig. 3.1. Fins provide additional radiating and conducting surfaces. Air cooling system is

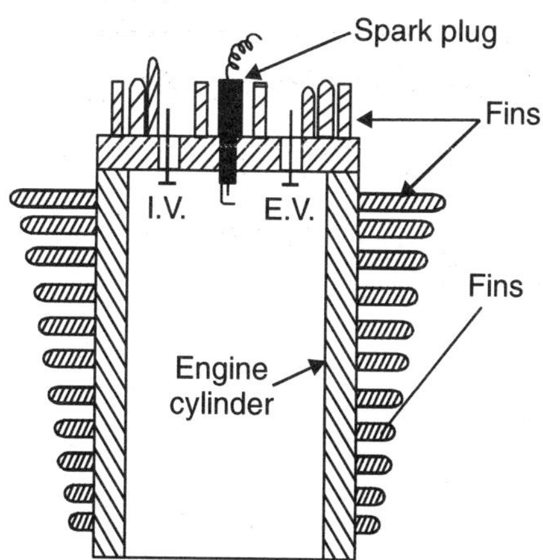

Fig. 3.1 Air cooled engine

simple, light in weight, cheap and no extra parts are required for cooling. For air-cooling to be effective, movement of the vehicle is essential, otherwise, an external air blower must be employed in stationary engine.

Water Cooling /Liquid Cooling: Almost invariably all the four wheelers employ a more efficient system of cooling, that is, water cooling. The water absorbs heat about four times more than air. In automobiles, a fixed amount of coolant or water is used and circulated for cooling upto very longer times. Atmospheric air is also needed for cooling one of its part-the RADIATOR. A pressure water cooling system with thermostat valve is shown in the Fig. 3.2. Referring to the figure, the liquid coolant or water is filled in the system through radiator cap. The coolant reaches the lower tank, from where the water is circulated in the water jacket by the water pump. The water pump shaft is driven by the crankshaft through pulley and belt system.

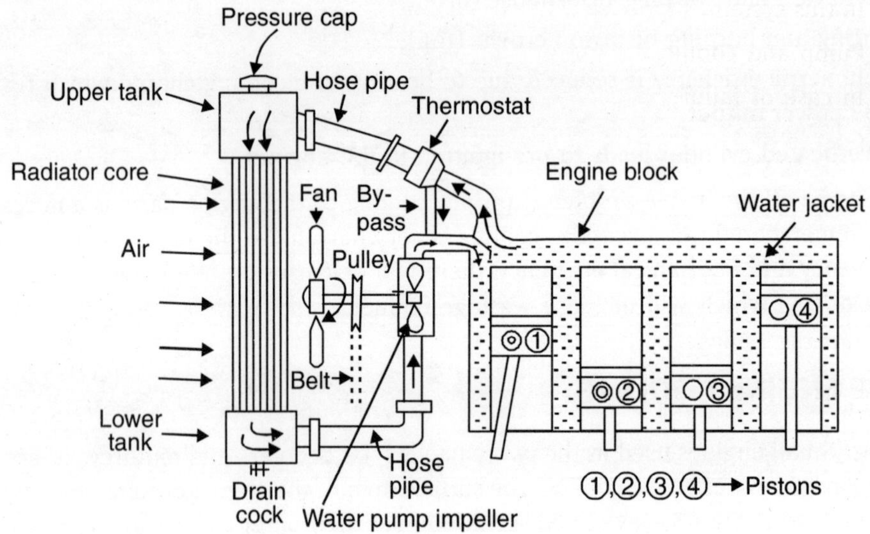

Fig. 3.2 Pressure cooling system with thermostat valve

On the other end of the shaft of the pump, a cooling fan is fitted which draws in the atmospheric air passing over the radiator core. The air cools the hot water flowing down in the radiator core. The cooling water after circulation and absorption of heat from the water jacket casted around the engine cylinders, passes on to the upper hose which leads it to the radiator's upper tank and then to the radiator core where it again gets cooled by air sucked by cooling fan. To avoid excessive cooling, particularly in cold weather, a thermostat is used. The thermostat valve is fitted inside the upper pipe where upper rubber hose is fitted. The thermostat valve does not allow cold water to circulate in the radiator but by-passes it back to the water jacket.

Advantages of Water Cooling

This system has got the following advantages:

- It allows a smaller radiator to be used, so to lower the bodyline and reduce the height of the vehicle.
- Since the water is continuously under pressure—it can be directed to the engine's very hot spots such as spark plug bosses and valve seats.

- Modern cars have a thermostat to control the engine temperature within limits, by restricting the water flow until the engine attains its normal working temperature. The thermostat allows the water to flow to radiator only when its temperature exceeds the pre-determined limit of about 80°–85°C. If the temperature of water is less than this limit then thermostat will send back the water to the pump through a by-pass passage.

This results in the improved fuel consumption *i.e.*, fuel economy, reduced cylinder bore wear and lesser engine oil dilution.

Disadvantages of Liquid Cooling

- In this system, supply of water or coolant is to be maintained at a required level always.
- Pump and cooling fan absorb lot of power.
- In case of failure of cooling system, the engine will be damaged badly.
- Cost is high.
- It requires regular check ups for various parts.
- Leveling up of the coolant/changing of coolant is required after certain intervals as prescribed by the manufacturer.

With the burning of air-fuel mixture in the engine cylinder, a large amount of heat is generated and the temperature in the cylinder rises up to 2500°C, about 35% of this heat escapes with exhaust gases and the remaining heat is sufficient enough to damage the engine parts, so some part of this heat is removed by the cooling system to save the engine from many troubles. The process of removing heat from the engine is called *engine cooling*.

Component of the Cooling System

1. Radiator.
2. Water pump.
3. Cooling fan.
4. Thermostat valve.
5. Flexible rubber hoses (pipes).
6. Temperature gauge.

Fig. 3.2 shows a line diagram of a pressure cooling system with thermostat.

***(a)* Radiator:** The function of radiator is to provide enough area of cooling surface so that downward flowing water in the radiator core-tubes is properly cooled by the air sucked by cooling fan. It also provides a storage tank for water or coolant to be circulated in the water jacket formed in the engine block. Refer to Fig. 3.3.

There are many types of constructions of radiators varying in design, partly depending upon the different working ideas of different engineers and partly in accordance with the limitations imposed on the radiator areas by the modern automobile body design.

As regards to their flow, the radiators of automobile consists of two general types:

- The vertical flow type.
- The cross-flow type.

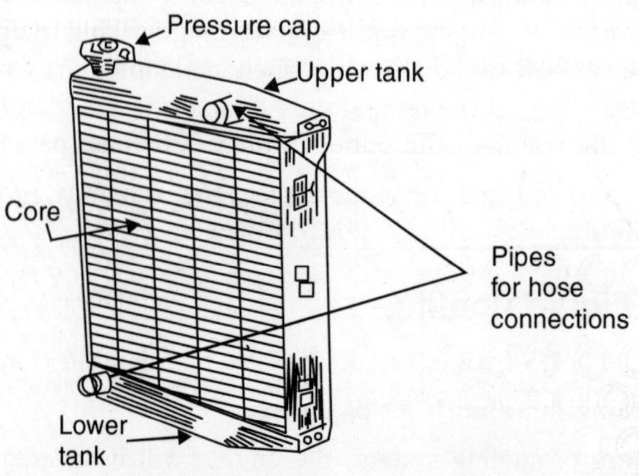

Fig. 3.3 A sealed radiator system

In the vertical type, thin water tubes to which air fins are attached are placed in vertically up position so that the water runs through them from the top to the bottom.

In cross-flow type, the water tubes are arranged in a horizontal plane to supply the water from a supply tank fitted separately.

Both these types of radiators consist essentially of two tanks known as *the upper tank* and *the lower tank.* The upper tank is the coolant filler tank. A radiating element also known as the *core* is placed in between these tanks to form the radiating portion. All radiator cores may be classed as either *tubular* or *cellular.* The upper tank of the radiator is connected by a radiator outlet hose to the water jacket inlet through water pump.

In a tubular radiator the upper and lower tanks are connected by the series of gilled tubes through which the water runs. A section of tubular type radiator core arranged in different manners is shown in the Fig. 3.4. Tubular radiators are widely used in all auto vehicles.

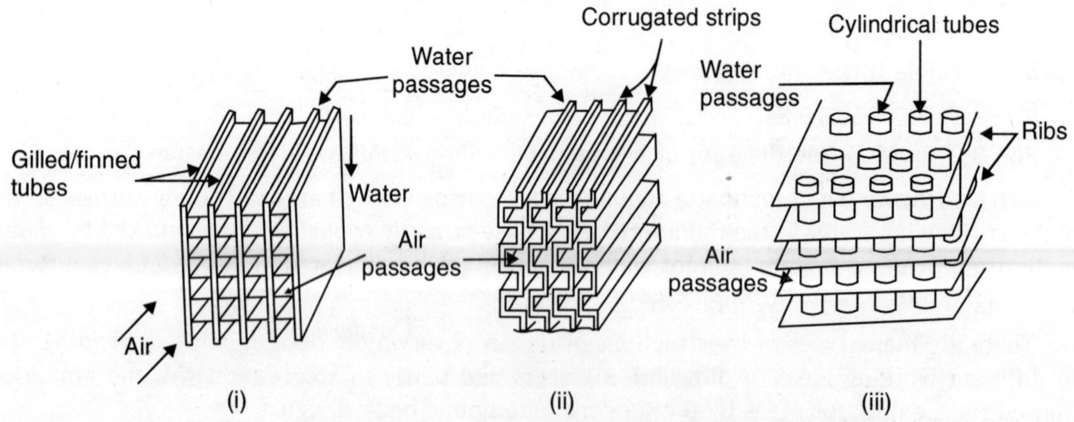

Fig. 3.4 Tubular radiator core sections

A cellular or 'Honey-Comb' radiator is composed of a large number of individual air cells which are surrounded by water. Its name is such given because of its appearances as 'Honey-Comb'.

The course of water cannot be confined to a definite vertical or angular-course especially when the air cells in front are of hexagonal form. Typical sections of cellular type core radiator are shown in the Fig. 3.5.

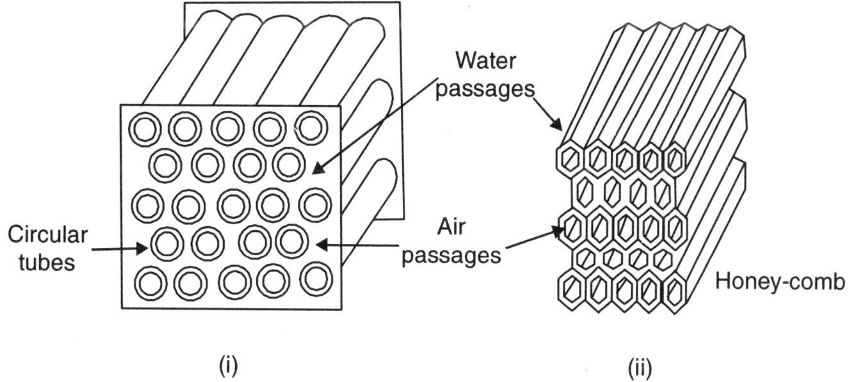

Fig. 3.5 Cellular radiator core sections

Materials used for radiators are brass and copper, the end supports are usually made of steel.

***(b)* Radiator Cap:** It provides a cover for the filler hole through which water/coolant is filled into the upper tank of the radiator. There are two types of caps in common use. One is the simple cap which acts only as a cover so that water may not flow out through the filler hole. The other type of cap contains a double check valve through which cooling contents go in or out to the overflow tank. This type of cap is known as *pressure cap*. By the use of this cap, cooling efficiency is increased; evaporation is prevented and the water losses due to surging when the vehicle is suddenly stopped are avoided. The pressure cap seals off the cooling system and does not allow escape of air or vapour.

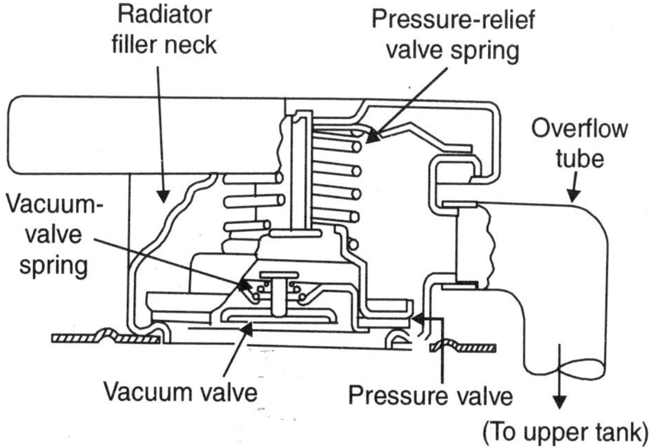

Fig. 3.6 Radiator pressure cap

At sea level, where atmospheric pressure is higher (about 1 bar), water boils at higher temperature (about 100°C). At higher altitudes, where atmospheric pressure is less, water will boil at lower temperature. Higher pressure increase the temperature required to boil the water whereas lower pressure lowers down boiling temperature. The use of pressure cap on the radiator increases air pressure (1.5 bar to 2.0 bar) within the cooling system and the water is circulated at higher temperatures without boiling. The water enters the radiator at a higher temperature and the difference between the cooling water and air temperature is greater. Heat is thus transferred from water to air more quickly which results in improvement in cooling efficiency and a smaller radiator is required for cooling under pressure.

The double check valve in the pressure cap consists of a blow-off valve or a pressure relief valve and a vacuum valve. The blow-off valve is held against the seat by mean of a calibrated spring. The spring keeps the valve held close so that pressure reaches the predetermined value for which the system is designed, the blow-off valve is raised off its seat and the excessive pressure is released through the overflow pipe. Sectional view of a pressure cap is shown in the Fig. 3.6. The vacuum valve prevents radiator tubes from collapsing when a vacuum is created as the system cools down.

(c) Thermostats: The thermostat is placed in the coolant passage between the cylinder head and the top of the radiator (Figs. 3.8 and 3.9). Its purpose is to close off this passage when the engine is cold. Now coolant circulation is restricted, causing the engine to reach normal operating temperature more quickly. This reduces the formation of acids, moisture, and sludge in an engine. Also, after warm-up, the thermostat keeps the engine running at a higher temperature than it would,

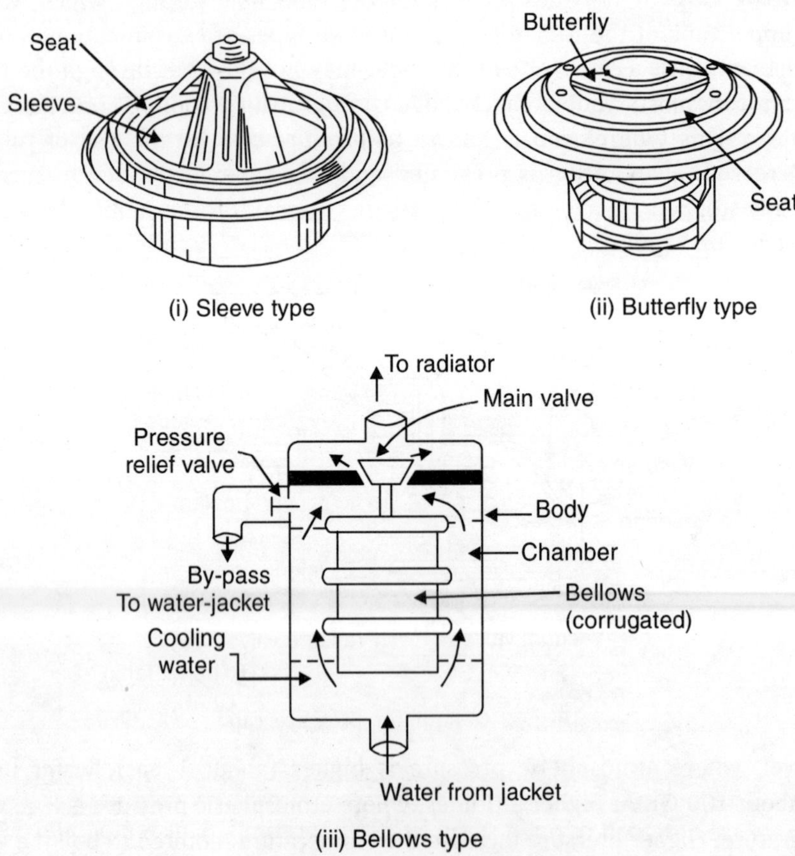

Fig. 3.7 Types of thermostat

for example, without a thermostat. The higher operating temperature improves engine efficiency and reduces exhaust emissions.

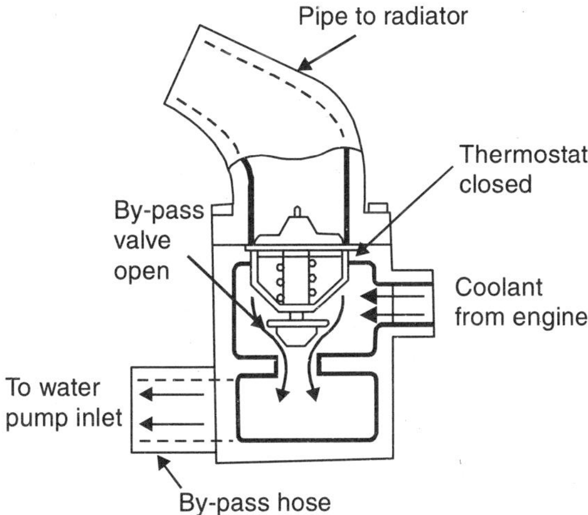

Fig. 3.8 Operation of blocking radiator pipe by closed thermostat

The thermostat consists of a thermostatic device and a valve (Figs. 3.8 and 3.9). Various valve arrangements and thermostatic devices have been used. Most thermostats today are operated by a wax pellet, which expands with increasing temperature to open a valve. The sleeve and the butterfly thermostats shown in the Fig. 3.7, both use the wax pellet.

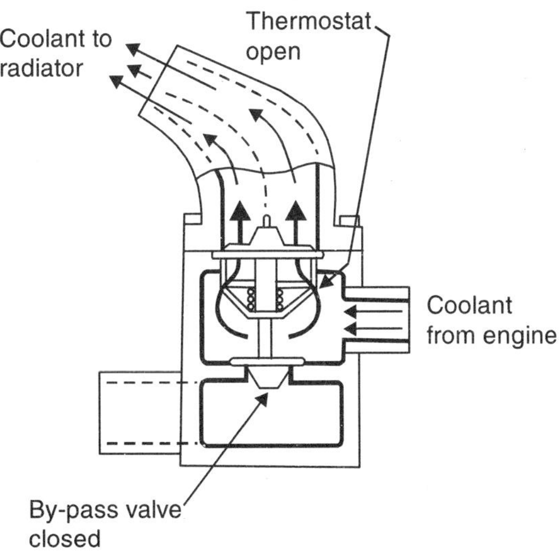

Fig. 3.9 (Thermostat opened), water going to radiator

Thermostats are designed to open at specific temperatures. This temperature is known as the *rating of the thermostat,* and may be stamped on it. Two frequently used thermostats are having ratings of 83°C and 91°C. Most thermostats begin to open at their rated temperature. They fully open about 11°C higher than its rating. For example, a thermostat with a rating of 91°C starts to open at that temperature. It is fully open at 102°C.

Most engines have a small coolant by-pass passage. It permits some coolant to circulate within the cylinder block and head when the engine is cold and the thermostat is closed. This provides equal warming of the cylinders and prevents hot spots. When the engine warms up, the by-pass must close or become restricted. Otherwise, the coolant would continue to circulate within the engine itself, and too little would go to the radiator for cooling.

The by-pass passage may be an internal passage, or an external by-pass hose. In Fig. 3.8, notice the small external by-pass hose at the top of the water pump.

One internal by-pass system uses a small, spring-loaded valve located in the back of the water pump. The valve is forced open by coolant pressure from the pump when the thermostat is closed. As the thermostat opens, the coolant pressure drops within the engine and the by-pass valve closes.

Another internal by-pass system has a blocking-by-pass thermostat (Fig. 3.9). This thermostat operates like those already described, but it also has a secondary, or by-pass, valve.

When the thermostat valve is closed, the circulation to the radiator is shut off. However, the by-pass valve is open, permitting coolant to circulate through the by-pass. As the thermostat valve opens, permitting coolant to flow to the radiator, the by-pass valve closes. This blocks off the engine by-pass passage.

(*d*) Bellows Type Thermostat: During running of engine, the hot water from engine enters into the bellow chamber and transfers the heat to the liquid filled in the bellow. This liquid converts into the vapour. The vapour exerts pressure in the vertical direction and opens the main valve. This water starts flowing into the radiator. Refer to Fig. 3.7.

(*e*) Cooling Fan: The fan is fitted between the engine and the radiator. It is driven by the belt and the pulley, drive is taken from the crankshaft. It serves the following purposes in the engine cooling system:

- It ensures regular flow of air through the radiator fins and thus increases its heat radiating ability.
- It throws fresh air over the outer surface of the engine due to which the heat conducted is taken away and thus it further increases efficiency of the cooling system.

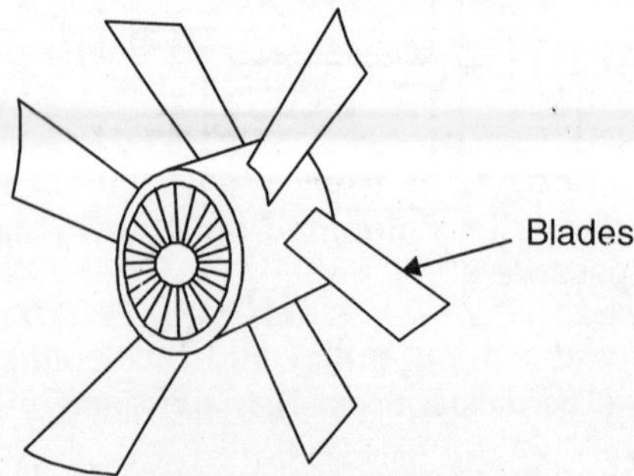

Fig. 3.10 Cooling fan

The fans vary in number of blades which range from two (02) to eight (08). In cold countries or during winter season, there is more problem of keeping the engine warm than to cool down. Fan is usually removed during extreme cold. Even a canvas sheet is covered over the radiator core or radiator shutters are employed to keep it warm. A simple cooling fan is shown in the Fig. 3.10. Removal and replacement of the fan leads to unnecessary headache in winter and summer there. In order to overcome this difficulty, *automatic fans* have been developed and employed in cars. Automatic fans are operated by the different ways, electromagnetically or thermostatically in cold countries' vehicles.

***(f)* Water Pump:** Centrifugal type water pumps are commonly used to circulate the cooling water through the cooling system. It is mounted in front of the engine block on a common shaft with the fan. The shaft is driven by the pulley-belt system. The power to drive the pump is taken from the crankshaft. A water pump is shown in the Fig. 3.11.

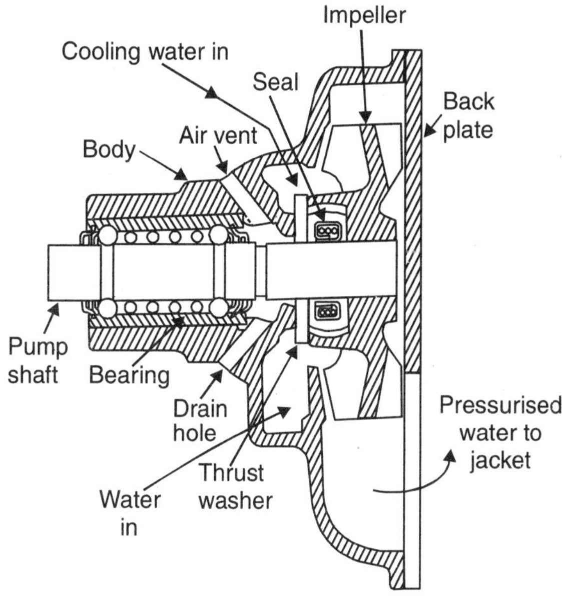

Fig. 3.11 Water pump for cooling system (driven by pulley-belt system from crankshaft)

***(g)* Drive Belts:** A fan belt drives the water pump and fan shaft and alternator or dynamo shaft. It fits in the V-groove of the pulleys mounted on the two shafts. A separate belt is used to drive Air Conditioner's compressor.

As shown in the Fig. 3.12, water pump pulley and the cooling fan are mounted on the same shaft. Fan draws in the air from atmosphere to cool the radiator tubes and engine block.

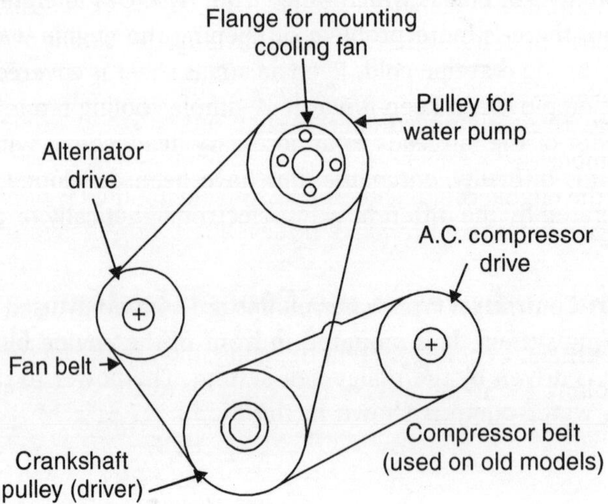

Fig. 3.12 Belt drives for water pump, fan and A.C. compressor and alternator

GRAVITY OR THERMO-SIPHON CIRCULATION SYSTEM OF COOLING

This system works on the principle that if a container of cold water is heated, the hot water will tend to rise (due to lowering of density at high temperature) and its place will be taken by cold water (convection method). When the water in water jacket gets heated, it becomes lighter and rises up. The hot water thus enters the radiator through upper pipe and flows down into radiator core, cooled by air-stream, becomes cold and accumulates in the bottom tank and then circulates through water jackets again without any pump.

It is a very simple system and its operation is automatic. But the water circulates very slowly which needs a large quantity of water and a big vessel for storage. This system is not used on automobiles these days. Previously in older days, it was used on some motor cars and on stationary engines of not more than 30 H.P.

AIR AND WATER COMBINED COOLING SYSTEM

These days some vehicles employ air cooling and liquid cooling both for their engines. Both the systems may work separately or simultaneously. This kind of cooling is known as *combined cooling system.*

The Volkswagen rear side mounted engine having two pairs of horizontally-opposed cylinders, incorporates oil cooling combined with the air cooling. An oil cooler is mounted in between the stream of air on this engine.

SEALED COOLING SYSTEM

In modern cars, the cooling system is entirely sealed-off from the surrounding and is filled with an anti-freeze fluid for the entire 'lifetime' of the engine. The fluid is blended with corrosion-inhibiting chemicals such as ethyl amines and methyl amines. An object of this method of cooling is to eliminate continued losses of coolant due to surge effects, vaporisation and others.

In this system, an expansion chamber is employed along with a relief valve. During rise in temperature of the engine, the cooling medium is pushed-out due to its expansion which enters back into the expansion chamber via a relief valve. The coolant is sucked-back again into the main circulating system when the engine cools. Thus any loss of cooling fluid is prevented. A sealed cooling system is almost maintenance free and requires no leveling up or change of cooling medium.

Use of Coolants and Additives in the Cooling System

Any auto engine's efficient cooling system depends upon many factors including the correct selection and use of coolant. An ideal and coolant must have the following properties:

- It must not cause corrosion to the parts of the cooling system.
- It must not form scales and deposits.
- It must have high heat transfer coefficient.
- It must not evaporate readily.
- It must have a wide operating range *i.e.,* lower freezing point and higher boiling point.
- It must not be toxic.
- It must be cheaper and easily available.
- It must be light in weight.
- It must have higher specific heat.

From various points of view, water is considered to be the best coolant. It has higher specific heat *i.e.,* more heat carrying capacity, low specific gravity, non-toxic property, easily and cheaply available. Distilled water and rain water are the purest forms of water and can be used as cooling medium.

During frosty and winter seasons, the cooling water is likely to freeze in the water jacket which may produce cracks in the radiator and the water jacket and harm the system. To prevent icing, anti-freezing solutions must be inevitably mixed with coolant. Such solutions lower the freezing temperature of the coolant. Glycerin, alcohol, washing soda, glycol-master, calcium chloride and ethylene glycol are used as anti-freezing agents. Some anti-freezing mixtures are not safe for the use in the engines with aluminium components. So they should be used according to the manufacturer's instructions. Most of the coolants are mixture of water and the cooling compounds. Maruti 800 car uses Maruti Golden (30:70) as coolant. Premier Diesel car and Tata-Cummins diesel engine utilise ethylene glycol (30 to 50%) with water.

Cooling of Exhaust Valve: When engine is running, the exhaust valve gets over-heated due to high temperature (about 2500°C) of burnt gases. To save the valve from pitting, distortion etc., it is very essential to have a proper cooling of the valve. To achieve this, the water of the water jacket of engine is allowed to circulate very closely to the exhaust passage through a bit wider channels.In Heavy duty engines, sodium-cooled exhaust valves are used (Fig. 3.13). Sodium is a very good conductor of heat and it melts at 105°C. Sodium is filled in the hollow stem of the valve up to

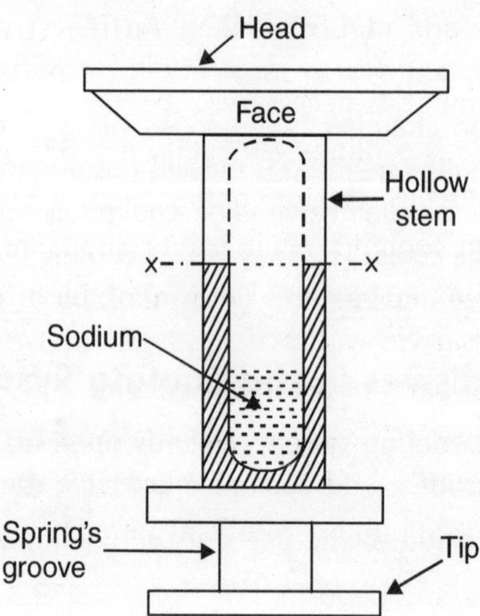

Fig. 3.13 Sodium-cooled exhaust valve

one-third of its length. When the valve gets heated, sodium melts and become liquid. The liquid sodium splashes up and down during lifting and sitting of valve from and at its seat. The liquid sodium takes away heat from the upper portion or face of the valve and transfers to the lower portion which remains at a lower temperature.

Servicing the Radiator

- Leakage are marked and repaired by applying suitable adhesive or by gas-welding. If it has several leakages, the radiator should be changed.
- Water pumps are generally sealed type and require no lubrication. Older type of pumps may require some lubrication. If pump is making noise and leakage are visible than replace it with a new one of the same size.
- After use for one to two years, the additives in anti-freeze compounds get expired and lot of rust begin to form. Therefore, a rust coloured anti freeze is an indication that cooling system service is now due. They cooling medium may also contain scales, grease, oil and some acids formed by exhaust gas leakage into the coolant.

First the dirty cooling liquid is drained out from radiator and fresh coolant is refilled after a plain water wash of the system. Sometimes a cooling system cleaner liquid is also used. To use the cleaner liquid, fill the mixture of water and cleaner into the radiator and run the engine idle for about 20 minutes and then stop the engine, drain out the water and cleaner mixture and refill the fresh coolant up to the mark.

Reverse flushing of radiator and water jacket is also a very effective method of cleaning the cooling system. In this method, a high pressure water stream is connected to the hose of the lower tank and this pressurized water is allowed to come out of the upper tank. The process is continued till the clear water starts coming out. After draining the water, the fresh anti freeze mixture is filled up to the level. Before reverse flushing, the thermostat must be taken out and re-installed after flushing.

QUESTIONNAIRE

1. Why cooling of I.C. engine is necessary?
2. Discuss in details various cooling systems used in I.C. engines.
3. Describe with neat sketches, the air cooling system. State its limitations.
4. What is thermo siphon cooling? What are its disadvantages?
5. Describe a water cooling system used in I.C. engines.
6. Why radiator is necessary in water cooling system? Describe its construction.
7. Describe the following parts of a cooling system:
   (*i*) Thermostat valve, (*ii*) Radiator cap,
   (*iii*) Cellular radiator and (*iv*) Overflow tank.
8. What is difference between honey-comb and tubular radiator?
9. Describe various types of thermostat valves.
10. What is the function of fan in cooling system? Why is it placed behind radiator?
11. What is an anti-freeze and why is it used?
12. Compare air cooling system with water cooling system.
13. What is reverse flushing of radiator?
14. What is the function of a 'Thermostatic valve' in the cooling system?
15. What are the systems of water circulation?
16. Describe briefly the advantages of a 'Centrifugal pump' system of water circulation?
17. What are the advantages of a 'Thermo syphon' system of water circulation?
18. What are the advantages of a 'Closed pressure' system of water circulation?
19. What are the precautions that will prevent a coolant from reaching its boiling point and its freezing points?
20. How does a cellular type radiator core differ from a tubular type?
21. How will you detect the cause of a coolant leak inside a engine?
22. What are the advantages of by-pass circulation?
23. What are the main causes of rust and scale formation in water jacket? How are they removed?
24. How does the water in the radiator get cooled?
25. What precautions are taken in removing the radiator cap in a pressure cooling system?

Chapter 4

Lubrication Systems

NEED OF A LUBRICATION SYSTEM

When two dry metallic surfaces move over each other, a lot of friction is experienced and the coefficient of friction varies from 0.15 to 0.5. If a film of lubricating oil is interposed between the two surfaces, the value of coefficient of friction falls below 0.1 that means lot of friction is reduced by lubricating the two surfaces which are sliding over each other. Therefore, to save energy, it is must to reduce friction by way of interposing a thin film of a lubricant between the two mating metallic surfaces of the moving parts.

TYPES OF LUBRICATIONS

There are four types of lubrications:

1. **Hydrodynamic Lubrication:** In this type, the lubricant between the two moving surfaces forms a wedge-shaped film. The moving part acts as a pump to form the oil into the narrow clearance.

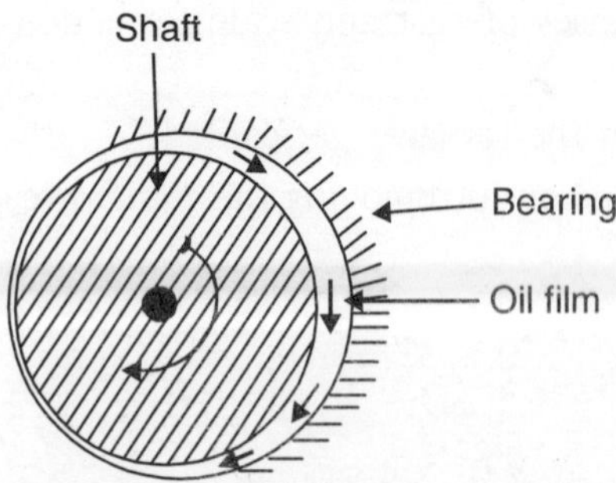

Fig. 4.1 Hydrodynamic lubrication

It occurs in main journals, cam shaft bearings, engine cylinder walls, valve guides, connecting rod ends bearings. Refer to Fig. 4.1.

2. **Elasto Hydrodynamic Lubrication:** When load on the moving part is very high, the material of the part itself deforms elastically against the pressure exerted by oil film. This type is now termed as *Elasto hydrodynamic lubrication.* It occurs between gear teeth, rolling bearings, cams and followers.

3. **Boundary Lubrication:** When moving surfaces break the oil film under heavy pressure, the irregularities of the moving surfaces come in contact with each other and make a metal to metal contact. Lot of wear and tear occurs due to insufficient supply of oil or poor quality of oil/lubricant.

4. **Hydrostatic Lubrication:** In this type, a thin film of lubricating oil resists instantaneous squeezing out under reversal of loads with relatively slow motions. The oil film acts as a cushion, if oil supply is quite sufficient the oil film thickness is restored before next reversal of load.

SOME IMPORTANT PROPERTIES OF LUBRICANTS

Viscosity: It is measure of oil's resistance to flow. It is also a measure of the ability of the oil film between two metallic moving parts to carry the load. It is measured in terms of *Saybolt Universal Seconds (SUS)*, which is the time required in seconds, for a given quantity of the oil to flow through a capillary tube under specified test conditions. It is also expressed in Centispokes, Centipose and Red wood seconds. Viscosity is usually expressed at –18°C and 99°C temperatures, oils with high viscosity can carry greater load but offer greater friction to sliding surfaces.

Flash Points: It is the lowest temperature which the oil flashes when a flame is brought near its surface. It should high for engines lubricants.

Fire Points: The temperature at which oil burns continuously is called *Fire Points.* It should also be high.

Stability: An oil must remain stable at high temperatures.

Film Strength: The thin film of oil should not break at high temperatures, loads and speeds.

MAIN ENGINE PARTS WHICH REQUIRE LUBRICATION

- Cylinder walls and piston assembly.
- All types of bearings.
- Timing gears.
- Crank pin and piston pin.
- Crankshaft bearings.
- Camshaft bearings and mountings.
- Valve operating mechanism.
- Connecting rod ends bearings.

Change of Engine Oil: When lubricating engine oil becomes too dark and thin, it is changed. It is generally changed after running every 10,000 kms. To follow manufacturer's instructions in this regard is advisable. Oil is always changed in warm-up condition by unscrewing the drain plug and after flushing out the oil, retight the drain plug. Fresh lubricant oil is poured or refilled from an opening at the top of crankcase, in the prescribed quantity for the particular model.

Mechanical Frictions Which Require Lubrication

Bearing Friction: For low friction in moving parts of engine, clearances between them should be more but in the case of automotive engines, low engine noise level is required. Then piston and bearing clearance must be minimised. Thus, requirements for low friction are in conflict with those for low noise level.

Investigations reveal that using roller cam followers instead of slider flat foot followers reduce frictional losses of energy.

Friction by Oil Ring: In I.C. engines, two types of rings are used. They are:

1. Compression rings designed to seal the cylinders against gas leakage, and
2. Oil rings designed to limit the passage of oil from the cylinder wall to the combustion space by scraping off the excess oil from the cylinder surfaces.

In the case of compression rings, the force exerted by the ring on the cylinder wall is due to two factors:

1. Elasticity of the ring.
2. Pressure of the gas which leaks into the clearance between the rings and the piston. Investigations reveal that the gas pressure in the top ring groove is nearly same as the cylinder pressure, in the second ring groove it is less than the cylinder pressure and it is very little in the third ring groove.

Oil ring grooves have holes drilled into the piston interior. Therefore, no gas pressure can build up in their grooves. In this case, the pressure exerted by the ring surface on the cylinder wall is entirely due to the elasticity of the ring.

The piston rings, because of the spring action, press against the cylinder walls at all times, that is, no period exists without load on the cylinder walls. This causes friction forces on the cylinder wall and hence wear of the cylinder as well as ring surfaces occur.

Cylinder Wear: Wear of the cylinder/cylinder liner and piston rings result in loss of compression, increased blow-by and oil consumption and reduced power output and thermal efficiency. Loss of compression may also result due to ring defect. The main causes of wear of the cylinder and piston rings are discussed below:

Corrosive wear: Corrosion, erosion and/or abrasion cause wear of cylinders, cylinder liner and piston rings. The degree to which each of these mechanisms influences wear depends upon to a large extent on the operating conditions. This aspect is particularly true with respect to corrosive wear.

Corrosion is a chemical attack on the cylinder liner surfaces and piston rings by the vapours or acids, formed by the products of combustion. The extent to which the corrosive materials are formed depends on the following:

1. Quality of the fuel used.
2. Completeness of combustion.
3. Jacket cooling water temperature.

4. Exhaust back pressure.
5. Quality of the lubricating oil.

Fuel impurities such as sulphur, chlorine and nitrogen components may form corrosive materials during combustion. These materials may produce corrosive acids in the presence of water vapour. The acids may condense on the cylinder walls when the jacket water temperatures are low during cold climates.

***Abrasive wear*:** This type of wear is mainly due to friction between two rubbing surfaces. This friction is due to surface roughness. Even the finest of surfaces, whether ground, lapped or polished are rough in molecular dimensions. Frictional force is the tangential force necessary to move the irregularities of one surface past those of the other when a normal force due to load is acting on them. The resistance to sliding is due to mechanical interlocking of the surface roughness, and the force of friction is that required to tear them apart. The load applied on the sliding metals is carried by a small fraction of the surface area. During sliding, there may be instantaneous temperature rise between the metal contacts reaching the melting point of the metals. Some portions of the surface are loaded beyond their yield point even though the applied load is sufficient only to deform the metal elastically. This plastic yielding causes molecular adhesion and a weld is formed. With subsequent sliding, the welded spots are torn apart and some portions of the metal are removed. This sometimes called *scouring*. Any foreign particle embedded between the sliding surfaces may also cause scouring of the surfaces.

To reduce abrasive wear, both intake air, fuel and lubricating oil must be kept free from abrasive materials. To achieve this end, filtering equipments must have adequate capacity and must be maintained at maximum efficiency.

***Scuffing wear by piston rings*:** During engine operation, scuffing wear occurs between the rings and the cylinder. This wear is due to momentary failure of the oil film, causing instantaneous temperature rise and plastic yielding of the skin surface under conditions of high load and speed. The frictional heat thus generated cannot be dissipated rapidly to the surrounding media. Consequently, an unstable system is set up, that leads to localized welding of the mating surfaces and their consequent rupture. If thin oil film condition prevails during scuffing, then, the degree of surface finish plays an important part in determining the load at which scuffing occurs.

A typical scuffed surface exhibits numerous torn particles and blobs or projections of transferred material. These are due to high temperatures produced at the points in contact. When speeds are low and loads are high, these blobs of transferred material were found to be large in size but few in number. At high speeds, when the loads are small, a great number of such blobs of comparatively smaller dimensions appear. At very high speeds, the blobs eventually spread out to cover the whole area of contact.

The wear of the cylinder/liner and piston rings are caused by one or more of the following reasons:

1. Improper filtration of air, fuel and lubricating oil.
2. Water in the fuel or oil.
3. Corrosive fuel being used.
4. Low viscosity oil being used.
5. Insufficient oil-feed to the cylinders.

6. Low cooling water (jacket) temperature.
7. Excessive blow-by past piston rings and cylinder walls.
8. Distorted piston or cylinders/liners.
9. Frequent cold starts.
10. Excessive piston clearance.
11. Excessive ring pressure on the cylinder walls.

Pitting of cylinder liners: In the case of wet liners, the exterior surface of the liner is in contact with the cooling water. This surface wears out by a process known as *pitting of cylinder liners.* The wear is a highly localised one in nature. The production of very turbulent coolant flow conditions adjacent to the liner surface by high speed vibration of the liner wall is responsible for pitting. In the extreme case, the pitted surface may cause hole formation in the liner and rushing of water through the liner into the cylinder and consequent stoppage of engine.

Blow-by Gases: The escape of gases from the top of the piston, past the piston rings into the crankcase during engine operation is called *blow-by.* The purpose of the compression rings is to prevent this gas leakage. The rate at which the harmful gases escape past the piston into the crankcase is an inverse measure of the efficiency of the compression rings. Blow-by results in the following defficiencies:

1. Loss of compression and hence reduction in compression pressure and power produced.
2. Undue heating of piston, rings and cylinder wall.
3. Carbonisation of oil in the ring grooves.
4. Sticking of ring in the ring groove of the piston.
5. Contamination of the lubricating oil and consequent sludge formation.

All piston rings installations have a critical speed (engine speed) at which **blow-by** increases abruptly. This is because at the critical speed, the ring begins to act as elastic vibrating member, and no longer remains in contact with the cylinder wall. This happens because of the uneven wear of the cylinder along its length (greater wear at the top than at the bottom).

Viscosity of the lubricating oil also affects the **blow-by** as the engine speed changes. Higher the viscosity of the oil, lesser will be the tendency for the breakage of the lubricating oil film that exists between the ring and the cylinder wall. This aspect reduces the **blow-by** when the higher viscosity oil is used. However, higher viscosity oil will cause greater fluid (film) friction loss. **Blow-by** contains mostly hydrocarbons, nitrogen, sulphur and moisture plus some amount of fuel vapours.

Blow-by can be reduced by the following means:

1. By adding compression rings. This reduces blow-by but increases frictional losses. This also increases the height of the ring belt and hence the height of the piston. As such adding compression rings is not desirable.
2. Reducing the width of the ring face (axial width) reduces blow-by and also shifts the point of critical speed towards higher speeds. Reducing the width of the ring face also reduces frictional losses.
3. Providing an inner spring ring behind the top compression ring increases ring tension *i.e.* the force exerted by the ring on the cylinder wall. This construction eliminates the critical speed condition and also reduces the **blow-by** at all engine speeds.

Purpose of Lubrication

The purpose of lubricating the various engine parts listed below are as follows:

1. To reduce friction between the moving parts.
2. To reduce wear of the moving parts.
3. To act as a seal and prevent leakage between the parts such as pistons, rings and cylinders.
4. To carry away much of the heat generated by friction, by flowing between the moving parts.
5. To keep down the temperature of the moving parts and thus prevent seizure.
6. To conserve power that would otherwise be wasted in overcoming excessive friction.
7. To wash away acidic accumulation and the abrasive metal worn from the friction surfaces.
8. Lubricants provide cushioning effect to bearings against cyclic shocks occuring in cylinders.

To effect the above an oil film of correct quality and thickness must be interposed between the rubbing (friction) surfaces. The actual working thickness is determined by the following:

- Heavy pressures acting to keep the surfaces into contact and in their movement.
- Viscosity of the oil used.
- Effect on the oil viscosity by the chemical additives added to the oil.
- Dilution effect of foreign matter.
- Working temperature of the oil.
- Quality of the oil.

Main parts of a Lubrication System: The following are the main parts of a lubrication system used in I.C. engines:

- Oil sump or Oil tank.
- Oil pump and Oil strainer.
- Oil filter.
- Oil dipstick (for indication of oil level).
- Oil pressure gauge.
- Oil cooler—used in heavy duty diesel engines only.
- Oil pressure relief valve—when oil pressure increases above the limit, this valve returns some of oil back to the sump.

METHODS OF LUBRICATIONS

The different parts of an engine, namely, pistons, cylinders, crankshaft and main bearings, crankpin and piston pin bearings, valve actuating mechanism may be lubricated by anyone of the following methods.

1. *Mist Lubrication or Petroil Method:* In small capacity engines (particularly 2-stroke engine) used in two-wheelers, 2 T lubricating oil (2%) is mixed with petrol directly, which lubricates cylinder walls and piston. In the process, some of the lubricant, is burnt with petrol which produces smoke and carbon deposits on piston and cylinder.

2. *Splash Lubrication or mechanical system*

3. *Pressure Lubrication Systems:* It may be classified as under:

- Wet sump.
- Dry sump.
- Semi pressure system.

Splash Lubrication System: In the splash lubrication system, the oil retained in the oil pan is churned and splashed up by the internal parts of the engine (connecting rod big and crankshaft) into a combination of liquid and mist. This oil is sprayed over the interior of the engine. This system can be seen in the Fig. 4.2.

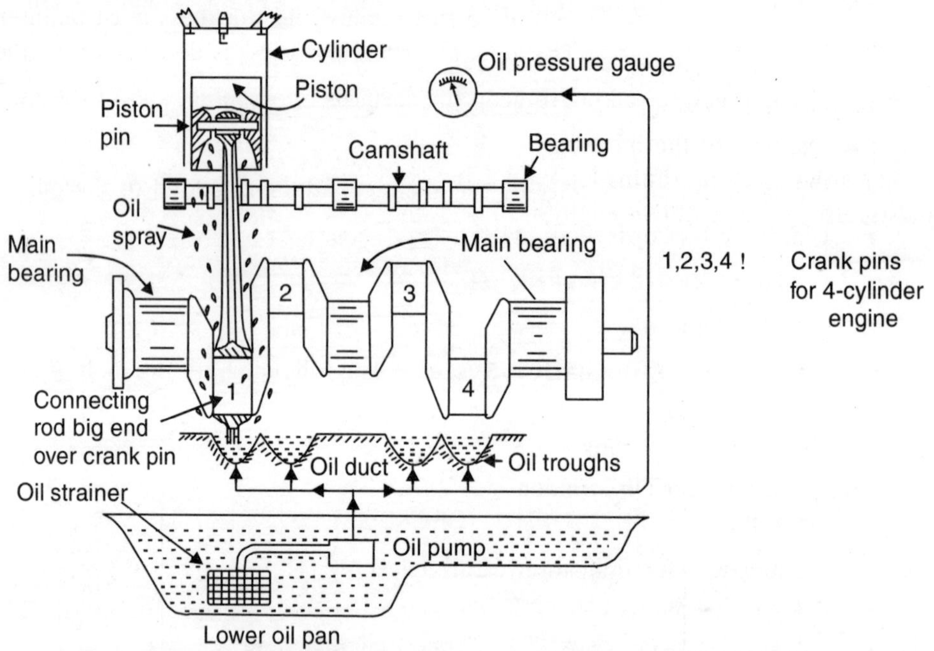

Fig. 4.2 Splash lubrication system with circulating oil pump

The big end of the connecting rod dips into the oil in the crankcase and splashes up the oil and thereby obtain lubrication. Some of the oil that is sprayed on the cylinder walls and the crankcase gets collected in pockets over the main bearings. The collected oil then flows through the bearings by means of grooves in the bearing surfaces.

In one of the designs, the connecting rod big end caps have scoops. These scoops pick up oil from the oil pan, during the lower ends of the connecting rod travel. The connecting rod caps have drilled-holes. Through these holes, part of the oil picked up by the scoop reaches the big end bearings. The remaining oil lubricates the rest of the engine parts by splash *i.e.*, the throwing of the oil by the connecting rods, crankshaft and other oiled moving parts.

Some of the engine components cannot be pressure fed in the same way as main bearings (*e.g.*, timing gears, timing chain etc.) while others should not fed with quantities of oil under pressure (*e.g.*, inlet valve stem). In practice, these parts are lubricated with splash and/or mist lubrication.

Pressure Lubrication System: In this system, the lube oil is pumped under pressure to the various engine bearings. The gear type oil pump delivers the oil into an oil gallery at a pressure of 1.3 to 4 kg/cm^2. The oil gallery distributes the oil to many engine parts and bearings. The gear pump is driven by the engine crankshaft. Figure 4.3 shows the pressure lubrication system. The main bearings (each one) of the crankshaft are pressure fed from the main oil gallery. Internally drilled ducts in the webs of the crankshaft carry oil from the main bearings to the connecting rod big end bearings. The oil from big end bearings, flow through the long duct drilled in the shank of the connecting rod to the piston pin and its supports. Flow of oil through bearings is controlled by maintaining limited clearance all around between a round shaft and its bearings (Fig. 4.1).

(*a*) Wet sump lubrication: In the wet sump lubrication system, the main oil supply is kept in the sump which is below the engine in the crankcase. The oil pump draws oil from the crankcase and forces it through lube oil filters to the various parts of the engine. In some engines, the oil pump is sub-merged in the oil sump. Where there is no oil hole in the connecting rod shank, the cylinder walls and piston pins are lubricated by the spray of oil that is forced out of the connecting rod bearings and thrown by the revolving cranks. This spray also lubricates many exposed internal parts. A separate oil line supplies lubricant to the accessories and their drive shafts, valve and rocker arms. Used oil drains back into the crankcase by gravity for recirculation. This method is mostly used in automotive engines. This system is shown in Fig. 4.3.

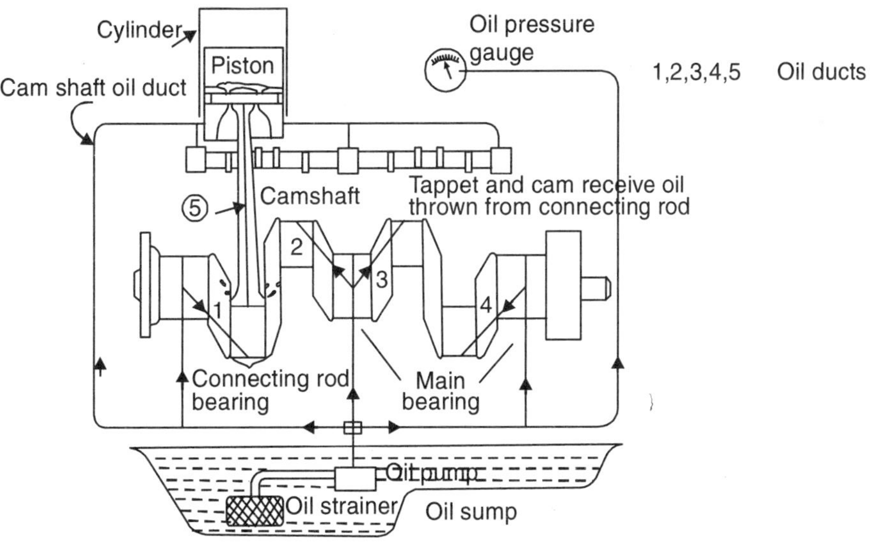

Fig. 4.3 Pressure lubrication system (wet sump type)

(*b*) Dry sump lubrication: In dry sump pressure system, there is no oil sump in the crankshaft chamber. In this system, the oil is kept either in a separate tank or in a separate reservoir, provided with the cooling fins. Two pumps are used in this system. One pump sucks the oil from the reservoir and forces it under high pressure, as in the wet sump system, to the various bearings. The other pump (also called *Scavenger oil pump)* is of larger capacity. This pump sucks up oil which drains down to the bottom of the crankshaft chamber, and returns it to the oil reservoir. The system is shown in the Fig. 4.4.

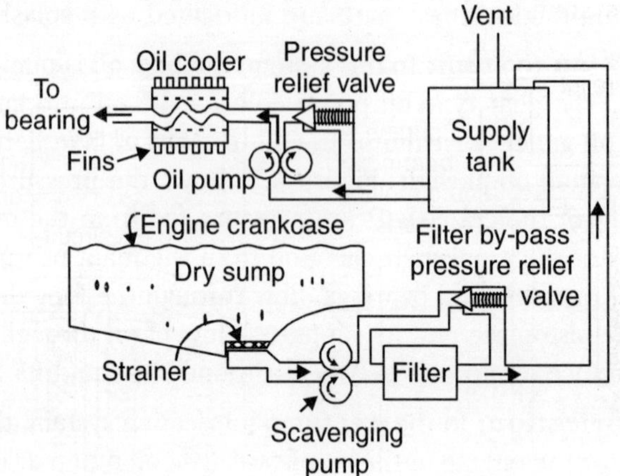

Fig. 4.4 Dry sump lubrication system

The main advantage of the dry sump lubrication system is that the oil is also cooled during its circulation. As such, the oil has better lubricating value. Dry sump system is suitable for engines that may have to work in inclined positions. Dry sump lubrication is used in more expensive luxury cars, and also on the many aircraft engines in which there is no oil sump in the crankshaft chamber.

(c) Semi-pressure lubrication system: The semi-pressure lubrication system is a combination of splash and pressure lubrication systems. Most engines use this system. This is simpler and less costly than the complete pressure lubrication system. This system also enables higher bearing loads and engine speeds to be employed than for the splash system. This system is shown in the Fig. 4.5. Some parts are lubricated by pressurised oil and some (exposed) parts are lubricated by splash of oil thrown up by crank webs and big-ends of connecting rods.

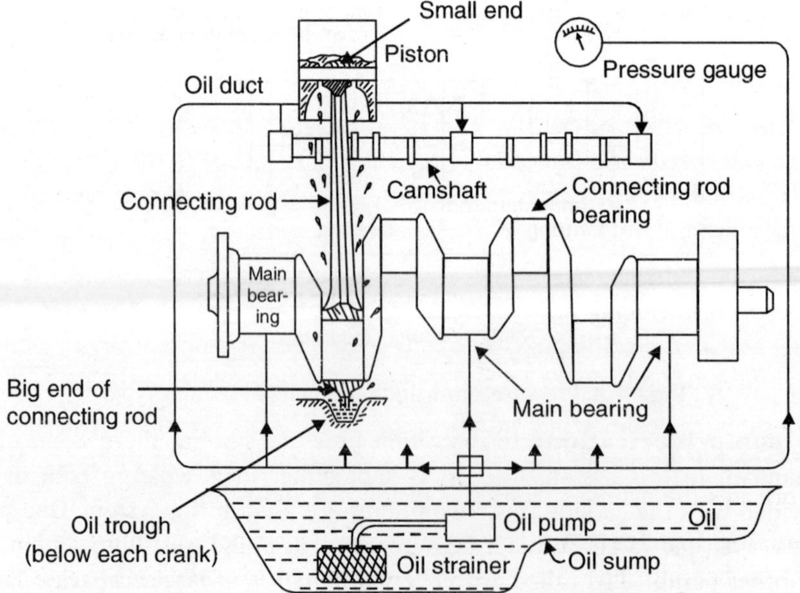

Fig. 4.5 Semi-pressure with splash lubrication system

Full flow and by-pass pressure systems: The partial and by-pass lubrication system and full flow lubrication system can be seen in the Fig. 4.6. Both these systems are the sub-types of the pressure lubrication system.

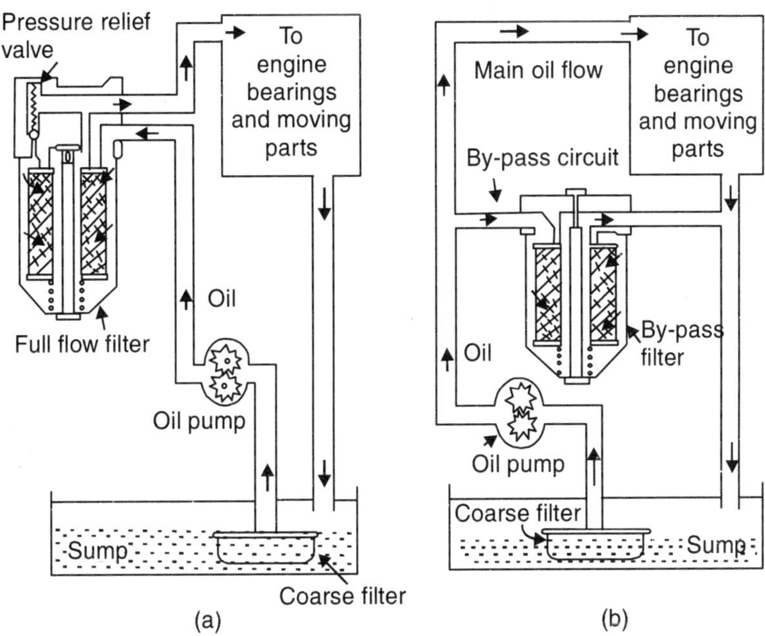

Fig. 4.6 (a) Full flow lubrication system and (b) partial (or by-pass) lubrication system

The full low lubrication system forces all of the oil through the oil filter before the oil reaches the parts of the engine. It is the most common type of lubrication system.

The by-pass lubrication system does not filter all of the oil that enters the engine bearings. It filters some of the extra oil not needed by the bearings. The by-pass lubrication system is not very common in auto engines.

Lubrication of cylinder walls: The splash lubrication system over-lubricates the engine cylinder wall at low speeds and under-lubricates at high speeds. On the other hand, the pressure lubrication system under-lubricates at low speeds, and over-lubricates at high speeds. Therefore, both these system are not very much satisfactory for the cylinder wall lubrication.

The more recent tendency is to provide positive lubrication of the cylinder walls by means of an oil jet or jets from the small holes drilled through the upper end of the big end bearing and a corresponding hole in the crank pin. These holes come into line once every revolution. When this occurs, the high pressure oil spurt outward/upward on the cylinder wall.

In some designs, the connecting rod big end cap has fins or an oil scoop at its bottom portion, This dashes on the oil surface in the crankcase and throw the oil on the cylinder wall and thus sufficiently lubricates the cylinder walls and piston pin and rings.

Main bearing lubrication: The main journal (shaft portion) must be slightly smaller in diameter than the bearing. The clearance between the two is called *oil clearance.* Oil circulates through this clearance. Lubricating system constantly feeds oil to the bearing. It enters through the oil hole, and the rotating journal carries it around to all parts of the bearing. From there, it is thrown off and this helps to lubricate other engine parts such as the cylinder walls, pistons and pistons rings.

As the oil moves across the faces of the bearings, it lubricates them. It also helps to cool them. The oil is relatively cool as it leaves the oil pan. It picks up heat in its passage through the bearing. The heat is then carried down to the oil pan and released to the air passing around the oil pan from outside.

The oil also flushes and cleans the bearings. It flushes out the particles of grit and dirt that may have sneaked into the bearing. The particles are carried to the oil pan by the circulating oil. They then tend to drop to the bottom of the oil pan or are removed from the oil by the oil screen or filter.

At the ends of the crankshaft oil seals are provided there to prevent the oil from leaking out at the front and rear of the engine. This may be single piece or two piece neoprene oil seal or rope or wick oil seal.

Crankpin and Piston pin Lubrication: To provide piston pin lubrication, most of connecting rods have an oil passage or oil duct drilled from the crankpin journal bearing to the piston pin bearing. Oil reaches the piston pin bearing by travelling through a specific path: from oil pump to oil lines in the cylinder block, from these lines to the main bearings, from the main bearings through the inclined oil passages drilled in the crankshaft to the big end bearings, from the big end bearings, through the connecting rod oil passages, to the piston pin bearings.

Main Parts of a Pressure Lubrication System

1. Oil Pump: It is used to pump and circulate the oil to various parts of the engines. It is driven by the camshaft by gearing arrangement. They are generally are of four types:

- Gear pump
- Rotary pump
- Plunger pump
- Vane pump

A Gear type pump is shown in the Fig. 4.7.

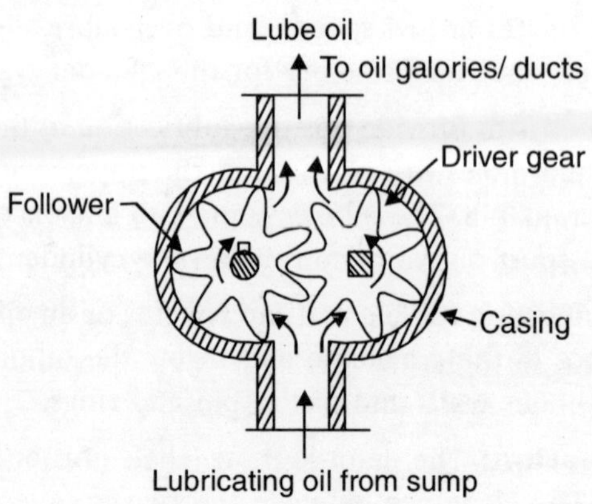

Fig. 4.7 Gear pump

It is widely used in auto engines. It has two gears enclosed in a casing. The driver gear is driven by cam shaft. The oil from sump is sucked into the casing and then pumped out at a pressure of 3.5 kg/sq. cm approx.

2. Oil Filter: A cartridge type oil filter has been shown in Fig. 4.8. The function of oil filter is to stop circulation of dust particles, metallic particles and other physical impurities in the lubrication system. The filter element is enclosed inside a M.S. sheet casing. The filter element may be made of cotton or a special porous paper.

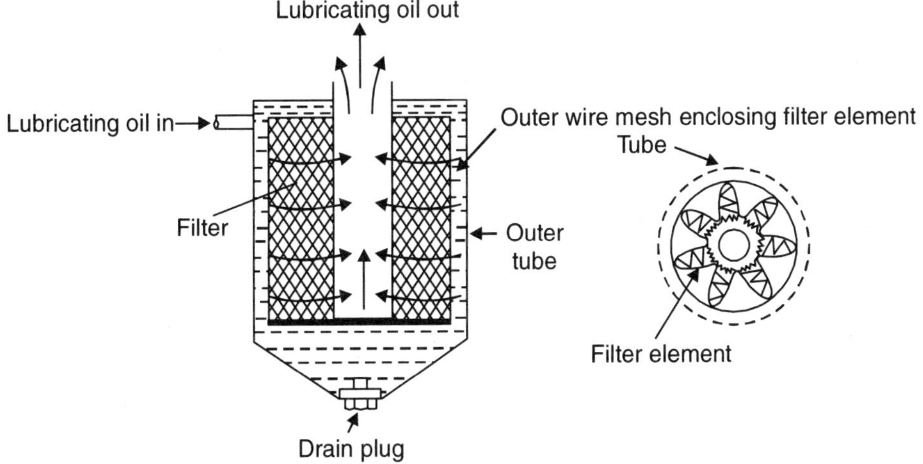

Fig. 4.8 Oil filter (cartridge type)

The oil leaves its all impurities at the filter element. The filter element are to be changed after some duration according to the instructions of the manufacturer. The oil filter is installed ahead of the oil pump. A wire mesh strainer contained in a bowl is also used before oil pump to stop bigger particles from entering into the pump.

3. Oil Ducts: From pump to the parts requiring lubrication, small diameter oil ducts or holes are drilled in the parts. The lube oil under pressure is fed to various parts through these ducts.

Vehicles are also provided oil level indicator or dip stick, oil pressure gauge, oil pressure warning light, and oil drain plug as a part of the lubrication system.

USES OF ADDITIVES IN LUBRICATING OILS

Mineral oils have most of the required properties of a good lubricant. But for I.C. engines operating in varying conditions require some other specific properties such as high viscosity index over a good range of temperatures, resistance to oxidation etc. In order to add above specific properties, some chemicals called additives are mixed with the mineral oils some of the additives are described below:

1. **Detergent:** It controls deposits of oil at high temperature, it also acts as a Acid Neutralizer.
2. **Antiwear:** It reduces wear, scoring, and seizure by not allowing the mating surfaces to come in direct metal to metal contact.

3. **Dispersants:** It controls formation of sludge due to blow-by gases and varnish deposits.
4. **V.I. Improver:** It increases velocity index of oil at a large range of temperature.
5. **Anti-rust:** It reduces rusting by acid neutralization or by forming a protective coating of oil over the metal surfaces.
6. **Anti-foam:** It reduces oil foaming due to presence of air in the crankcase.
7. **Anti-oxidants:** It reduces oxidation of oil which protects metallic parts and bearings against corrosion.
8. **Pour point depressant:** It reduces pour point of oil by not allowing wax crystals to grow and stick together.

Some other additives are also used to prevent acid formation, rust and dirt. These are called—metal deactivator, water repellents, color stabilizers, foam inhibitors, emulsifiers, dyes and odour control agents.

CLASSIFICATION OF LUBRICATING OILS ON THE BASIS OF VISCOSITY

The lube oils are classified according to their viscosity. SAE (Society of Automotive Engineers) has given a method of assigning numbers to different oils whose viscosity is maintained at a given temp. ranges. There are two temp, ranges (a) –18 degree centigrade and 99 degree centigrade for monograde oils such as SAE 5W, 10W and 20W. These are suitable for cold climates whereas SAE 20,30,40 and 60 monogrades of 99 degree centigrade are suitable for ordinary and hot climate. Some multigrade oils (such as SAE 20W/60) has a viscosity equal to SAE 20W oil at –18 degree centigrade and a viscosity equal to SAE 60 at 99 degree centigrade. The multi-grade oils are used on vehicles running on cold hilly areas and planes as well.

Another Classification of Lubricants

Lubricants are also classified on the basis of their state:

(*i*) *Solid Lubricants.* Such as graphite, soapstone, talc, mica, molybdenum sulphide.

(*ii*) *Liquid Lubes.* Such as mineral oils (petroleum products) and vegetable oils and animal oils; the last two oils easily get oxidised, hence not suitable for engines.

(*iii*) *Semi-solids.* Such as grease, Al-paste, Lithium paste and Petroleum jelly etc.

(*iv*) *Emulsions.* They are water mixed oils, cutting fluids, soap water etc.

CRANKCASE VENTILATION

Combustion chamber products contain mostly hydro-carbons such as CO_2, CO, and some amount of sulphur and water vapour. A lot of nitrogen gas is also present. These products continuously leak into the crankcase past cylinder walls and piston rings; they contaminate the engine lubricating oil and form a sludge. These gas vapours containing lube oil particles corrode the metal parts. Sulphur forms acid traces with moisture, which is also injurious to metal. Therefore, it is essential to remove the harmful combustion products from the crankcase. For this purpose, air is circulated in the crank case. The removal of combustion products from crankcase is termed as *Crankcase*

Ventilation. It also does not allow build-up of any pressure in the crankcase which reduces leakage of oil through seals and gaskets. If any pressure builds up in the lubrication system, it is exposed to atmosphere by a oil-breather valve provided in the oil-filter tube or at the air-cleaner pipe to crankcase. Thus atmospheric pressure is maintained inside the crankcase.

Types of Crankcase Ventilations

1. *Open System*. A stream of air is allowed to enter the crankcase through a breather cap in the passage from air-cleaner. The air passes through the crankcase and drives out the harmful gases, fuel vapours and moisture to the atmosphere. The cooling fan also helps in providing flow to the air-stream. This system increases air-pollution and this is type of ventilation is very poor at slow speeds due to slower flow of air stream. This system is also known as 'Road Draft System'. It is now out of use.

2. *Positive Crankcase Ventilation (PCV)*. In this system, the cleaned air is taken in from the inlet manifold through a breather valve. This air is circulated inside the crankcase and the air collects harmful **blow-by** gases, fuel vapours and moisture and recycles them into the air-intake manifold. Thus crankcase emissions or pollutants are fully eliminated by recycling the blow-by gases from the crankcase to the air intake system by means of positive crankcase ventilation valve. The PCV system has been made mandatory on all vehicles in many countries to control pollution.

The blow-by gases fed to intake manifold are sucked by the engine for reburning. Thus the circulation of air takes place in a closed circuit, therefore, it is also called "Closed Ventilation System (CVS)" or "Manifold Suction System (MSS)". The flow of air into the crankcase is regulated by a regulating valve fitted between inlet manifold and crankcase. The line diagram of the PCV is shown in the Fig. 4.9 below.

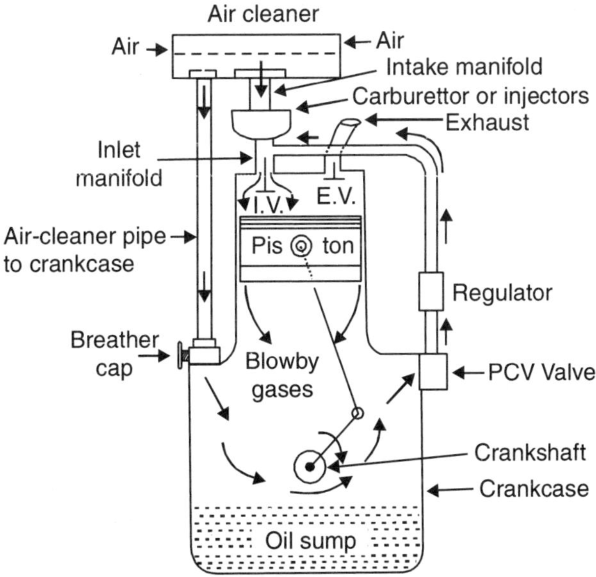

Fig. 4.9 Positive or closed crankcase ventilation

CHASSIS LUBRICATION

Modern car chassis of the present period are mostly lubrication-proof, *i.e.*, they don't need lubrication. The joints are filled with high quality lubricant and sealed. On other models, there are about 21 to 45 such joints which require lubrication. The following points/parts need lubrication at specified intervals:

(*i*) Wheel hub bearings.

(*ii*) Universal joints.

(*iii*) Stub axle swivel, ball joints.

(*iv*) Steering drag links joints, ball joints.

(*v*) Steering track rod joints.

(*vi*) Cooling water pump bearings.

(*vii*) Steering centre pivots.

(*viii*) Front suspension and links.

(*ix*) Brake cables

(*x*) Rear suspension points.

(*xi*) Rear axle bearings.

(*xii*) Suspension springs, leaf springs, shackles bushes.

(*xiii*) Clutch pedal pivots, gear box, steering box and column, starter motor and generator bearings, A.C. compressor bearings, door hinges, break linkages etc.

(*xiv*) Gear box lubrication—Oil is changed and refilled after prescribed periods.

Some parts of chassis which take up load and have some movement also require greasing such as leaf springs and shackles. In LCV and HCV, the greasing is done by a greasing machine or by a Grease-Gun. A hand Grease-Gun is shown in Fig. 4.10.

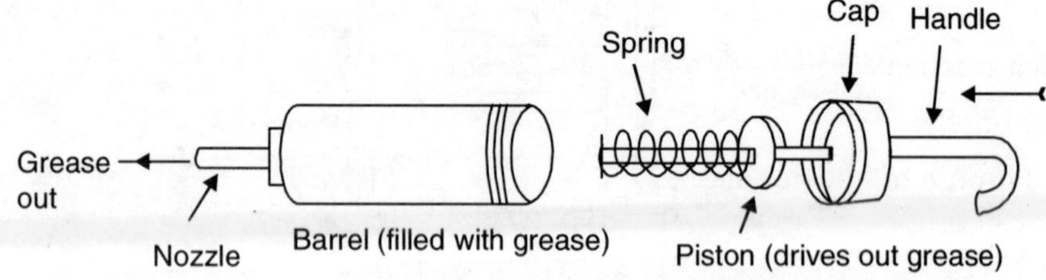

Fig. 4.10 A hand grease gun

QUESTIONNAIRE

1. Classify lubricants. On what basis are they categorised?
2. Discuss the utility of a lubricating oil in an engine.

3. Define:
   (*i*) Hydrodynamic lubrication.
   (*ii*) Elasto hydrodynamic lubrication.
   (*iii*) Boundary lubrication.
4. What is the significance of pour point, flash and fire points of a lubricating oil?
5. Describe with a neat sketch the dry sump lubrication system. In which engine is it used?
6. Describe with a neat sketch the wet sump method of lubrication. Where is it used?
7. Describe with sketches the different types of pumps used in lubrication system.
8. What are additives and why are they used in lubricating oils?
9. What is the importance of viscosity in the selection/grading of lubricants?
10. (*i*) What is the purpose of lubrication?
    (*ii*) Which type of lubrication is used in two-stroke petrol engines?
11. Differentiate between splash system and pressure lubrication system.
12. Describe working of a Hand Grease Gun, with a neat sketch.
13. What are the points in a car chassis which require lubrication?
14. Give reasons for changing lubricating oil at specified periods.
15. What is crank case ventilation? How is it done?
16. Write short notes on:
    (*i*) Grading of lubricating oils
    (*ii*) Positive crank case ventilation
    (*iii*) Oil filter
17. What functions a lubricating oil has to perform? What properties it must have in order to perform these functions?
18. What is pressure lubrication? Explain the types of pumps used in pressure lubrication.
19. Which mechanical parts of engine require lubrication for removing friction?
20. Write briefly:
    (*i*) Purpose of lubrication (*ii*) Blowby gases
    (*iii*) Additives.
21. Describe briefly:
    (*i*) Gear type oil pump (*ii*) Oil filter
    (*iii*) A hand grease gun.
22. (*i*) What is the difference between open system and closed system of crankcase ventilation?
    (*ii*) Name the parts of the chasis which require lubrication.

CHAPTER 5

Transmission System

1. CLUTCHES

AUTOMOTIVE CLUTCHES

Clutch plays a very important role in safe and continuous running of the automobiles. It allows to couple the engine shaft to the gear box and uncouple the engine shaft, when desired, from gear box shaft. It is placed between the engine and the gear box as shown in layout of the transmission system in Fig. 5.1.

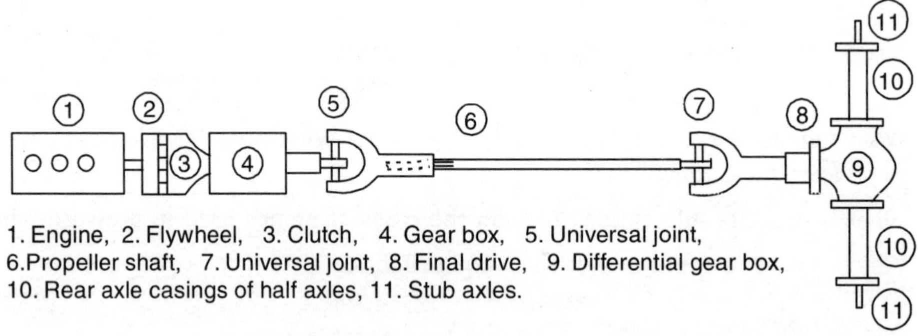

Fig. 5.1 Layout of transmission system of a rear axle driven vehicle

The layout also shows the various parts of the transmission system. The engine power transmitted to road wheels through clutch, gear box, propeller shaft and differential gear box in case of rear axle driven heavy vehicles. Most of the small and medium segment cars are now front wheel driven. Clutch may also be defined as a mechanism by means of which engine's main shaft is connected or disconnected from rest of the transmission. The clutch allows the rotary motion and power of the engine shaft to be cut-off without stopping engine, from the gear box shaft. In normal running, it always remains engaged and the engine power/torque continues to flow to the rest of the transmission system. The need of de-clutching is arised when gears are to be changed or braking is required. The clutch disconnects the engine from gear box shaft at the will of the driver to facilitate the braking and/or gear shifting.

Requirements of a Good Clutch

1. Gradual and smooth engagement without noise and jerks. (Between engine crankshaft and gear-box shaft).
2. Effortless operation for disengagement and re-coupling, it saves driver from fatigue.
3. Smallest possible size and weights.

   Moment of inertia, $i = mk^2$

   where m = mass of the clutch

   k = radius of gyration of rotating parts of the clutch about its axis of rotation.

   Lesser inertia increases smoothness in gear shifting.
4. Full transmission of power and torque: Clutches are designed taking into consideration more than 125% of the engine torque so that they may transmit engine torque in full to the transmission without any damage.
5. Dissipation of heat: The rubbing surfaces must have enough mass and area with passages for ventilation in dry plate clutches so as to absorb the heat produced during engagement. In multi-plate clutches, oil cooling is used.
6. The clutch must be dynamically balanced for smooth and noiseless running at high speeds.
7. Suitable provisions should be incorporated in the clutch plates for vibration and noise damping.

TYPES OF CLUTCHES

1. **Friction or mechanical clutches:** They are of many types such as:
   (*a*) Cone clutch
   (*b*) Single plate dry clutch
   (*c*) Multiplate wet clutch
   (*d*) Diaphragm clutch
   (*e*) Semi-centrifugal clutch
   (*f*) Centrifugal or Automatic clutch
2. **Fluid flywheel clutch**
3. **Electromagnetic clutch**
4. **Power (Booster) clutch**
5. **Fluid filled Electro-Rheological clutch**

Cone Clutch

The cone clutch is a very basic type of clutch. In this type the contacting friction surfaces are in the form of cones as shown in the Fig. 5.2. In its engaged position, the sliding cone remains in full contact inside the cone mounted on the engine shaft. The clutch spring keeps the sliding cone pressed over the engine shaft cone during running.

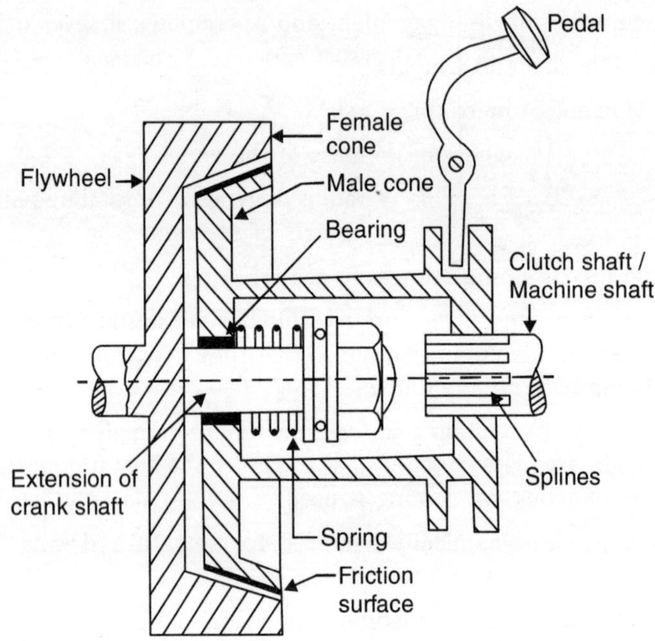

Fig. 5.2 Cone clutch

When clutch pedal is pressed, the sliding cone assembly slides back at splines cut on the clutch shaft and thus the clutch is dis-engaged. When the driver releases the clutch pedal, the spring pulls back the sliding cone assembly till the two cones get engaged. It is being used on no-gear small 2-wheelers. The normal force acting on the friction surfaces is larger than the axial force as compared to single plate clutch. Small wear of contact surfaces causes 'slip' between the two cones.

Single Plate Dry Clutch

Single plate dry clutch is used on almost all the Indian vehicles. A simplified diagram of the clutch is shown in the Fig. 5.3(*a*). It is mounted on the outer surface of the flywheel of the engine.

The clutch plate having friction lining on both sides is held between the flywheel and the pressure plate by means of clutch springs arranged circumferentially. The friction plate or the clutch plate is mounted on the splined portion of the clutch shaft. The clutch plate is also splined at inner side of its hub. The splined end of the clutch shaft rests at bearing fitted in the circular groove cut in the flywheel centre. It also consists of a clutch cover, pressure springs, release levers (fingers) and throw-out bearings. The cover, pressure plate and clutch springs form one unit. The whole unit is fixed to the revolving flywheel. The friction or clutch plate is the driven member. The clutch plate, held tightly between pressure plate and the flywheel, rotates the clutch shaft because of their splined compling. The clutch shaft transmits the motion and torque to the gear box shaft. View of clutch plate with one side lining removed is also shown in the Fig. 5.3 (b).

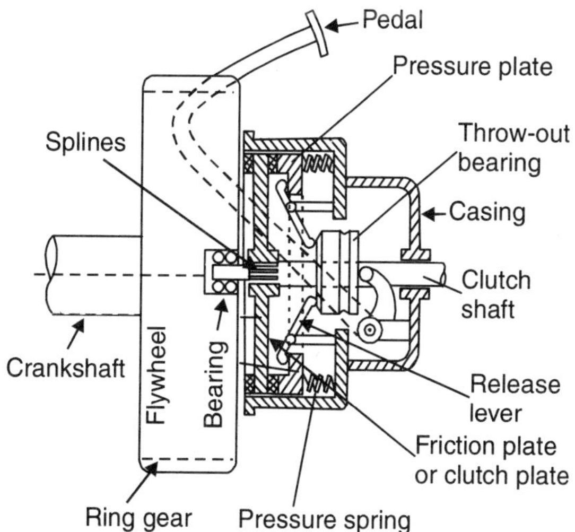

Fig. 5.3 (a) Single plate dry clutch

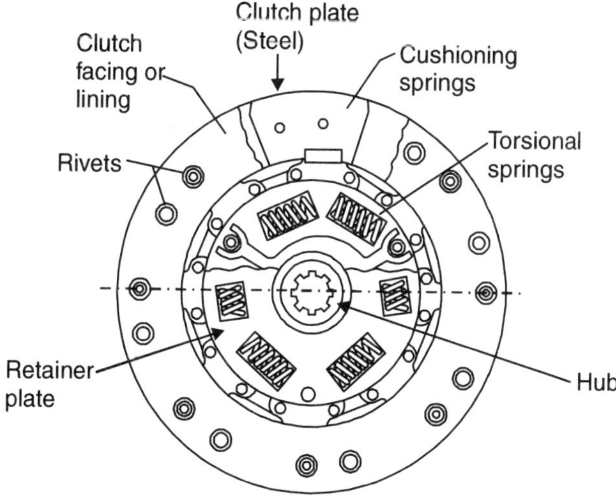

Fig. 5.3 (b) Construction details of friction plate

To disengage the clutch, the clutch pedal is pressed, a linkage pushes the throw-out bearing towards pressure plate. The throw-out bearing presses release levers which move back the pressure against the springs force thereby releasing the clutch plate. The clutch plate stops rotating and such the clutch shaft also stops rotating and the rest of the transmissions system is totally disconnected from the engine. When pedal is released, the spring force again holds the pressure plate pressing against the clutch plate and flywheel, allowing the power transmission again. However, it is possible to have the clutch plate partially engaged, allowing the clutch plate to slip by partial pressing of the clutch pedal. As a result, only a friction of the power from the engine reaches the drive or gear box shaft. Running of vehicle at very slow speed is required while passing through crowded roads. Slipping of clutch is also termed as *feathering* the clutch.

Riding the Clutch: It is the practice of some drivers to needlessly keep the foot on clutch pedal which may partially disengage the clutch. It causes (*i*) premature wear of clutch lining and (*ii*) the release or throw-out bearing also remains spinning which leads to premature bearing failure. To minimise this trouble, a small amount of free travel/play of clutch pedal is provided. During the Free Clutch Pedal Play (or travel) the clutch does not disengage.

Adjustment of Clutch: Free pedal play requires adjustment after sometime. Free pedal play is necessary due to wear of friction lining of clutch plate or due to wear of throw-out bearing's carbon ring which wears out due to the clutch-riding by driver. The wear of friction lining decreases the free pedal play, whereas the wear of carbon ring increases it. If free pedal play is more, clutch will not disengage fully. If free pedal play is decreased then clutch cannot engage fully. An adjustment nut is provided at the lower end of the clutch lever in Ambassador cars. Other vehicles have different systems for adjustment. Free pedal play has been illustrated in Fig. 5.4.

Refer to Fig. 5.4.

Pedal position (1) — Clutch fully engaged

Pedal position (2) — Free pedal play (about 25 mm)

Pedal position (3) — Clutch fully disengaged.

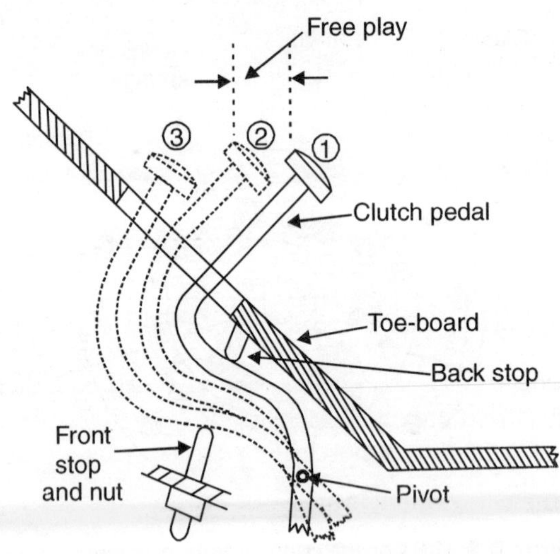

Fig. 5.4 Free pedal play

Multiplate Wet Clutch

Multiplate clutch operates more smoothly than single plate dry clutch. It consists of a number of small size frictions plates arranged between the pressure plate and the flywheel. The number of plates mounted on flywheel are more by one number than the number of clutch plates on pressure plate. The increase in the number of friction surfaces increases capacity of the clutch to transmit torque. The overall diameter of the clutch is reduced for the same torque transmission as compared to a single plate dry clutch. Multiplate clutch is used on racing cars for high torque transmission and invariably used on scooters and motor cycles due to limitation of available space. These clutches run in oil bath. When clutch operates, the oil is gradually squeezed out from the gaps between the plates, this results in smooth take up. The plates mounted on flywheel's inner splines, have grooves on its outer periphery whereas the plates mounted on pressure plate hub have internal grooves on their hubs which slide on the splines cut on the pressure plate hub. The internal splines of pressure plate hub drive the clutch shaft on whose surface external splines are cut. The clutch plates remain engaged by spring force. When pedal is pressed, the pressure plate moves back releasing flywheel plates which results in disengagement.

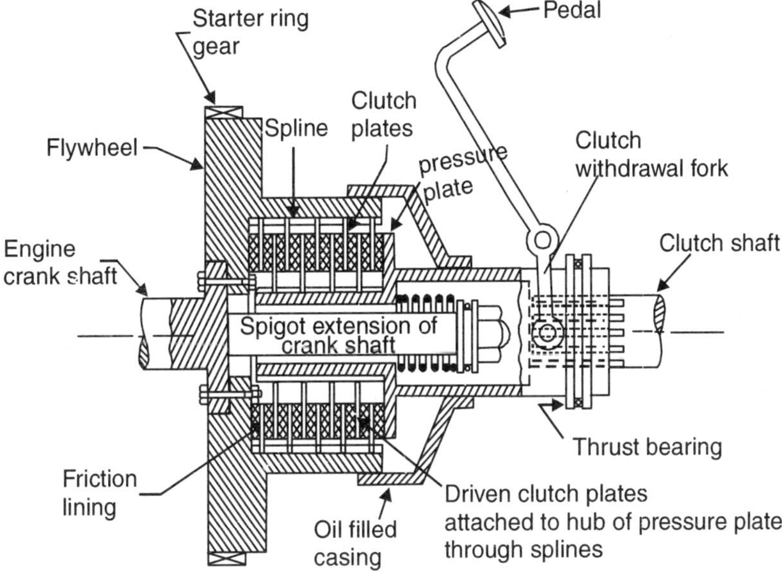

Fig. 5.5 Multiplate wet clutch

Diaphragm Clutch

In this type of clutch, a diaphragm is used in place of coil type clutch springs. It does not require release levers which makes this clutch light and simple.

In engaged position, the clutch plate is pressed between the flywheel and pressure plate under pressure by almost flat diaphragm spring. To disengage the clutch, the clutch pedal moves the throw-out bearing with slider pad towards pressure plate. The diaphragm spring when pressed in the centre takes the shape of a saucer and pushes back the pressure plate by its outer periphery and consequently clutch plate is released and starts rotating freely on clutch shaft splines.

This clutch is more compact than other clutches. It is easier to balance for rotation and it less subjected to centrifugal forces at high speeds. The diaphragm clutch or Lay Cock—Hausermann clutch is shown in Fig. 5.6. This clutch is fast becoming popular on auto vehicles due to its many favourable aspects.

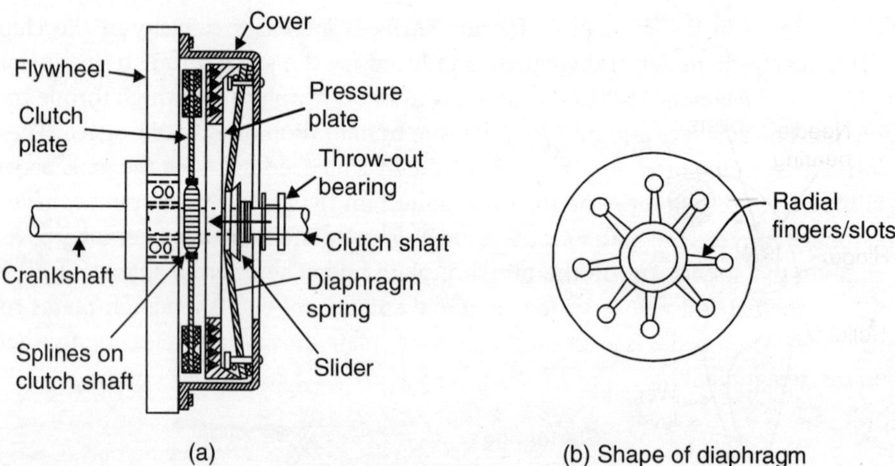

Fig. 5.6 (a) Diaphragm spring clutch in disengaged position diagram (b) Shows the shape of the diaphragm used in the clutch

Semi-Centrifugal Clutch

There is a little difference in a single plate dry clutch and semi-centrifugal clutch. In centrifugal clutch the shape of operating fingers or release levers is different. An extra weight A is provided at each of three weighted levers which acts on the pressure plate. When pressure plate rotates, centrifugal forces through the weighted arm of the levers, exert on extra pressure in addition to the pressure exerted by compression springs on the clutch plate via pressure plate. This force is sufficient/enough to prevent 'slip' at full engine speed. The centrifugal force is proportional to square of the speed ($F_c \propto N^2$). Vauxhall car employs such type of clutch. It is also used in racing cars. This clutch is shown in Fig. 5.7 in which only one compression ring and one weighted lever have been shown for simplicity. Rest of the working, *i.e.*, disengagement process is same as that of a single dry plate clutch.

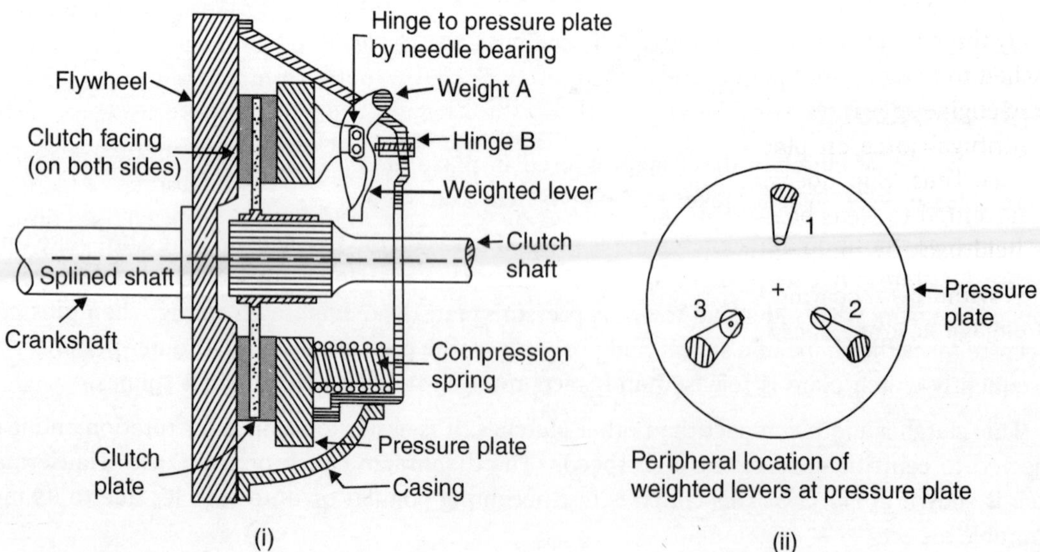

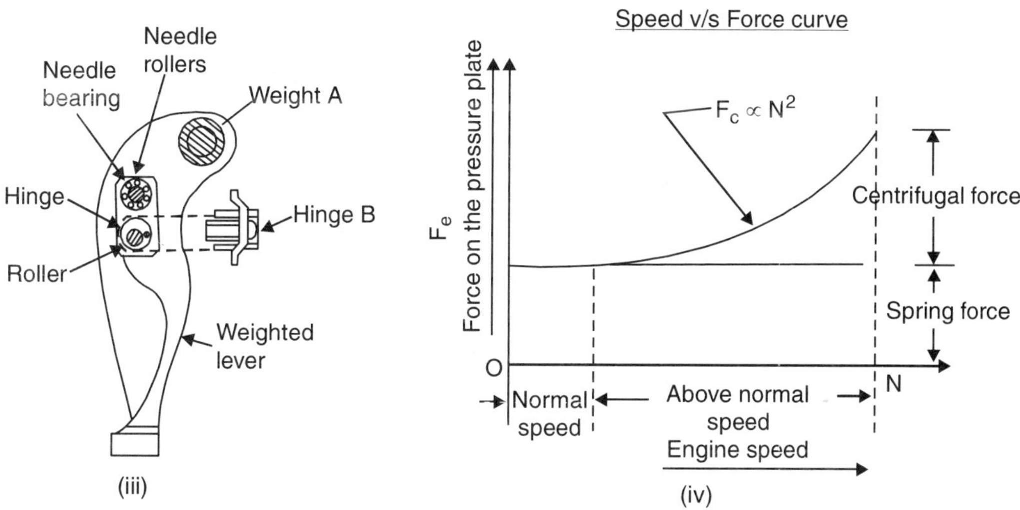

Fig. 5.7 SEMI-CENTRIFUGAL CLUTCH (i) constructional details, (ii) peripheral location of weighted levers, (iii) details of a weighted lever, and (iv) forces acting on the pressure plate at various engine speeds

Fully Centrifugal Clutch

This is an automatic type clutch. It needs no clutch pedal. The clutch automatically operates by change in speeds obtained by accelerator pedal. The clutch disengages automatically when engine runs at less than 600 r.p.m.

A Newton type fully centrifugal clutch is shown in Fig. 5.8. It consists of a number of clutch plates held between the flywheel and the pressure plate as usual. The pressure on the pressure plate is applied through a centrifugally controlled plate A and circumferentially placed and spaced equally three inner compression springs. This spring force is increased by similar number of weights attached to flywheel and plate A by pivoted levers alongwith outer springs as shown in the figure. As the engine speed increases the weights fly outwards compressing the outer springs and exerting a centrifugal force on plate A and pressure plate to hold the clutch plates more firmly at the flywheel. Thus, 'Slip' does not occur at high speeds with full loads and the full torque of the engine is transmitted to Gear box. When the clutch is disengaged due to low speed of engine, the plate A is held back by inner springs.

A withdrawal bearing is also provided to disengage the clutch at the times of sudden braking and engage at lower speeds.

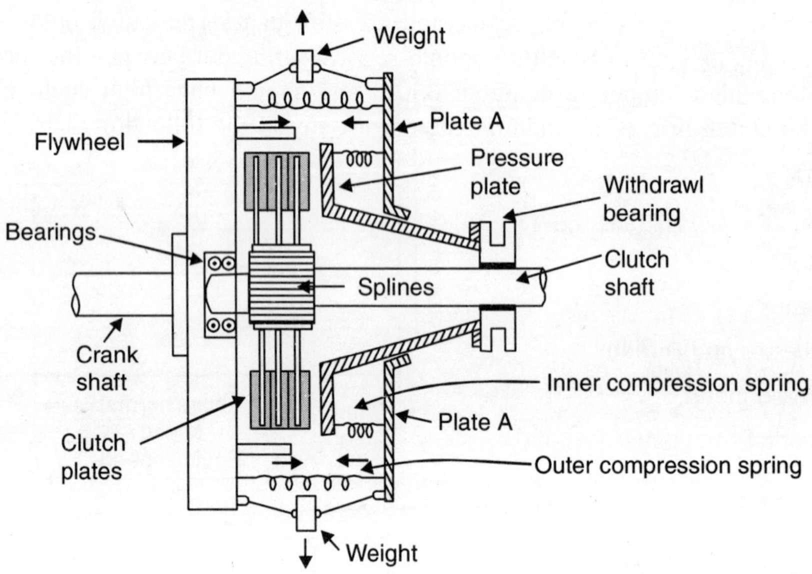

Fig. 5.8 Working of fully centrifugal clutch (disengaged)

Clutch for Front Wheels Drive Vehicles

In front wheel-drive cars, a transaxle is attached to the engine. Trans axle is a combined transmission and differential, with a clutch on manually shifted transmission.

A simplified diagram of power flow from the engine crankshaft, through clutch, to the gear box, differential assembly and then to wheels by half-axles, is shown in Fig. 5.9. The clutch in most of the small cars with transaxle is a single disc diaphragm spring clutch. The clutch, instead of being contained within a clutch housing, is fitted into the transaxle assembly as shown in the Fig. 5.9.

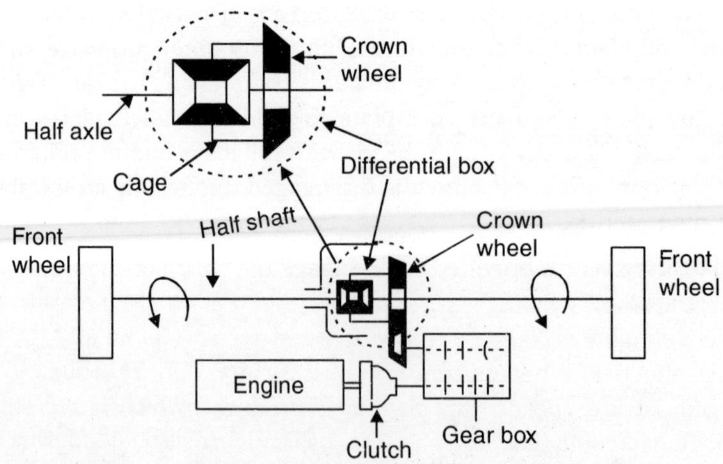

Fig. 5.9 Transaxle drive in front wheel drive vehicles

Clutch Lining: Clutch facing or lining is rivetted on both sides of the clutch plate for slipless frictional grip between the flywheel surface and pressure plate. The clutch lining should have a longer life and must be light in weight. It should resist wearing out between the rubbing surfaces, able to withstand high temperatures about 350°C and should have high coefficient of friction (0.4–0.75). The clutch linings are manufactured with any of the following materials:

(*i*) Fabric

(*ii*) Cork

(*iii*) Leather

(*iv*) Asbestos

(*v*) Reybestos and Ferodo

(*vi*) Sintered metal

(*vii*) Organic-fibre composites and

(*viii*) Some particulate matter such as metal-ceramic combination. Out of above materials, asbestos and reybestos are widely used.

Size of a Friction Plates

If r_1 = outer radius of friction plate in cm.

r_2 = inner radius of friction plate in cm. This is the radius just below the friction lining.

then effective radius $= \frac{r_1 - r_2}{2}$, say it is equal to r_e

The frictional area of the clutch $= \pi r_e^2$ (cm)^2

Some Indian vehicles has following outer diameters ($2r_1$) and frictional areas:

Fiat 1100—18.4 cm dia and about 365 cm^2 friction area.

Ambassador (Diesel DX)—21.6 cms dia and about 500 cm^2 friction area.

Tata LPS 1612/32 Tractor—31.0 cms dia and about 1030 cm^2 friction area.

Electromagnetic Clutch

Few models of Renault cars of France employ an electromagnetic type clutch which is shown in Fig. 5.10.

It contains coil windings housed in the flywheel rim side. The current to the windings is provided by Dynamo. When windings get energised they attract the pressure plate towards the flywheel and the clutch plate is clamped firmly between pressure plate and the flywheel and such the clutch becomes effective. When current is cut-off by a switch, controlled by clutch pedal, the clutch is disengaged. At low speeds, low current is produced which is not sufficient enough to engage the clutch. To overcome this problem, 3 to 4 springs are also added. Due to this defficiency, the electromagnetic clutches are not popular. Moreover, they also increase load on engine due to use of dynamo supply at flywheel rim.

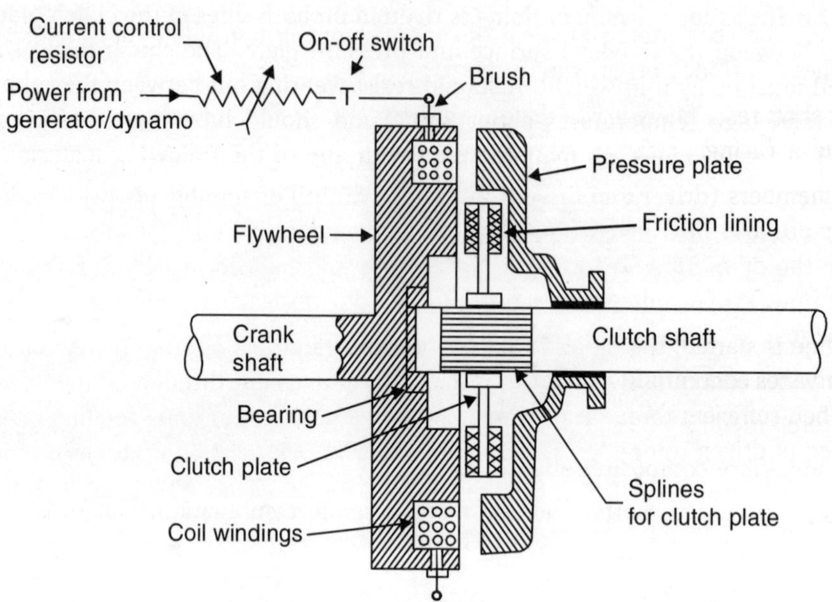

Fig. 5.10 Electromagnetic clutch

FLUID FLYWHEEL

In the vehicles using automatic transmission, a hydraulic coupling or a Fluid Flywheel is used in place of conventional friction clutch.

A simplified diagram of fluid flywheel is shown in Fig. 5.11. It has two main members or rotors:

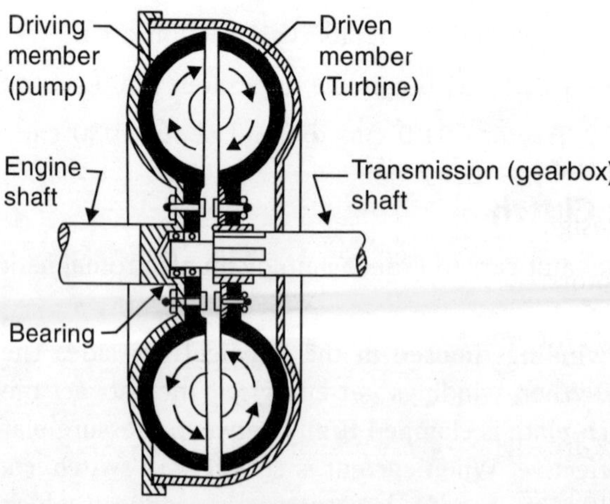

Fig. 5.11 Fluid flywheel

(*i*) A driving member attached to the engine shaft just after the Flywheel (called Pump or Driving Torus).

(*ii*) A driven member attached to gear box input shaft through splines at its end. The member is called "Turbine" or Driven Torus.

The input shaft rests freely inside the engine shaft end cavity on bearings. The whole assembly is placed inside a casing.

Both the members (driver and driven) are filled with oil of suitable viscosity under pressure. The two rotors are also provided with radial ribs to form a large number of oil passages. These passages guide the oil to flow in the desired direction and help to avoid the formation of eddy currents of oil flow. Proper oil seals are fitted to prevent oil leakage from the system.

When engine is started, the oil also rotates with torus and this oil from driver rotor strikes the stationary rotor vanes continuously and attains a circular movement through all the vanes as shown in Fig. 5.12. When sufficient torque is developed, the driven rotor also starts rotating but at a slower speed. The speed of driven rotor goes on increasing till it becomes equal to the speed of driver rotor. At this stage, engine torque is fully transmitted to the gear box shaft through the such formed oil coupling. At engine shaft r.p.m. below 500, the power transmission became ineffective due to failure of fluid coupling.

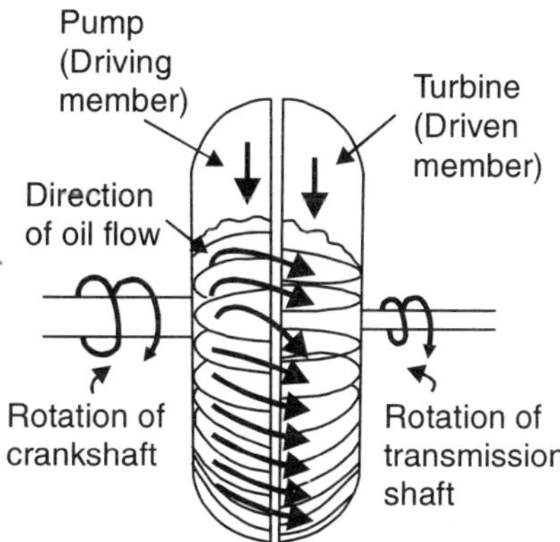

Fig. 5.12 Oil flow in a fluid coupling

The fluid flywheel has certain advantages over friction clutches:

1. Very simple design.
2. No wear of rotors.
3. No adjustments for wears.
4. Maintenance free.
5. Noise and jerk free when gears are changed.
6. Easy to operate.
7. Vehicle can be stopped in gear and may also be moved by simply pressing accelerator pedal.

Disadvantage

1. Levelling-up of oil is required. Some energy is lost in overcoming friction of oil.
2. Below 500 r.p.m. of engine, the slip is 100% and there comes a drag on the gear box shaft but epicyclic gear box takes it easily.

$$\% \text{ Slip} = \frac{N_1 - N_2}{N_1} \times 100$$

where N_1 = speed of driving member

N_2 = speed of driven member

100% slip means fully disengaged clutch, where 0% slip means fully engaged clutch.

CAUSES OF SLIP IN FRICTION CLUTCHES

(*i*) Adjustment of various links is not proper.

(*ii*) Free pedal play is insufficient or excessive.

(*iii*) Friction lining has worn out or become greasy or oily due to leakage of oil from gear box.

(*iv*) Clutch springs are weak or broken and worn out fingers.

Rattling of Clutch. It may be due to defect in pedal return spring or damaged clutch plate or loose/worn out release bearings.

Pulsations in Clutch Pedal. The clutch pedal pulsates due to clutch misalignment with flywheel or an unbalanced flywheel.

Disengagement become Difficult. When there is excessive free pedal play, warped pressure plate, buckled clutch plate, fingers broken, seized clutch plate in splines of the clutch shaft.

2. TRANSMISSION OR GEAR BOX

PURPOSE OF GEAR BOX

The purpose/functions of the gear box are:

(*i*) To change the speed of vehicle according to load, road and traffic demands.

(*ii*) To run the vehicles in reverse direction.

(*iii*) To provide a Neutral position/Neutral gear so that the engine power does not transmit to the rest of the transmission line upto the wheels even when the clutch is in the engaged position.

(*iv*) To provide a means to change the leverage or torque ratio between the engine and the road wheels and when the need arises to meet the different conditions of running.

There are many resistances which oppose the motion of vehicle. These resistances are:

(*i*) Self weights of vehicle including weight of passengers and luggage.

(*ii*) Wind resistance opposing vehicle movement.

(*iii*) Internal frictional resistances of the transmission system.

(*iv*) Resistances due to deformation of tyres and road conditions.

(*v*) Gradient resistance due to road gradient (uphill resistance).

In order to keep the vehicle running at an uniform speed, a tractive effort or a driving force exceeding the total value of all resistances has to be applied to it. If the tractive effort is less than the sum of all resistances, the vehicle speed will lower down. To provide adequate leverage to the vehicle against all resisting forces, a gear box is must for smooth movement of the vehicle.

The gear box transmits the desired torque under varying conditions of load and speed. The gear box itself does not increase or decrease the engine's tractive power but it changes the torque and speed for the same power output of the engine. At lower speeds, gear box increases torque and at higher speeds, it provides the lower but the desired torque. To move the vehicle from rest a very high initial torque at slow speed is needed to overcome high rolling resistance. Similarly more torque is needed when the vehicle is climbing up a hill. This is possible only with a gear box. We know that, an engine's Brake Power (B.P.) is determined by the following formula:

$$\text{B.P.} = \frac{2\pi\,NT}{60} \text{ watts.}$$

where N = speed in r.p.m.

T = Torque in N-m.

Since $\left(\frac{2\pi}{60}\right)$ is a constant

therefore, B.P. $\propto N \times T$

which means that for the same brake power, if N is increased, T will be decreased and if N is decreased, T will increase keeping the B.P. of the engine at the same value. Generally, 4 to 6 such combinations of N and T are provided by the gear boxes of the automobile vehicles.

TYPES OF GEAR BOXES

Many types of gear boxes are being employed on automobile vehicles. In the modern vehicles synchromesh type gear boxes are widely used. Some vehicles especially foreign models, use epicyclic gearings which are automatic type gear boxes.

On Indian vehicles, following three types of gear boxes are widely used:

1. Sliding mesh gear box
2. Constant mesh gear box
3. Synchromesh gear box.

Sliding Mesh Gear Box

It is also called as crash mesh type gear box. It is a very earlier type of gear box in which gears are shifted or meshed by sliding one on to the other. Due to difficulty in engaging the gears, lot of noise is produced in sliding and meshing them and sometimes teeth also got damaged during meshing. But it is very simple in design and is cheap.

A sliding mesh gear box is shown in Fig. 5.13. In this type of gear box, straight spur gears are used which are cheaper in cost but make noise in slide meshing.

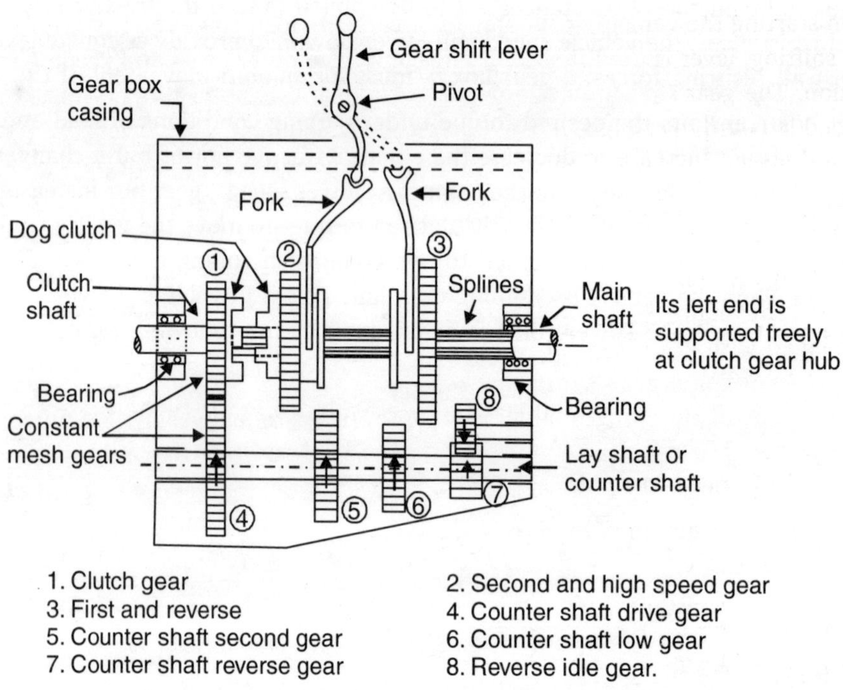

Fig. 5.13 Sliding mesh gear box with gears in neutral

While changing the gears, clutch must be fully disengaged and when gear shifting is over, the clutch should be slowly engaged to avoid jerk.

Referring to the above figure, gear number 1 is the clutch gear mounted on the clutch shaft. This gear, *i.e.*, clutch gear is always remains in mesh with the lay or counter shaft gear number 4 and all the gear's (4, 5, 6 and 7) are firmly mounted on lay shaft and remain rotating at the same speed as long as the clutch is in engaged position during engine running condition. A small pinion called reverse gear number 8 (idler) is also attached with lay shaft reverse gear No.7. In alignment with the clutch shaft, a main shaft is placed in the gear box, the outer end of which is connected to the final drive through propeller shaft in case of rear wheel drive or directly to final drive and differential in case of front wheel drive. Two gears (numbers 2 and 3) are mounted on the main shaft splines cut on its inner portion. These two gears may be shifted back and forth on the splines of the main shaft. These gears too have inner splines on their hubs. The gears are slided by means of a gear shifting mechanism having a lever and forks. A dog clutch is attached to clutch gear and on main shaft gear number 2 for direct meshing to transmit power in top gear at high speed. The gear box casing is filled with suitable grade oil upto half level for lubrication of gears.

1. When dog gears are meshed by sliding gear number 2 towards clutch shaft, the power from clutch shaft is directly transmitted to the gear box main shaft and then onwards.
2. When gear number 2 is slide back and engaged with layshaft gear number 5, the power is transmitted from clutch shaft to counter shaft gear number 5 to gear number 2 on main shaft. The main shaft to rotates at medium speed with more load. It is called second gear. When stopping the vehicle the lever should brought in Neutral position.
3. When starting the vehicle or running on a very crowded road, the first gear is used. The gear shifting, lever is used first to disengage it from other gears and keeping it in neutral position. The gear shift lever is now placed in gear number 3's fork. The gear number 3 is slided leftward to engage with gear number 6 on layshaft. The power is now transmitted through layshaft gear number 6 to main shaft gear number 3 which rotates the main shaft. This is the lowest or first gear. In this matching, the speed is very slow but torque is high enough to propel the vehicle from rest and at very slow speed.
4. For moving the vehicle in reverse direction, *i.e.*, to move back the vehicle, the reverse gear is used. The main shaft gear number 3 is further slided to right slide, it first gets disengaged with gear number 6 and takes Neutral position. It is further slided to mesh with reverse pinion gear number 8. The pinion acts as idler gear and has no effect on gear-ratio. It simply changes the direction of rotation of main shaft gear number 3 which moves the vehicle in reverse direction. Figure 5.14 shows how a reverse speed is achieved by interposing an idler gear between the driver and driven gears.

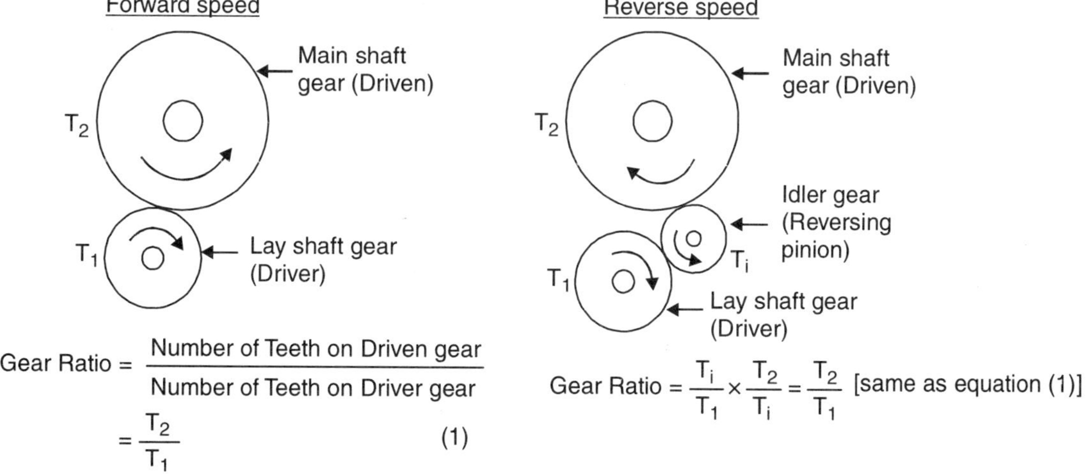

Fig. 5.14 Forward and reverse speed gears

It is to be remembered that to move vehicle backwards, it must be brought to rest first and then reverse gear should be applied, otherwise, gears may be damaged and vehicle will experience a serious jerk.

A Sliding Mesh Gear Box may have following gear/speed ratios on various gears:

Gear Ratio = (Speed of clutch shaft : speed of main shaft of gear box)

Top Gear 1.0 : 1

Second Gear 1.6 : 1

First Gear 2.7 : 1

Reverse Gear 3.5 : 1

Constant Mesh Gear Box

A constant mesh gear box is shown in Fig. 5.15. In this type of gear box, all the gears of main shaft remain in constant mesh with corresponding gears on the lay shaft. The gears on main shaft rotate freely on bearings mounted between their hubs and main shaft. The remaining portion of main shaft is splined over which two Dog Clutches 1 and 2, as shown in the figure, rotate with the shaft and can slide to and fro to engage and disengage with Dog teeth made on constant mesh gear's side faces. The gears mounted on layshaft are fixed on it and rotate with the layshaft. The power to layshaft gears flow from the clutch gear. All the gears are fitted in a gear boxcasing which is half filled with lubricating oil of suitable grade.

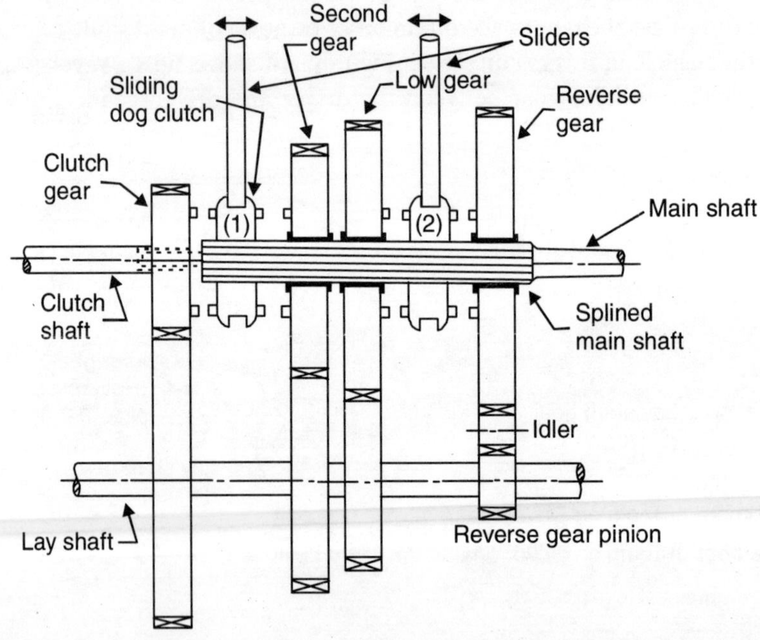

Fig. 5.15 Constant mesh gear box (3 forward and 1 reverse gears)

The dog-clutches are shifted by means of a selector mechanism.

1. When left dog clutch number 1 is slid to left side, its teeth get engaged with teeth on clutch gear and the top gear speed is transmitted to main shaft of gear box.

2. When left dog clutch number 1 is slid to right side, its teeth get engaged with teeth on second gear, the second gear becomes effective and a medium speed is obtained.
3. When right dog-clutch number 2 is slid to left side, its teeth get engaged with teeth on low gear, the slow speed in low gear is obtained.
4. When right dog-clutch number 2 is slide to right side, its teeth get engaged with teeth on reverse gear and the reverse speed is obtained.

The teeth on constant mesh gears are helical type which run quite noiselessly. All the teeth on dog clutches engage simultaneously, therefore, wear is reduced. The dog teeth are easier to engage than engaging the teeth of the gear wheels. The engagement of sliding gear wheels teeth takes place in between only two or three teeth at a time which causes lot of wear and tear and noise and sometimes breakage due to unequal speeds.

Constant mesh gearboxes employed on 2 and 3 wheelers very successfully. A six speed gearbox on Sitibus of Ashok Leyland and a 5-speed gearbox on Suzuki Shaolui are in use.

A two-wheeler uses the following gear ratios:

First gear—1:12.52, Second gear—1:8.76, Third gear—1:6.30, Fourth gear—1:4.82

Double De-clutching in Constant Mesh Gear Box

For smooth engagement of dog clutches, it is necessary that the speed of main shaft gear and sliding dog clutch must be the same. To obtain a lower gear, the speed of the layshaft and main shaft should be increased. This is done by double de-clutching to avoid 'clash' of teeth during engagement.

The clutch (main) is disengaged and the gear is brought to neutral. Again the clutch is engaged and accelerator pedal is pressed to increase the speed of the layshaft and main shaft gears. Then clutch is disengaged again, the dog-clutch is moved into the required lower gear and after engaging the dog-clutch, the main clutch is re-engaged and thus power through lower gear starts flowing. Since in this process, the main clutch is disengaged twice, the operation is called Double De-clutching.

For moving to higher gear from lower gear, reverse process is desired. The driver brings the gear in Neutral, waits for sometimes till the main shaft speed is decreased enough to engage higher gear smoothly.

Advantages of Constant Mesh Gear Box

1. Less noisy due to use of helical gears and running in constant mesh.
2. Wear of teeth is lesser.
3. Compact in size.

Synchromesh Gear Box

Synchromesh gear box is similar to constant mesh gear box (Fig. 5.16). The synchromesh gear box has synchromesh devices which avoid the double de-clutching every time the gears are to be changed. The synchronising cones first come in frictional contact and equalise the speeds of engaging gears, which facilitates the smooth engagement of gears thereafter. On small cars synchromesh devices are generally not fitted on low and reverse gears as the speeds in these gears are not so high to need synchronising. This also reduces the cost of the gear box. The low and reverse gears are sliding mesh type. The low or the first gear and reverse gears are applied when the vehicle is at rest for starting or reversing back respectively.

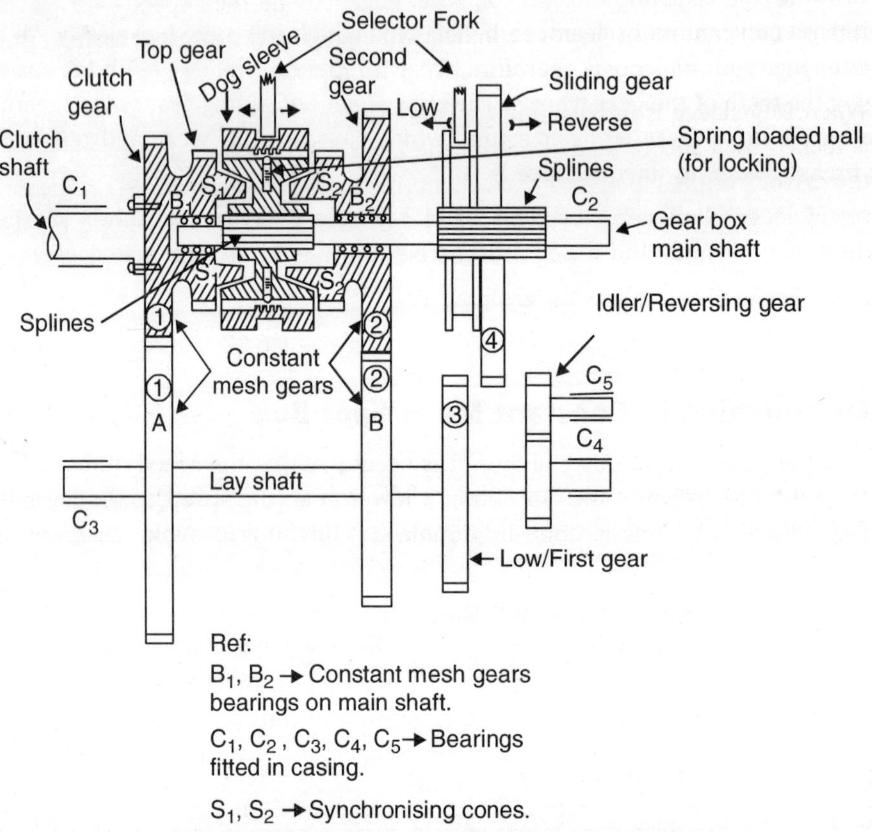

Fig. 5.16 Synchromesh gears with sliding gears

The use of synchronising cones an high speed gears reduces ratling noise, wear and tear in the gear teeth and saves time wasted in double de-clutching. Zen, Wagon R, Alto, Esteem, Baleno, Lancer, Matiz, Spark, Santro, Indica etc., have synchronising devices on all the gears. Maruti 800, Ambassador Mark II, Fiat 1000, Jeep CJ-3 have gears with synchronising devices at high speed gears and sliding system in low or first gear and reverse gear.

Both sets (1 + 1) and (2 + 2) of constant mesh gears (B_1 and B_2) on main shaft are free to rotate on it. They are in constant mesh with layshaft gear (1) and (2). All the gears on lay shaft are firmly keyed to it. The synchronising cones S_1 and S_2 can slide on the main shaft splines and transmit power.

1. When top gear is to be engaged the synchronising cone is slided towards left side with help of selector fork. First, the conical frictional surfaces come in contact and the speed

of synchronising cone becomes equal to the speed of top gear cone attached with clutch gear. Further movement of the selector fork, slides the dog sleeve gear with internal teeth on to the top gear teeth. Since top gear and dog-sleeve were already rotating at the same speed or at the synchronised speeds, the engagement of top gear with dog-sleeve teeth (internal) is very smooth and noiseless.

2. When second gear is to be engaged, the selector fork moves the synchronising unit towards right side and the two cones get engaged and their speeds are synchronised then further movement of selector fork, helps in engaging the internal teeth of dog sleeve gear with the teeth of second gear attached with constant mesh gear 2.

   When Dog-sleeve is in between the two constant mesh gears, *i.e.*, in Neutral Position, it is locked by a spring loaded ball in the centre. When Dog sleeve is moved right or left the driver's effort on selector fork is sufficient enough to overcome resistance offered by spring loaded ball. When top gear or second gear is firmly engaged by sliding Dog sleeve, the spring loaded ball (placed in a slot cut in the synchronising unit) locks the dog sleeve by setting its half portion in the half-round groove made below the Dog sleeve. In Fig. 5.16, the spring loaded ball is shown locked in Neutral position.

3. The low or first gear is engaged by sliding gear number 4 on layshaft gear number 3.

4. The reverse gear is obtained by sliding gear number 4 towards right side to engage with idler gear or reverse gear.

In most of the vehicles, all the gears are synchromesh type and sliding gears are not used. For synchronising low and reverse gears, similar type of synchronising unit (as shown for top and second gear) will be employed in place of sliding gears. (number 4).

Gear Ratios for Cars

Car	**First gear**	**Second gear**	**Third gear**	**Fourth gear**	**Top gear**	**Reverse gear**
Maruti 800	3.583	2.166	1.333	—	0.999	3.363
Jeep CJ-3	2.798	1.151	—	—	1.0	3.798
Zen, Wagon R, and Alto, Esteem	3.416	1.894	1.280	0.914	0.757	3.416
Santro	3.833	2.105	1.310	0.919	0.784	4.00
Indica	3.420	1.950	1.360	0.950	0.770	3.58
Ambassador Mark-II (Over all ratio)	18.559	10.983	7.392	4.875	1.0	18.559

The cones formed on gear hubs or bosses are made of special bronze and those on synchronising unit are made of case hardened steel, having radial slots for giving passages to the lubricating oil for cooling and lubrication.

Gear Selector Mechanism

The mechanism used for selecting the desired gear and sliding the same to engage with the corresponding gear on layshaft is called gear selector mechanism. They are made in two types:

1. Gear shift lever is mounted on the top of the gear box.
2. Gear shift lever is mounted on the steering column—It involves a complicated linkage but saves the space. Hindustan Ambassador Mark-II uses steering column shifting mechanism.

Most of the vehicles are provided with gear shift lever on top of the gear box because of its ease to operate and good efficiency.

A gear shifting mechanism with gear lever on the top of the box is shown in Fig. 5.17.

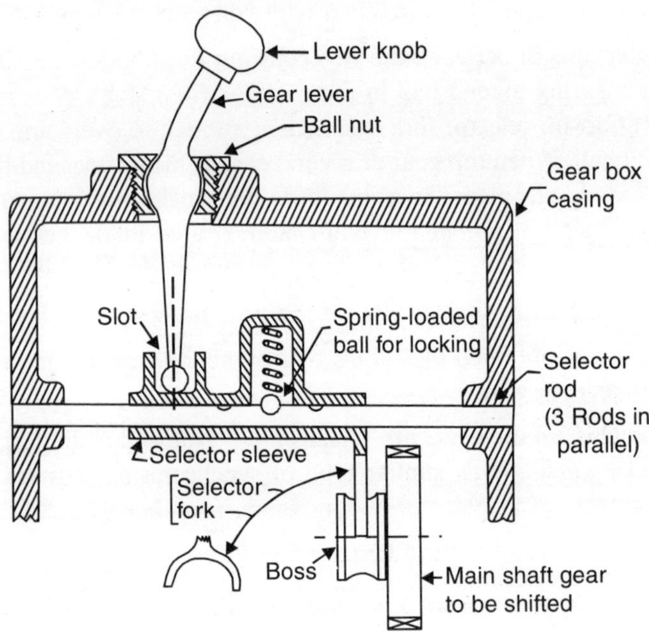

Fig. 5.17 Selector mechanism with gear lever on top of the gearbox

The gear lever is ball mounted in the gear box cover top. The ball joint facilitates the movement of lever in any direction. The lower end of the lever rests into a slot made in the spring loaded selector sleeve. There are forks mounted on the selector sleeves on three separate parallel selector rods which are supported in the gear box casing (only one selector rod is shown in the figure). Sleeves are provided with spring loaded balls. These balls resist the movement of the forks until some force is applied to gear lever to overcome their resistance. The ball helps in avoiding unwanted engagement of gears and lock the sleeve. Grooves are provided on gear bosses where selector forks can fit in. The cross-wise motions of the lever selects the proper fork which is to be engaged and longwise movement of lever slides the fork to engage the desired gear. Various gear positions are marked on the gear lever knob itself, which help driver to select the required gear (Refer to Fig. 5.18). An interlocking mechanism is also provided on the middle selector rod to ensure only one gear engagement at a time by any rod.

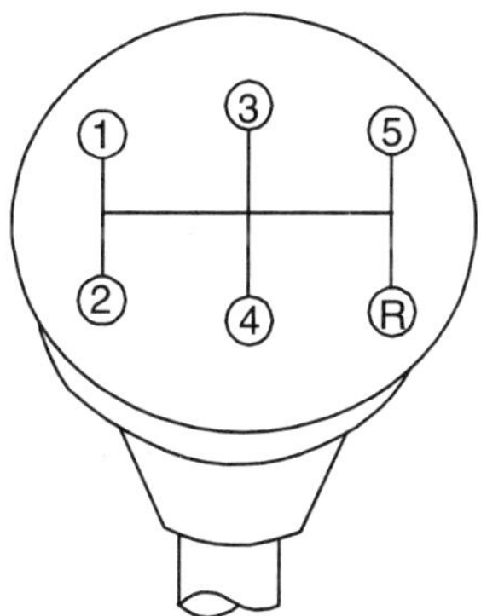

Fig. 5.18 Gear lever knob

TRANSFER CASE

In all the four-wheel drive vehicles, a transfer case is provided, which enables the driver to either drive in two-wheels drive on plain roads or shift to four-wheels drive in high or low gear for cross-country operation; when all the wheels require a greater tractive force. Four wheel drive is also required when one or two wheels of the car are plugged in muddy pit or when the vehicle is climbing a hill. The transfer cases are generally used on military vehicles and jeeps. Many passenger cars are also being provided the transfer cases for efficient running in all conditions.

The transfer case is shown in Fig. 5.19. It has two output shafts and one input shaft. The input shaft gets power from the main shaft of the main gear box. The front output shaft is connected to front axles through universal joint and front differential. The rear output shaft is connected to rear axles through a universal joint, propeller shaft and rear differential as shown in 5.19.

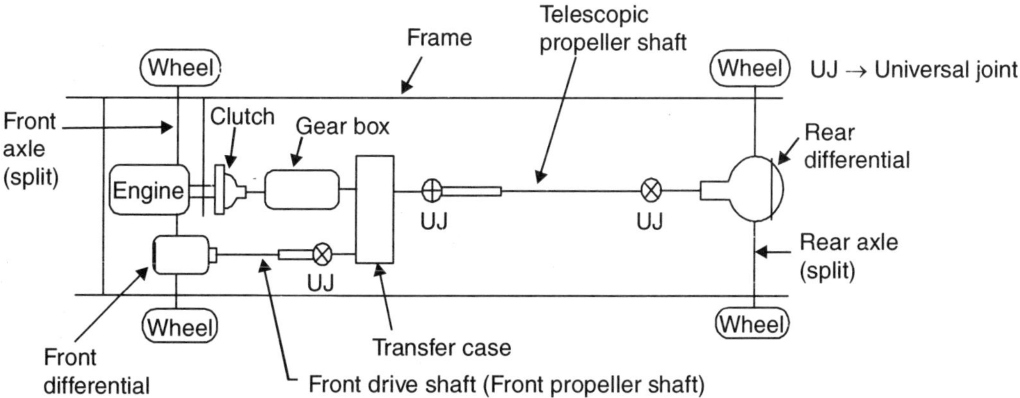

Fig. 5.19 Lay-out of a 4-wheel drive vehicle. The transfer case gives choice to driver to select rear-wheels drive or all the four wheels drive (2×2) or (FWD)

A detailed construction of the transfer case and its working has been shown in Fig. 5.20.

The sliding dog clutch-1 (Rear wheels) on the splines of the main shaft may be locked with low range gear A or high range gear B. A dog-clutch-2 on front axle drive gear (C) is also used to rotate or lock the front axle. In normal running, rear wheels act as driving wheels.

1. When sliding dog clutch (Rear wheels) is locked with high-range (speed) gear B and the dog-clutch on front axle is kept in Neutral, the power transmits to rear wheels only and high speed 2-wheel drive (rear) is obtained. (Fig. 5.20 (1))
2. When rear dog clutch-1 is engaged with gear B and front axle clutch-2 with gear C, a high-speed range four wheel drive is obtained. (Fig. 5.20 (2))
3. When sliding rear clutch is locked with gear A and front axle clutch-2 with gear C, a low-speed 4-wheel drive is obtained (Fig. 5.20 (3). This combination gives maximum torque to wheels, about 2.5 times the normal torque available in 2-wheel drive. Wheels A, B and C are bearing-mounted on their shafts and rotate freely. They transmit power only when they are engaged with dog clutches. A simple transfer has been in shown Fig. 5.20, with various combinations for different drives. The epicyclic gear transfer cases are used with automatic transmissions.

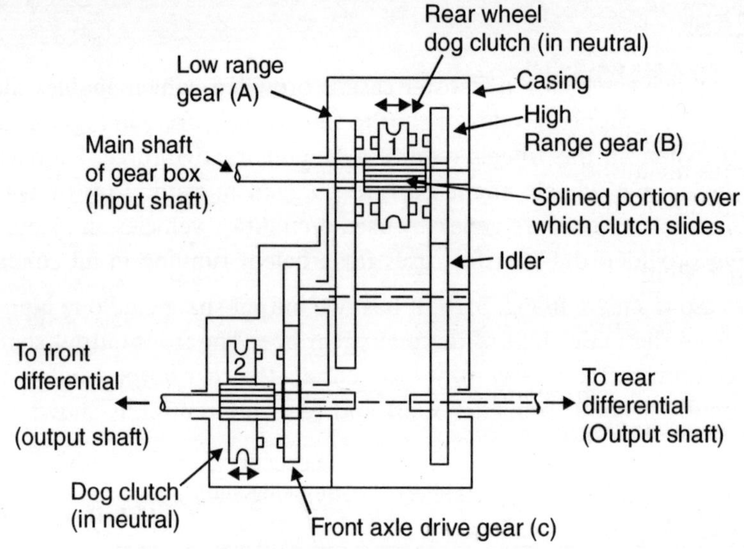

Transfer case gears in neutral

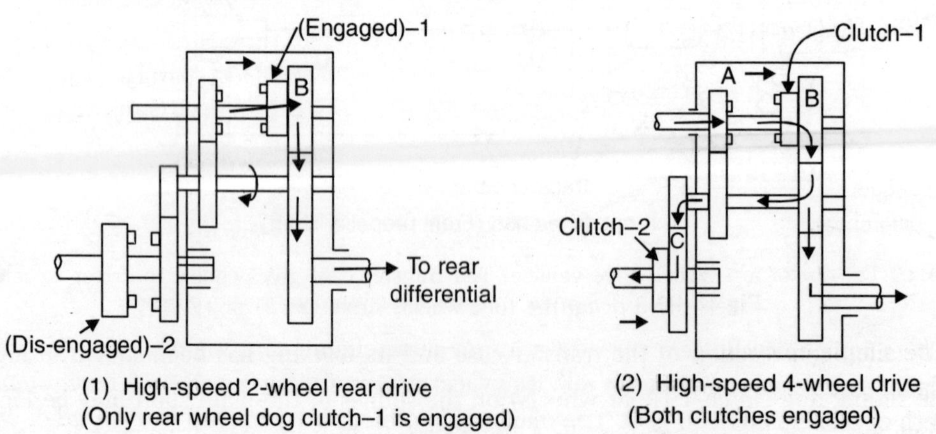

(1) High-speed 2-wheel rear drive
(Only rear wheel dog clutch–1 is engaged)

(2) High-speed 4-wheel drive
(Both clutches engaged)

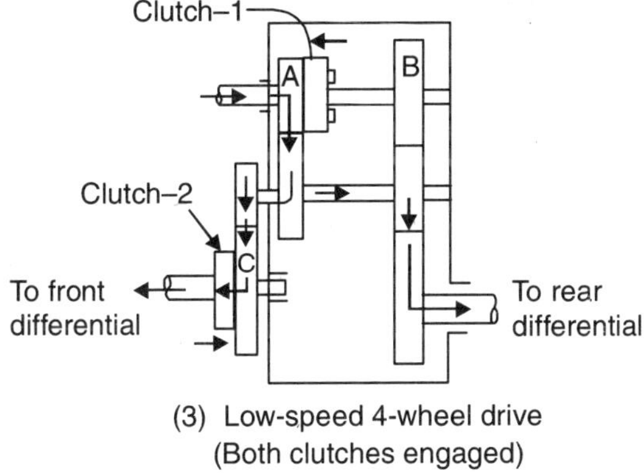

Fig. 5.20 Various speed combinations in transfer case

AUTOMATIC TRANSMISSION/GEAR BOX

Automatic transmissions are operated by controlling speed and engine load. The accelerator pedal's movement actuates the automatic system of epicyclic gearing and according to load variations, the speed is automatically increased or decreased as per the driver's need.

Epicyclic Gear Box

An automatic transmission utilises epicyclic gearing system. The principle of working of epicyclic gears may be well be understood by the simple epicyclic gear set shown in Fig. 5.21 (a). In actual practice, Wilson automatic gear box is used, whose construction is very complex. It is very difficult to draw its line diagram.

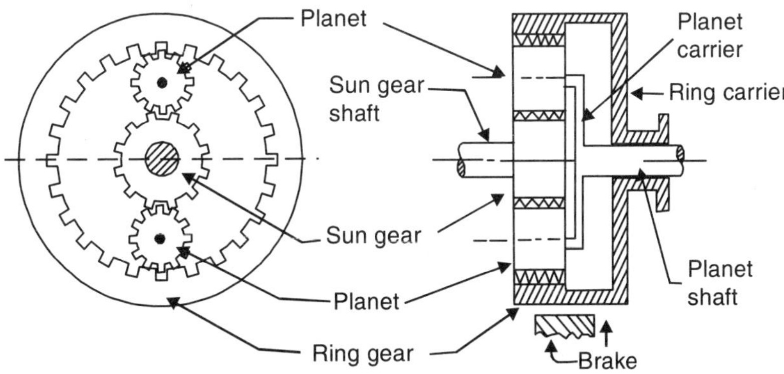

Fig. 5.21 (a) Simple epicyclic gear automatic transmission

The simple epicyclic gear set shown in the above figure, consists of a sun gear, about which two planet gears turn round inside the Ring Gear or Annulus having internal teeth matching with the teeth of planets and sun gear. The planet gear pinions are attached to each other by a carrier which is also attached to a shaft, called *planet shaft*.

Different Torque Ratios or Velocity Speed Ratios may be obtained by braking/locking anyone of the parts, *i.e.*, sun gear or planet gear or ring gear. By locking two parts with each other, a direct drive gearing is obtained. The epicyclic gear shown in the figure above, may have the following speeds—4 forward and 2 reverse:

(*a*) Lock the sun gear, planet carrier is driver and ring gear is driven—we get fast forward speed.

(*b*) Lock the ring gear, planet is driver and sun gear is driven—we obtain very fast forward speed.

(*c*) Lock the sun gear, ring gear is driver, planet carrier is driven—a slow forward speed is obtained.

(*d*) Lock ring gear, sun gear is driver and planet carrier is driven—a very slow output is available.

(*e*) Lock the planet carrier, make ring gear driving and sun gear driven—a fast reverse speed is obtained.

(*f*) Lock the planet carrier, sun gear is made driver and ring gear driven—a very slow reverse speed will be obtained.

In practice, all the above settings cannot be obtained due to complexity of the construction. Some of the settings are avoided to make the design easy to fabricate the gear box.

The above six settings may be summarised as given in the following table: S—Sun gear, P—Planet carrier, R—Ring gear.

Speed combination	**Fast forward**	**Very fast forward**	**Slow forward**	**Very slow forward**	**Fast reverse**	**Slow reverse**
Braked	S	R	S	R	P	P
Driver	P	P	R	S	R	S
Driven	R	S	P	P	S	R

THREE SPEED AUTOMATIC TRANSMISSION

A schematic diagram of a 3-speed automatic transmission using torque converter is shown in Fig. 5.21(b). It will be seen that the turbine unit of the converter drives the drive plates of the front and rear clutches through the input shaft. Both the clutches are operated with the help of hydraulic controls. In the planetary gear system, there is a small and large sun-gear as well as a small and long planet pinion. The small sun-gear becomes the driving member of the system when the front clutch is engaged. When the rear clutch is engaged, the large sun-gear becomes the driving member of the system. The ring gear is the driven member of the planetary gear system and transmits power to the output shaft. Let us now see how the different speeds are obtained with this set-up.

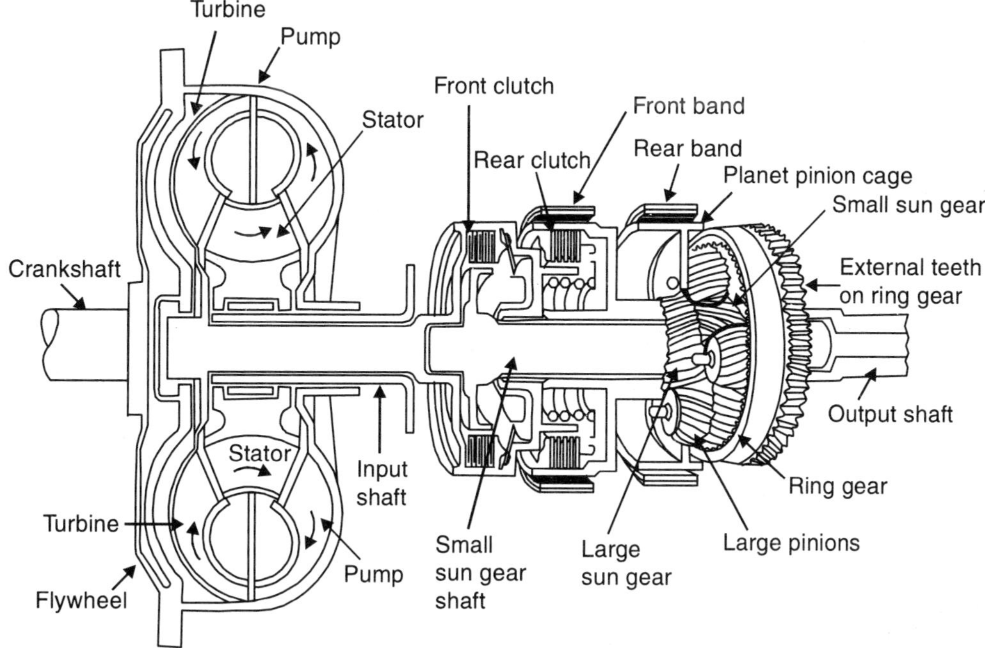

Fig. 5.21 (b) Cutsection of three-speed automatic transmission with torque converter

First Gear

To engage first gear, the front clutch is engaged and the band brake is applied to the rear drum thus holding the planet cage stationary. Under the condition, the power flows from engine crankshaft to the pump, from there to the turbine, to the input shaft, to the front clutch, to the small sun-gear, to the small planet pinion, to the long planet pinion, to the ring gear and then to the output shaft.

Second Gear

Second gear is obtained by applying the front clutch and also applying the band brake on the front drum. The application of the front clutch makes the small sun-gear the driving member while the application of the band brake holds the larger sun-gear stationary. Under this condition, the power from the converter flows to the small sun-gear through the front clutch and input shaft, to the small planet pinions and then to the long planet pinions which revolve around the stationary large sun-gear thus causing the planet pinion cage to rotate, and thereby driving the ring gear. The ring gear in turn drives the output shaft.

Third Gear

To obtain third gear, both the clutches are engaged whereas both the band brakes are released. Under this condition, the entire unit is locked and power is transmitted directly from the converter to the output shaft. It is also called the direct drive since no reduction takes place.

Reverse Gear

Reverse gear is obtained by engaging the rear clutch and also by applying the band brake to the rear drum thus holding the planet pinion cage stationary. Under this condition, the power flows from the converter to the larger sun-gear through the input shaft and from there to the long planet pinions and then to the ring gear. In turn, the ring gear transmits power to the output shaft. The output shaft rotates in the reverse direction since the planet cage is stationary and the long planet pinions act as reverse idlers.

Neutral Gear

In this position, both the clutches are released thereby the power from the converter does not get transmitted beyond the drive plates of the clutches.

Parking Brake

The ring gear is provided with external teeth. In this position both the clutches are released and the parking pawl is inserted in the external teeth of the ring gear thereby locking the ring gear and the output shaft to the casing of the transmission. Under this condition, the rear wheels of the vehicle cannot move.

ALGEBRAIC METHOD OF DETERMINING VELOCITY—RATIOS IN AUTOMATIC TRANSMISSION

The principle of this method may be described as under. 'The gear ratio of a pair of mating gear wheels with respect to the link carrying the axes of the gears is always the same whether the link carrying the axes is moving or fixed'.

Referring to Fig. 5.22, the two gears A and B are mating with each other and are connected by a link 'L'.

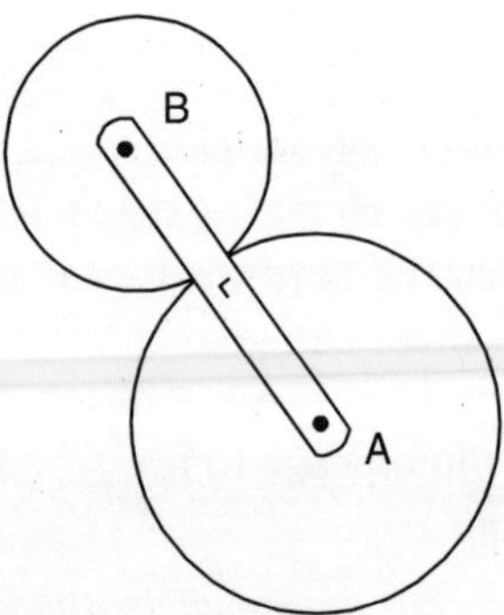

Fig. 5.22 Algebraic method for determining velocity ratio

According to algebraic method, the speed ratio:

$$\frac{N_A}{N_B} = (-)\frac{T_B}{T_A} = \frac{\text{Speed of gear A w.r.t. Link 'L'}}{\text{Speed of gear B w.r.t. Link 'L'}}$$

where T_A = Number of teeth on gear A; N_A = R.P.M. of gear A

T_B = Number of teeth on gear B; N_B = R.P.M. of gear B.

In the above formula, $\left(\frac{T_B}{T_A}\right)$ is unaffected by the condition in which the link 'L' is either fixed or moving.

Advantages of Epicyclic Gears

1. Change from one velocity ratio to another can be made without loosing any traction effort.
2. There is no engagement or disengagement of gear teeth or dog-clutches which make lot of noise. So it is quiter in operation.
3. It is very compact compared to the conventional gear box.
4. The loads coming on epicyclic gear teeth are lesser because the load is shared by at least two parts with 2 or 3 teeth sharing the load at a time.
5. By employing several planet gears the load on bearings can be balanced perfectly, whereas the shaft bearings in a conventional gear box are heavily loaded.
6. The gear changing processes take place by accelerator pedal movement and thus driver is saved from frequently moving gear shift lever for various running conditions.

GEAR BOX LUBRICATION

The lubricating oil used for gear boxes is generally SAE–90. The oil is completely changed at the intervals specified by the manufacturer. As a rough estimate, it is changed at every 25000 kms. Proper oil seals are fitted to check the leaks past the input shaft to the clutch, oil seal is also fitted at bearings of the output shaft. A breather hole is also provided at the top of the gear box casing to relieve any excess pressure caused by oil vapours. The oil is filled up to the prescribed level or the prescribed quantity of oil is filled up. The actual grade/name of oil is prescribed by the vehicle manufacturer.

DESIGN OF A SLIDING MESH GEAR BOX

Gear speeds/ratios help in the designing of a gear box.

SOLVED EXAMPLES

Example 1: *A sliding mesh gear box is shown in the Fig. 5.23 with 3-forwards speeds and one reverse speed. If teeth on gear A are 15, B has 20, C has 14 and D has 28 teeth. Calculate the speed of propeller shaft when low gear is engaged and the speed of engine is 2400 r.p.m.*

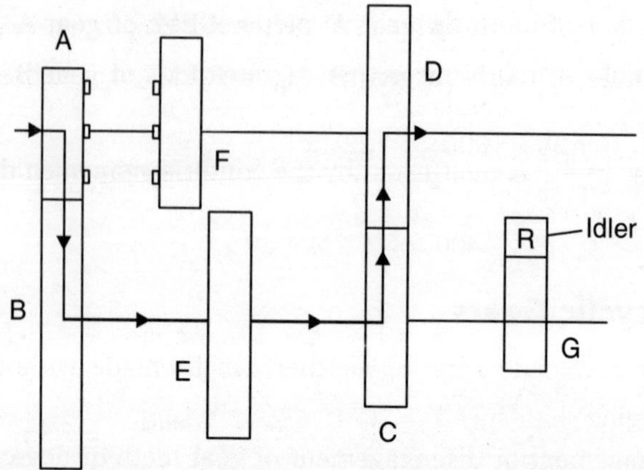

Fig. 5.23 Three forward and one reverse speed gear box

Solution: Gear ratio between gears A and B $= \frac{20}{15} = \frac{4}{3} = \frac{T_B}{T_A}$ ...(*i*)

Gear ratio between gear C and D $\frac{T_D}{T_C} = \frac{28}{14} = \frac{2}{1}$ ...(*ii*)

Multiplying equations (*i*) and (*ii*)

$$\text{Combined Gear Ratio} = \frac{T_B}{T_A} \times \frac{T_D}{T_C} = \frac{4}{3} \times \frac{2}{1} = \frac{8}{3}$$

$$\therefore \quad \text{Speed of propeller shaft} = \frac{\text{Speed of engine shaft}}{\text{Combined gear ratio}}$$

$$= \frac{2400}{\frac{8}{3}} = 900 \text{ r.p.m.}$$ **Ans.**

Example 2: *A clutch gear pinion of a gear box has 17 teeth it meshes with lay shaft gear having 32 teeth. A gear mounted on layshaft having 21 teeth is engaged with a 27 teeth gear on main shaft. Differential gear ratio is 4.82 : 1 in a rear wheels driven vehicle with wheels having an effective diameter of 73 cms. If the vehicle is running at velocity of 24 kms per hour, determine (i) the gear ratio and (ii) engine shaft speed.*

Solution: (*i*)

$$\text{Gear ratio} = \frac{\text{Teeth of driven gears multiplied}}{\text{Teeth of driver gears multiplied}}$$

$$= \frac{32 \times 27}{17 \times 21}$$

$$= 2.42 : 1$$ ...(*i*)

If r.p.m. of wheels are N and diameter is D meter. Given : D = 0.73 m

Then velocity of vehicle = π D N meter per minute.

$$\text{Then } 24 \text{ km/hr} = \frac{24 \times 1000}{60} = \frac{22}{7} \times 0.73 \times N \text{ m/mt}$$

or $$400 = 2.3 \text{ N}$$

∴ $$N = \frac{400}{2.3} = 174 \text{ r.p.m.}$$

$$\text{Engine shaft speed} = 174 \times 2.42 \times 4.82$$

$$\simeq 2030 \text{ r.p.m.}$$ **Ans.**

Example 3: *An engine develops 10 kg-m torque at 2500 r.p.m. If gear ratio is 2.1 : 1 and speed ratio at differential is 5 : 1. Find torque at wheels and speed of propeller shaft.*

Solution: (*i*) Speed of propeller shaft $= \frac{2500}{2.1} = 1190.5$ r.p.m.

(*ii*) Torque at rear wheels = 10 × 2.1 × 5 = 105 kg.m **Ans.**

Example 4: *A 3-forward speed and one reverse speed gear box (sliding mesh type) has to be designed. The gear box should have the following gear ratio:*

Top gear — 1 : 1, second gear—1.5, third gear—2.5 and reverse gear—3.9.

The distance between the axes of main shaft and layshaft is to be kept at 80 mm. The clutch gear should have 16 teeth with a diametral pitch of 3.20 mm.

Solution: Refer to Fig. 5.23 of Example 1.

(*i*) **Top Gear** ratio is 1 : 1, the dog-teeth on both gears A and F will be equal, which may be decided as per gear sizes.

(*ii*) **Second Gear:** The power flow will be from gears A–B–E–F.

$$G_2 \text{ Gear ratio} = 1.5 = \frac{T_B \times T_F}{T_A \times T_E} \quad \text{...(i)}$$

(*iii*) **Third Gear:** The power flow will be from gears A–B–C–D

$$G_3 = 2.5 = \frac{T_B \times T_D}{T_A \times T_C} \quad \text{...(ii)}$$

(*iv*) **Reverse Gear:** The power transmission will be from gears A–B–G–R–D. As the Idler gear has no effect on gear ratio.

$$G_4 = 3.9 = \frac{T_B}{T_A} \times \frac{T_R}{T_G} \times \frac{T_D}{T_R}$$

$$= \frac{T_B}{T_A} \times \frac{T_D}{T_G} \quad \text{...(iii)}$$

Since Dia. of gear A + Dia. of gear B = 2 × 80 = 160 mm. (where 80 mm is the distance between main shaft and layshaft).

$$\therefore \quad T_A + T_B = T_E + T_F = T_C + T_D = \frac{160}{3.20} = 50$$

(where 3.20 mm is the diametral pitch)

Since $\quad T_A = 16$ teeth

$$\therefore \quad T_B = 50 - 16 = 34 \text{ teeth}$$

Substituting in equation (*i*) for second gear:

$$1.5 = \frac{T_B}{T_A} \times \frac{T_F}{T_E} = \frac{34}{16} \times \frac{T_F}{T_E} \text{ or } \frac{T_F}{T_E} = 0.70$$

$$\therefore \quad T_F = 0.70\, T_E$$

Since $\quad T_F + T_E = 180$

or $\quad 0.70\, T_E + T_E = 50$

$$1.70\, T_E = 50$$

$$T_E = \frac{50}{1.70} = 29 \text{ teeth}$$

$$T_F = 50 - 29 = 21 \text{ teeth}$$

$$\text{Third gear ratio} = 2.5 = \frac{T_B \times T_D}{T_A \times T_C} = \frac{34}{16} \times \frac{T_D}{T_C}$$

or $\quad \dfrac{T_D}{T_C} = 1.178$

$$\therefore \quad T_D = 1.178\, T_C$$

Since $\quad T_C + T_D = 50$

or $\quad T_C + 1.178\, T_C = 50$

or $\quad T_C = \dfrac{50}{2.178} = 23$ teeth

$$\therefore \quad T_D = 50 - 23 = 27 \text{ teeth}$$

$$\text{Reverse gear ratio} = 3.9 = \frac{T_B}{T_A} \times \frac{T_D}{T_G} = \frac{34}{16} \times \frac{27}{T_G}$$

or $\quad T_G = \dfrac{34 \times 27}{16 \times 3.9} = 15$ teeth.

Idler pinion size and teeth may be chosen according to the space available for fitting it over the gear G. The size of idler has no effect on gear ratio.

Now, actual gear ratios may be calculated:

1. Top gear — It will be as it is, *i.e.*, 1 : 1

2. Second gear $= \dfrac{T_B}{T_A} \times \dfrac{T_F}{T_E} = \dfrac{34}{16} \times \dfrac{21}{29} = 1.538 : 1$

3. Third gear $= \dfrac{T_B}{T_A} \times \dfrac{T_D}{T_C} = \dfrac{34}{16} \times \dfrac{27}{23} = 2.495 : 1$

4. Reverse gear $= \dfrac{T_B}{T_A} \times \dfrac{T_D}{T_G} = \dfrac{34}{16} \times \dfrac{27}{15} = 3.825 : 1$ **Ans.**

OVERDRIVES

In most of the advanced foreign countries where roads are plain and horizontal, on which cars can run in top gears for long hours and long distances; a device called overdrive, is used to step up gear ratio, generally, in top gear only. In Europe and America, overdrives are very popular on luxury and sports cars. The overdrive enables a high cruising speed (long journey at higher speeds) about (20 to 25%) higher speed than the maximum speed of the engine. Sports cars employ overdrive in each gear. Overdrive may be fitted in the gear box itself or outside of the gearbox at output shaft of the gear box.

Advantages

(*i*) Lesser wear of engine parts as engine runs at lower speeds.

(*ii*) Low noise and less vibrations.

(*iii*) Fuel saving due to low friction losses.

(*iv*) Higher gear ratio than top gear ratio; car speed is more than engine speed.

(*v*) Lesser running cost.

(*vi*) Longer maintenance periods.

The overdrives are operated either manually or automatically at a pre-fixed speed.

Working of Overdrive

It is an epicyclic gear train, in which sun gear is free to rotate over the bearings on the input shaft (see Fig. 5.24). Planet carrier can move on the splines cut on the input shaft. A free wheel clutch is also fitted on its splines. The ring gear is connected to the output shaft.

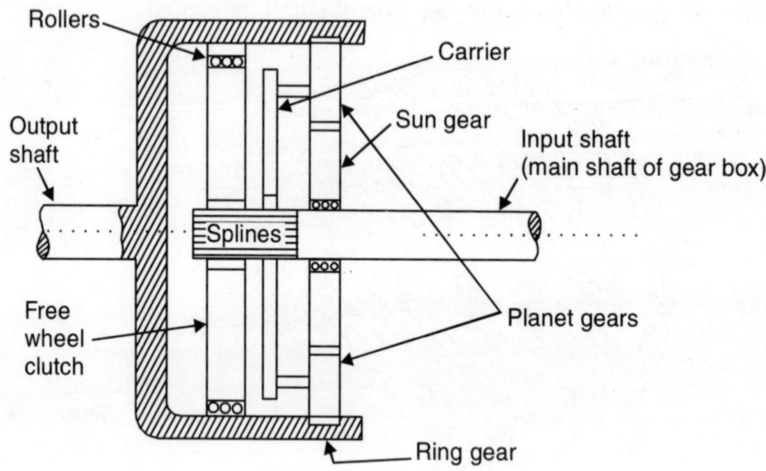

Fig. 5.24 The overdrive

On locking the sun gear with casing, it becomes stationary and the speed of output shaft is increased, which means, overdrive has become effective.

If sun gear is locked to the carrier or to the ring gear (annulus), the normal drive is obtained.

When sun gear is rotating freely on the input shaft, there is direct drive through the freewheel clutch when engine is running fast and developing power. When engine is simply idling, the output shaft will try to back overdrive the input shaft. Now the rollers of the freewheel clutch no longer remain wedged in the clutch and allow the output shaft to free wheel *i.e.*, rorate freely without affecting the input shaft. Due to free wheeling like this way, the driver has only to lift his foot off the accelerator pedal to change the desired gear without operating main clutch.

Overdrives on Indian Vehicles

Overdrive ratio in Tata Sumo (Deluxe) in 5th gear → 0.8 : 1

Overdrive ratio Ambassador diesel car 5th gear → 0.775 : 1

Overdrive ratio Contessa classic 1.8 GL car 5th gear → 0.775 : 1

Free Wheel Clutch

It is also called "One way clutch". The vehicles using overdrive and/or epicyclic gear train transmission are provided with a freewheel clutch. It is very much similar in operation to the free wheel used on bicycles. The free wheel clutch is shown in Fig. 5.25. The inner rotating member (3-lobes cam) is mounted on the splines of the main gear box shaft or input shaft.

The inner lobes are fitted with three spring loaded rollers in their steps. When inner member is rotating clockwise as shown in the figure, the outer ring (driven) will also rotate in the same direction due to firm wedging action by rollers. But when the outer ring (fitted inside Annulus case) becomes the driver (it happens when car is moving down hill with engine stopped or clutch disengaged) the inner member will not rotate with outer one because the wedging action by rollers is relieved. Thus gear box and engine are isolated from driving wheels which also results in saving in fuel consumption. In reverse drive, the free wheel clutch is locked.

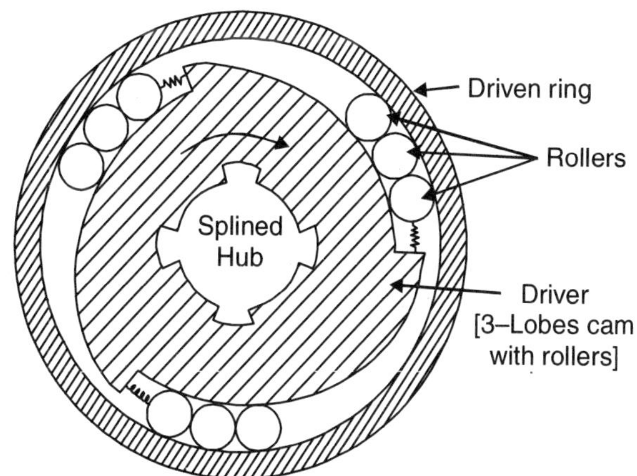

Fig. 5.25 Free-wheel clutch

There are some advantages of using a free wheel clutch:

1. The engine can idle without bringing the gears in Neutral and disengaging the main clutch.
2. Wear is reduced in transmission because during free wheeling engine and gear box are disconnected from propeller shaft.
3. While driving on long distances on downhills, the engine may be stopped which saves about 20% fuel. The brakes must be more effective to be used on downhill journey as freewheeling car or vehicle has less friction.

Freewheel is also employed alongwith torque converter in modern automatic transmissions.

TORQUE CONVERTER

The torque converter is a device to increase the engine torque two to three times. In construction, it is similar to a fluid flywheel. It employs an additional stationary member, called stator, fitted with vanes or blades.

Figure 5.26 shows the working of a torque converter. It consists of three main parts:

(*a*) Driving member or impeller— it attached to input shaft of the engine.

(*b*) Driven member or turbine—it transmits torque to gear box for onward flow to wheels through propeller shaft and final drive.

(*c*) A stater—it is fixed to the casing through a free wheel.

The whole assembly is filled with proper oil at a high pressure of 200 to 1200 kPa with the help of an oil pump.

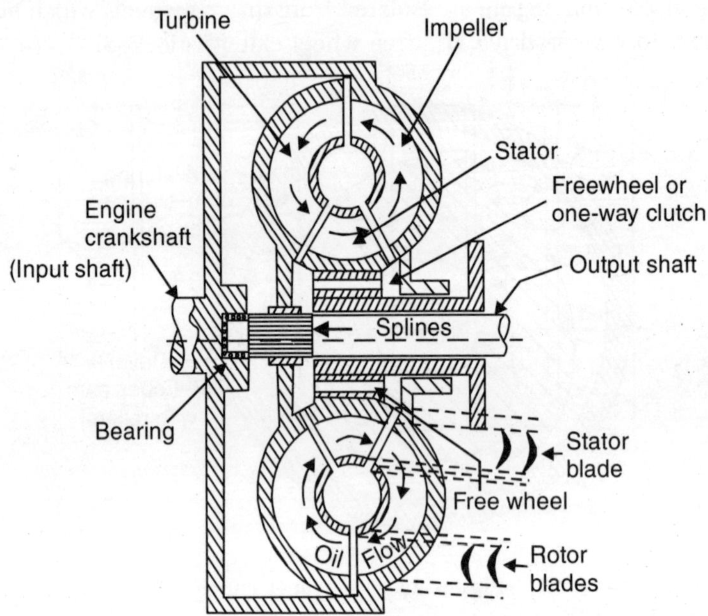

Fig. 5.26 Simple diagram of a three element torque converter

When engine starts, the impeller rotates with input shaft. The oil from the impeller is pushed into the turbine due to high centrifugal force created by rotation of impeller. The oil from impeller, gains kinetic energy and strikes the blades of the turbine and forces it to cause rotation. When this force and flow of oil is sufficiently increased with increase in the engine speed, the turbine starts rotating and thereby moves the vehicle. The use of torque converter is limited at low and medium speeds only.

The angle of turbine blades is made such that it changes the direction of oil flow. When it oozes out of the turbine blades, its direction is such that it smoothly enters the stationary stator blades, which again change its direction to the impeller blades in such a manner that the oil strikes the impeller blades in the same direction in which impeller is already, rotating (Fig. 5.27). The impeller again pushes the oil into turbine and the cycle goes on repeating. This continuous and repeated pushing of turbine blades causes the torque on turbine to increase. It is called torque multiplication. The maximum torque-multiplication (2 to 3 times) occurs when turbine is stationary and impeller is running at the engine speed. This is called "Stall". As the turbine starts rotating, the torque on turbine gradually decreases and it imparts the same torque as that at impeller when speeds of turbine and impeller become equal and at this stage it acts as direct drive.

When near equal speed is gained by turbine, the oil from turbine leaves in forward direction—hitting the back of the stator blades and thus stator blades act as a hindrance in oil flow to impeller. To avoid this, the stator is mounted over a free wheel clutch which allows it to rotate in the direction of turbine or impeller only and does not allow to rotate in the opposite-direction. Thus, the stator adjusts its desired position by rotating in the direction of turbine or impeller and allows the oil flow without hindrance. At higher speeds, torque converter behaves like a fluid flywheel and gives direct drive.

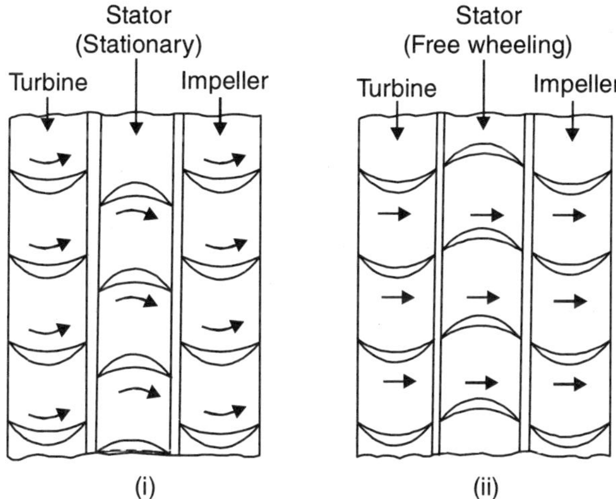

Fig. 5.27 Flow of oil in torque converter (i) At low speed, the stator deflects the fluid flow on to the turbine, (ii) At high speed the stator free wheels and oil flows straight without deflection, striking at the back of stator vanes

MANUAL GEAR BOXES TROUBLE SHOOTING

1. Gear Slips:
   (*i*) Adjust gear shifting mechanism.
   (*ii*) Shifting lever needs lubrication.
   (*iii*) Locking springs may be weak—replace them.
   (*iv*) Bearings worn—replace.
2. No power through gear box:
   (*i*) Adjust clutch slipping.
   (*ii*) Gear teeth stripped—replace.
   (*iii*) Gears shaft broken—replace.
3. Hard gear shifting or sticking gears:
   (*i*) Adjust gear shifting mechanism and lubricate.
   (*ii*) Excessive clutch pedal free play—adjust.
   (*iii*) Shifting fork bent—replace.
   (*iv*) Synchronising unit damaged—replace.
   (*v*) Locking balls stuck—free them and lubricate.
4. Noisy in neutral
   (*i*) Gears worn or teeth broken/chipped—replace.
   (*ii*) Bearings worn—replace.
   (*iii*) Gear box misaligned with engine shaft—align.

5. Noisy in gears
   (*i*) Clutch plate defective—replace.
   (*ii*) Insufficient lubricant—correct the level of oil.
   (*iii*) Synchromesh units worn—replace.
6. Gears clash during shifting
   (*i*) Synchronising cones defective—replace.
   (*ii*) Clutch not disengaging due to incorrect free play of pedal—adjust freeplay.
   (*iii*) Insufficient oil—fill the oil to the level.
   (*iv*) Idle speed excessive—adjust it.
7. Gears make noise in reverse direction
   (*i*) Reverse gear pinion/idler/bushings defective—replace.
   (*ii*) Reverse gear on main shaft damaged—replace.
   (*iii*) Counter gear worn or defective—replace.
8. Oil leakage
   (*i*) Oil level too high—correct the level.
   (*ii*) Gaskets broken—replace.
   (*iii*) Oil-seals defective—replace.
   (*iv*) Casing leaking—weld it.
   (*v*) Drain plug loose or washer broken—Tight the plug, change the washer.

QUESTIONNAIRE ON CLUTCHES

1. What are the functions of a clutch?
2. Give a classification of various types of clutches.
3. Describe a Single Plate Dry Clutch with the help of a neat sketch.
4. Describe the construction of a clutch plate. What materials are used for its lining?
5. Explain the working of a multiplate wet clutch.
6. Describe a cone clutch. How many types of friction clutches do you know?
7. Explain the working of an automatic clutch with suitable sketches.
8. How a diaphragm clutch works? How it differs from other clutches?
9. Explain the working of a Fluid Flywheel with a neat sketch.
10. What are the advantages of Fluid Flywheel? Also state its disadvantages.
11. Where are freewheel clutches used? Describe their working with suitable sketches.
12. What do you understand by 'Free Pedal Play'? Why is it desirable?

13. Where is the clutch located? Can it be placed before differential gear box? If 'No' or 'yes'—give the reasons.
14. Why is clutch required in auto vehicles? State its desirable characteristics.
15. Describe the working of an electromagnetic clutch. What are its shortcomings?
16. Describe the construction of a friction plate. How is it differ from pressure plate?
17. Write short notes on:
    (*a*) Riding the clutch
    (*b*) Clutch adjustment
    (*c*) Free pedal play
    (*d*) Clutch lining.
18. Explain the working of a diaphargm clutch with the helf a neat sketch.
19. Describe a semi-centrifugal clutch.
20. Explain the working of a Fully centrifugal clutch with a neat sketch.
21. How a front wheels driven car gets power from its engine?

QUESTIONNAIRE ON GEAR BOXES

1. Describe the functions of a gear box.
2. Describe a synchromesh gear box with the help of a neat sketch.
3. Describe the working of a 3-forward and one reverse speed sliding mesh type gear box.
4. What are the advantages of constant mesh gear box over sliding mesh gear box? Describe a simple constant mesh gear box with a neat sketch.
5. Enlist various types of gear boxes. What are the advantages of an epicyclic gear box.
6. What do you understand by the word synchronising? How is it achieved in a synchromesh gear box?
7. What is the use of a transfer case? Where is it used and what are its advantages?
8. Explain the working of a epicyclic gear train with neat sketches.
9. What is the need of a gear box? On which principle it works?
10. Explain the working of a selector mechanism in a manual gear box.
11. In a car gear box, the clutch gear pinion has 14 teeth, the main shaft low gear is having 32 teeth and its rating layshaft gear has 18 teeth. The lay shaftdrive gear, in constant mesh with clutch pinion, has 36 teeth. The differential gear ratio is 3.5 : 1 and effective radius of wheels is 35 cms. If engine speed is 2500 r.p.m., determine the speed of the car in kms per hour. [**Ans.** 41.25 kms/per]
12. What is a freewheel clutch? Why is it needed in automatic transmission? Describe it with a neat sketch.

13. Describe the working of a torque converter with suitable sketches.
14. What is overdrive? Describe its working with a neat sketch.
15. Describe the principle of working of an automatic transmission.
16. What are advantages of overdrive?
17. Explain how the various speeds are achieved in an epicyclic gear system.
18. What are the advantages of freewheeling? What arrangement is made for reversing the vehicle which is fitted with a freewheel clutch?
19. Explain how the oil flows in the rotors of a torque converter. Why the oil is filled under high pressure?
20. Which of the following part is responsible for increasing torque in an epicyclic gear train?

    (*i*) Ring gear (*ii*) Sun gear

    (*iii*) Sun gear shaft (*iv*) Planet carrier. [**Ans.** (*iv*)]
21. Draw a neat sketch of freewheel clutch and describe its working.

CHAPTER 6

Electrical System

TYPES OF ELECTRICAL SYSTEM

Electrical systems of automobiles play very important roles, without which the auto vehicles cannot run properly. S.I. engines need an electrical/electronic system for producing spark. The diesel engines, not vehicles, can run without any electrical system, but vehicles fitted with diesel engines do need an electrical systems for the efficient driving on the roads. An Automobile vehicle consists of the following electrical systems:

1. *Starting System.* It includes battery, starting motor, switches, drive assembly and wiring.
2. *Charging System.* It consists of a Dynamo or Alternator, regulator with rectifiers and wiring.
3. *Ignition System.* Employed only in Petrol engines and Gas engines. It has a Battery, Ignition coil, Distributor, Condenser, Spark plugs, low and high tension wirings and switches. The ignition system has already been described in details in the Chapter 2 of Ignition Systems in S.I. Engines.
4. *A System of Accessories.* It consists of various electrical appliances such as lighting system, horns, instrument panel, warning devices, heating and cooling systems, music system, indicator lights, dashboard instruments such as speedometer, temperature meter of cooling water, fuel gauge, warning lights, cigarette lighter, mobile charger, DVD player, fuses box etc.
5. *Wiring System.* In majority of vehicles, single wire earth-return type wiring is used especially in passenger cars. This type of wiring is called *Single-Pole Wiring*. The car body, frame, engine block serve as earth or return-wire, generally positive pole of battery is earthed. This type of wiring requires only half as much wiring as in double pole wiring. Another system of auto vehicle wiring is *Double-Pole Wiring*. In this type, two wires are taken from battery—one positive and other negative. The both wires are connected to the lighting system and other devices just like domestic wiring. The double pole wiring is done only on vehicles transporting inflammable materials such as petrol, diesel oil, kerosine, spirit, alcohol, gas and explosives to save them from short-circuiting through body. Wiring

diagrams are provided by vehicle manufacturers. These diagrams help in electrical-faults finding at the vehicle.

A third wiring system, called '*3-wire system*' is also in use on 24-Volts battery employed by public transport vehicles, city buses etc. In this system, a third wire (Neutral) is also attached to the centre of the battery and the load is equally divided on two sides. In a perfectly balanced load condition, no current flows through Neutral wire.

Wiring Colour Code

Manufacturers have made a wiring colour code for different wires in various circuits. The colour of the wire helps in repair work:

(*i*) *White coloured wires* are used for number plate light, instrument penal lights, interior lights, bonnet light, boot light, side light, reversing light.

(*ii*) *Milky white wires*—Head lights.

(*iii*) *Yellow wires*—Fog light, parking light (amber colour also), front spot light, direction indicators (Amber colour).

(*iv*) *Red wires*—Rear tail light, rear spot light, fare meter light in taxis.

On some vehicles green, black, pink, grey coloured wires are also used for different circuits. The detailed information is provided by the manufacturer of the vehicle.

Wiring Harness

Insulated wires of different colours and circuits run together upto a certain junction point. All such wires are placed in one covering of insulating rubber, just like a thick cable housing a number of wires. The covered multi-wire cable is termed as wiring harness. It provides extra safety of already insulated wires and keeps them to run like a single wire upto the junction. From junction point, they are distributed individually up to their appliances/devices/bulbs etc.

A simple electrical system with +ve earthing and showing various components, is shown in Fig. 6.1.

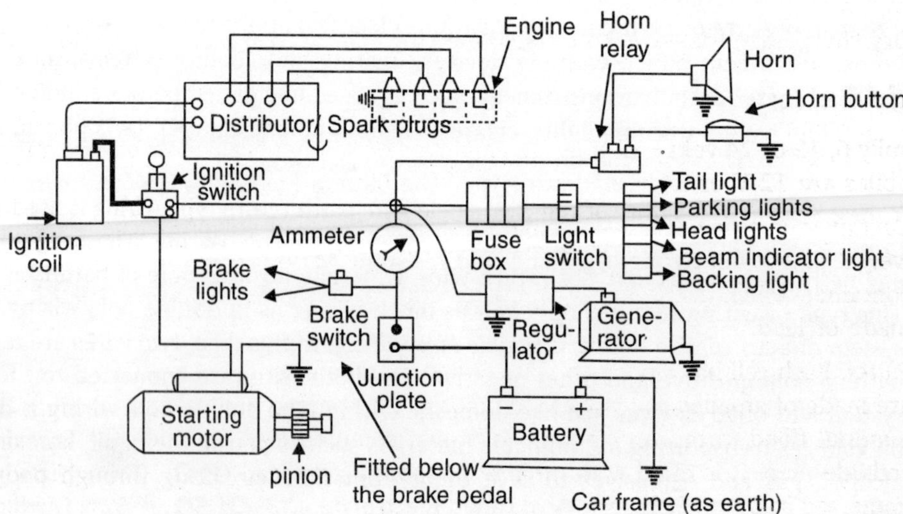

Fig. 6.1 Simplified electrical system of a car

The electrical system of an automobile serves the following purposes:

1. To supply power to starting motor for cranking.
2. To generate electricity for charging of battery and regulation of voltage.
3. To store power in battery and then supply it to various units.
4. To supply current to lighting system.
5. To supply current to Ignition coil in Otto cycle engines.
6. To supply electricity to dashboards instruments, gauges, horn etc.
7. To supply power to air-conditioning and music systems.
8. To supply power for other accessories such as cigar lighter, mobile charger, automatic windows opening and closing, anti-theft devices etc.

THE BATTERY

Battery is an electro-chemical device for storing energy in chemical form which can be released in the form of electricity. The battery gets electrical energy from Alternator or Dynamo for charging. This electrical energy is converted into chemical energy and stored in the battery. When engine is not running, the battery supplies its stored energy in the form of electric current to the starter motor, ignition system, lighting system and horn. When engine is started, the Alternator/Dynamo also starts and develops electrical energy. When sufficient speed is attained by crankshaft, the alternator not only supplies current to electrical systems but also charges the battery.

Types of Batteries

Generally two types of batteries are employed in automobiles:

1. Lead-acid battery.
2. Alkaline battery (Mostly Nickel-Iron battery) with Potassium Hydroxide as electrolyte are in use on heavy vehicles and also on some motor cycles.
3. Dry charge batteries—without liquid electrolyte—used on 2 and 3 wheelers.

Lead–Acid Battery

Generally 6, 12 or 24 volts batteries are employed on automobiles. Most of the batteries fitted on automobiles are 12 V or 24 V lead-acid type. The battery has the following parts:

(*i*) *Container or casing*. It is a single piece container made of polypropylene—which is a light weight plastic but strong, rigid and acid resistant. Various partitions are made inside the container, which divide it into compartments for different cells. Each cell has two plates made of lead.

(*ii*) *Plates*. Each cell has a positive plate (Anode) and a negative plate (Cathode). The plates are made of an alloy of lead and antimony (5%) or lead and calcium (0.1%). The active material filled in the Anode plate is brown coloured lead peroxide ($Pb.O_2$) and in the cathode plate it is filled with litharge or spongy grey lead (PbO). These plate grids are immersed in Dilute Sulphuric Acid called Electrolyte. (38% H_2SO_4 + 62% Distilled Water by weights). The level of electrolyte is kept about 10 mm above the plates.

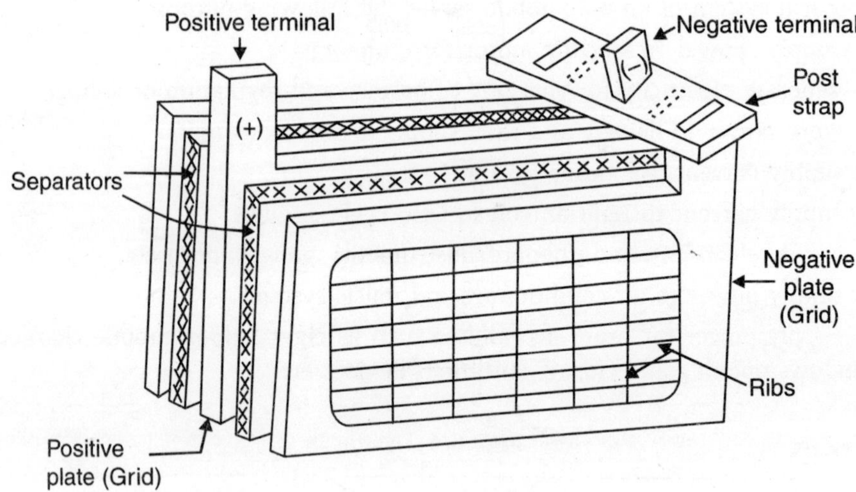

Fig. 6.2 Construction of a battery cell

The (+ve) and (–ve) plates, are kept apart by interposing separators as shown in Fig. 6.2. These separators are made of a non-conducting porous material such as fibre glass, resin-bonded cellulose and the most widely used sintered P.V.C. (Poly Vinyl Chloride). The separators prevent short circuiting between the two plates. In some batteries, the number of negative plates are two in each cell as the positive plate is to be surrounded by negative plates from both sides. Each group of (–ve) plates is held by post straps. The plates are suspended from the top of the cover and are kept slightly above the bottom of container and assembled together to form a high current yielding battery for vehicles. Voltage per cell is 2^+ volts. A 12-volts battery contains 6 cells connected in series as shown in Fig. 6.3. Each cell is sealed with a cover of hard rubber through which +ve and –ve lugs project out. Each cell has opening for adding distilled water and an air–vent for gases and vapours to escape into atmosphere.

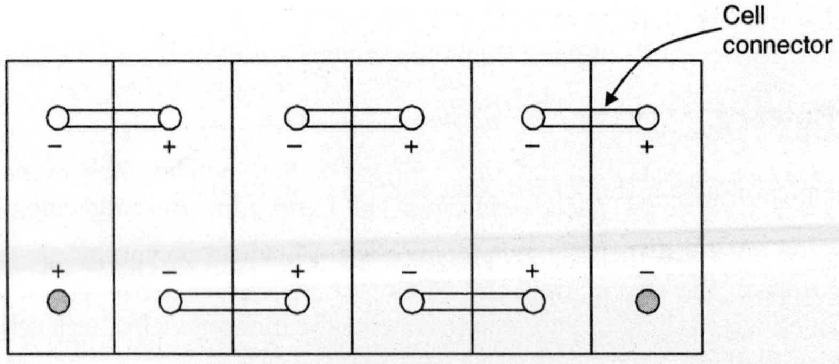

Fig. 6.3 Connections in a 12-V battery

Electrolyte

In the assembled condition of lead-acid battery, the battery is filled with electrolyte (1 part of sulphuric acid to 2 parts of distilled water by volume). Specific gravity of the electrolyte is of great importance. It is determined by a Hydrometer test. The hydrometer is shown in Fig. 6.4.

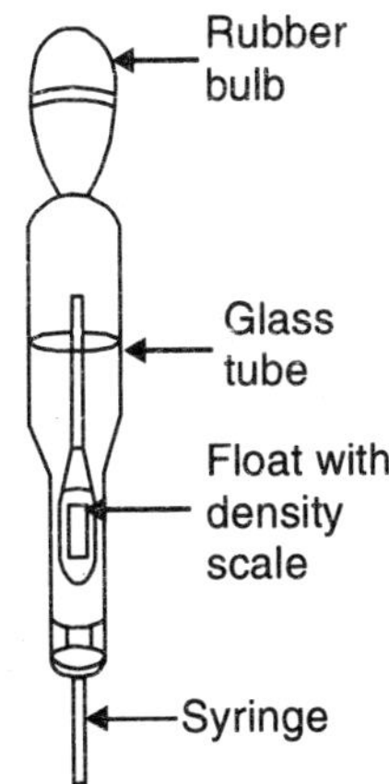

Fig. 6.4 Hydrometer

Specific Gravity–at 26°C	Battery Condition
1.2810 gm/c.c.	Fully charged
1.240	3/4 charged
1.200	Half charged
1.160	1/4 charged
1.120	Discharged

Specific gravity increases with decrease of temperature, fully charged battery has a voltage of 2.1 volts per cell.

Working of a Lead-Acid Battery

The (+ve) plates contain chocolate colour or brown colour lead per oxide ($Pb.O_2$) and (–ve) plates contain spongy lead of grey colour. The chemical changes that take place during discharging and recharging processes are given by the following equations:

$$\underset{\text{!+ve plate}}{Pb.O_2} + \underset{\text{!Electrolyte}}{2H_2SO_4} + \underset{\text{!–ve plate}}{Pb} \underset{\text{Recharge}}{\overset{\text{Discharge}}{\rightleftharpoons}} \underset{\text{!+ve plate}}{PbSO_4} + \underset{\text{!Electrolyte}}{2H_2O} + \underset{\text{!–ve plate}}{PbSO_4}$$

The chemical reaction that takes place during charging is as follows:

At positive plate # # $\underset{\text{!white colour}}{PbSO_4} + 2H_2O = Pb.O_2 + 2H_2SO_4$

At negative plate %# # $\underset{\text{!white colour}}{2PbSO_4} + 2H_2O = 2Pb + 2H_2SO_4 + O_2$

Water is continuously split up into oxygen and hydrogen gases during charging process which decreases its quantity, therefore, topping up with distilled water is necessary. Specific gravity of electrolyte is increased during charging process.

Battery Capacity

The capacity of a battery is the amount of current it can deliver or the ability to store electrical energy:

(*a*) **Twenty hour rate (Ampere-hour capacity):** It shows lasting capacity of the battery on small load. It is the rate of current a battery will deliver continuously for 20 hours with the condition that cell voltage should not drop below 1.75 volts at 26°C.

(*b*) **Ampere-hour (Ah) rating:** A battery of A.44-Ah is capable of delivery a current of 1 Amp. for 44 hours. The same battery will give 2 Amps for 22 hours.

(*c*) **C.C.A. rating:** It is based on engine displacement volume. Cold Cranking Amperes (C.C.A.) is also given for automobiles batteries. 1 Ampere of C.C.A. is required for 16 cubic centimeter of engine displacement. Thus, a 800 CC engine must have a battery with

C.C.A. rating of $\frac{800}{16} = 50$ Ampere.

Battery Charging from A.C. Mains or Outside Charging

For recharging new or fully discharged batteries, the outside battery charging is needed. The A.C. supply is first converted into D.C. supply by battery charger, in which rectifiers are used for this purpose. The battery charger first steps down the A.C. voltage to the required voltage of battery by the transformer. This stepped down A.C. supply is now converted into D.C. by a valve or a metal rectifier. If the ampere-hour capacity of the battery is 100 and charging current is 5A, than approximate charging time will be $100 \div 5 = 20$ hours.

External Charging

Normally, battery gets continuously charged by car generator/dynamo during running. But under certain conditions the battery needs external charging:

1. When vehicle is parked for many days.
2. When lights/parking lights are left 'on' for long periods of parking.
3. When A.C., heater, blower, car, radio etc. are used under engine's stopped condition.
4. When the self-starter is frequently used. The self starter draws heavy current which discharges the battery quickly.

Only D.C. (Direct Current) is used for battery charging. To convert A.C. supply into D.C., a Battery charger or the equipments for converting A.C. to D.C. is used. The charger is the rectifier-transformer combination equipments with controls for current and voltage etc. In a "slow" or normal charging method, the battery is supplied with a small current at 1 ampere per positive plate or 7% of the Ampere-hour rating of the battery. It takes lot of time depending upon its discharged condition. The process is shown in Fig. 6.5(a). In a Quick Charge Method, the battery is supplied with a high direct current (about 50 amperes for 12–V battery) for a short period. This method is used in emergency only.

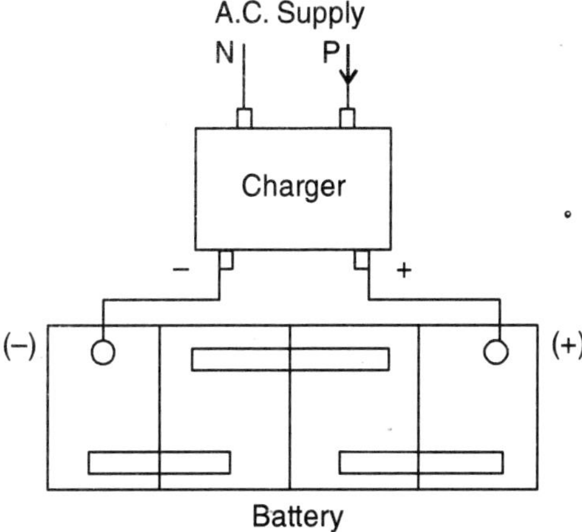

Fig. 6.5(a) Battery charging circuit from external source

Battery Charging from Dynamo or Generator

An internal charging circuit is shown in Fig. 6.5(b). In automobiles, outside charging from A.C. mains supply is seldom required. The generator constantly remains charging the battery with a combined current and voltage regulator and cut out placed in between generator and battery.

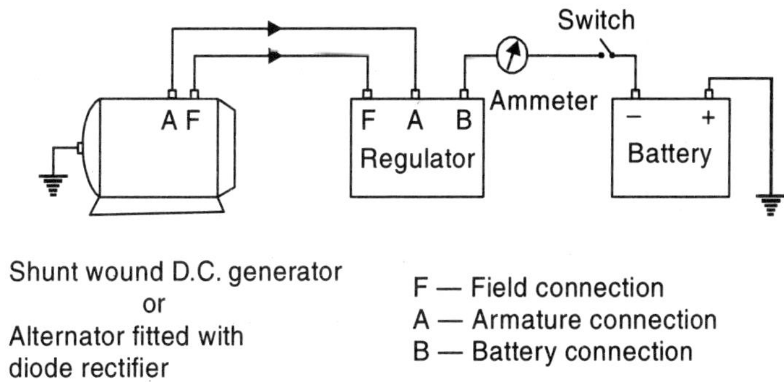

Fig. 6.5(b) Battery charging circuit (internal)

Maintenance Free Batteries

Due to liberation of hydrogen and oxygen gases at negative plates during the process of charging as the battery approaches the full charge condition, the water is consumed and its level in the battery falls. This needs a periodic water level checking and topping up to the marked level with distilled water in order to compensate for the loss of water. The liberation of H_2 and O_2 is called 'Gassing'.

To overcome this problem, maintenance free batteries are now available in the market. These batteries use envelope-type separators which completely enclose each plate, enabling the battery plates to rest on the bottom of the battery case, instead of keeping them hanging a little above the bottom. This system provides additional electrolyte volume and less gassing, thereby more useful life without maintenance is achieved.

SELF STARTER OR STARTING MOTOR OF A CAR

The starting motor causes the flywheel of the engine to rotate, which in turn rotates the crankshaft which activates the processes of suction, compression, ignition, and expansion in the engine cylinders. As soon as the engine is started the starting motor is shut-off. A cranking speed of about 100 r.p.m. is enough to start the engine. A simple line diagram of a starting circuit has been shown in Fig. 6.6.

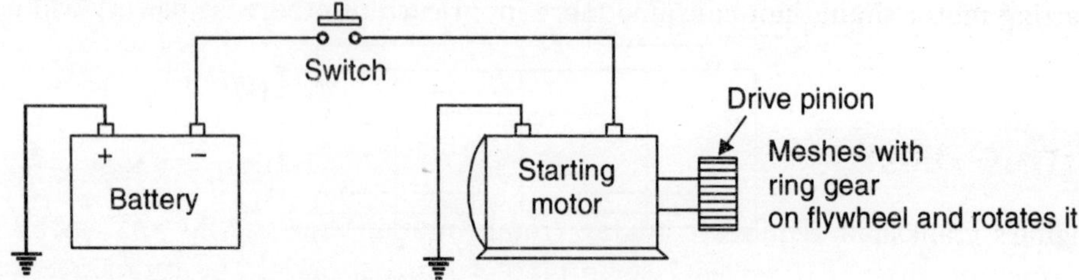

Fig. 6.6 Starting system of a car

Types of Starting Motors

The starting or cranking motor is a compact D.C. motor. It draws current from battery. It is mounted in the close vicinity of the flywheel. The output shaft of the motor is fitted with a drive system for rotating the flywheel of the engine. The starting motor should have sufficient power to rotate the heavy flywheel from its rest position. For that, it should produce a high initial torque at low speeds. For light and medium cars initial torque should be 10 to 30 N-m and for larger cars its value ranges from 50 to 150 N-m, the starting current may be high up to 500 Amperes. The starting motor used on automobiles are classified on the basis of field windings and magnetic poles:

A. On the basis of field winding:

1. Series-wound motor—widely used.
2. Series-shunt wound motor—rarely used.

B. On the basis of poles used:

1. 2-pole motor
2. 4-pole motor

The series wound motor is commonly employed for starting purposes. It is a variable speed motor and produces sufficiently high torque at low speeds with full load, when the starting pinion engages with the ring gear of the flywheel on taking 'self' or switching on the starter switch.

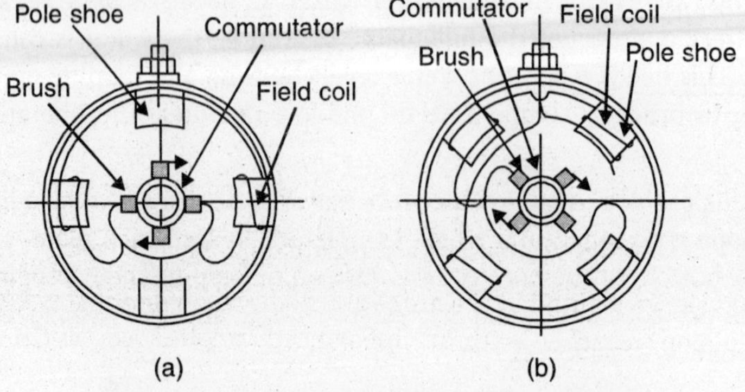

Fig. 6.7 (a) 2-pole (b) 4-pole starting motors

The series wound starter motor may have either 2-poles winding or 4-poles as shown in Fig. 6.7. On 2-poles motor, the opposite field coils are wound in the same directions, so that the both poles have the same polarity. The remaining two intermediate unwound poles then acquire opposite polarity by induction. In 4-poles motors the field coils are wound in opposite directions on the adjacent poles so that they acquire strong opposite polarity. A 2-pole starter motor is best suited for small and medium cars whereas a 4-pole motors are required for starting heavy vehicles. The starting motor should not consume more input current otherwise battery will get discharged sooner.

STARTING DRIVES

I.C. engine's crankshaft requires cranking (rotary motion) for starting the engine. The starting motor armature shaft is fitted with some type of starting drive. The rotation of starting motor activates the starting drive to give rotary motion to the flywheel, which being mounted on the crankshaft, rotates the crankshaft. The rotation of crankshaft is transmitted to piston to reciprocate inside the engine cylinder to start the working cycle.

Main types of starting devices used on vehicles are:

(*i*) Bendix Drives

(*ii*) Over Running Clutch

(*iii*) Dyer Drive

(i) Bendix Drives

These are the inertia type of drives. The starting pinion is moved to engage or disengage with the toothed ring gear made on the periphery of the flywheel.

A standard Bendix drive (Inboard Type) is shown in Fig. 6.8.

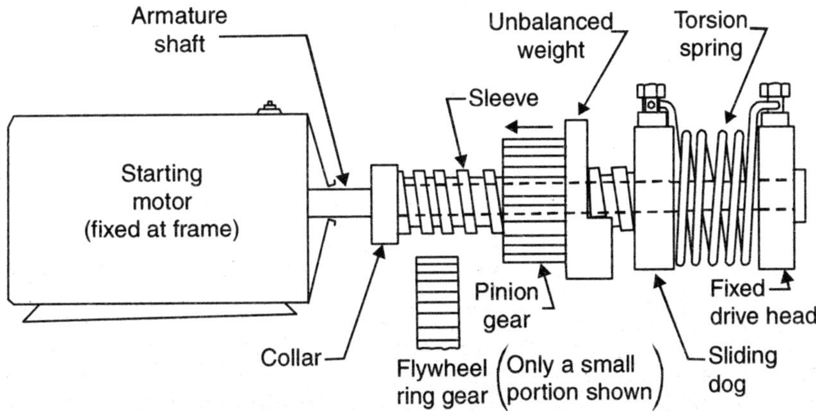

Fig. 6.8 A standard bendix drive (inboard type)

The threaded sleeve is fitted at the armature shaft, when motor is started, the armature shaft rotates but the pinion, attached with an unbalanced weight, does not rotate but slides axially towards the motor due to effect of unbalanced weight, till it engages with ring gear of flywheel and further it is stopped by the collar. As soon as it touches the collar, it starts rotating and also forces the flywheel to rotate. When the engine is started the starting switch is turned off and the flywheel runs under its own energy. When it reaches at a speed greater than that of starting motor, the pinion is spun back along the threaded sleeve and goes out of engagement. The torsional spring bears the shock of engagement and return of pinion, thereby, giving a smooth start. Some Bendix drives employ a compression spring on a splined armature shaft in place of torsion spring as shown in Fig. 6.9 (a). Auxiliary spring gets compressed at starting and pushes back the pinion fastly after starting.

Another type of Bendix drive is 'Folo-Thru Drive'. It is similar to the standard bendix drive. The pinion is connected to the pinion barrel and the armature is attached to threaded sleeve through a spring, lock-pin and an anti-drift pin. A folo-thru drive is shown in Fig. 6.9 (b).

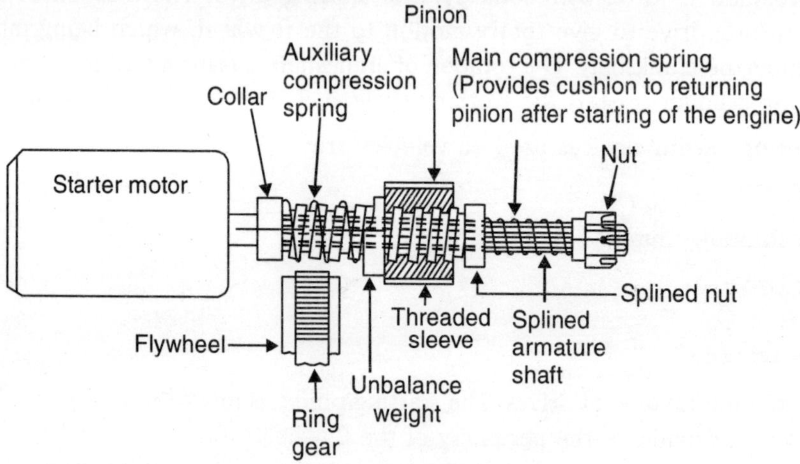

Fig. 6.9 (a) Bendix drive with compression spring

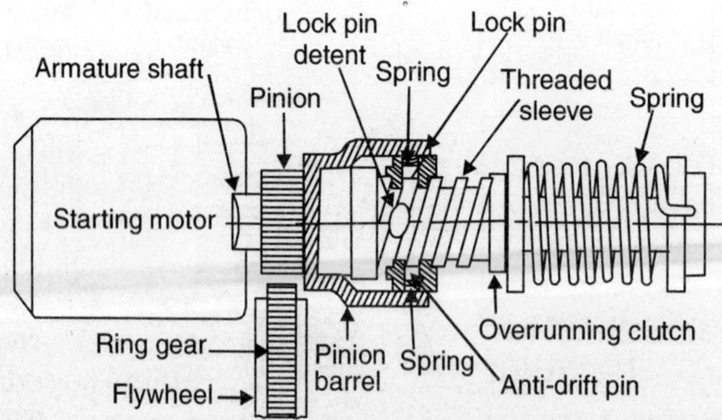

Fig. 6.9 (b) Construction of a folo-thru drive (pinion engaged with ring gear)

The pinion meshes with ring gear of flywheel in the same manner as in case of a standard Bendix Drive. But, at the end of the pinion travel, the lock pin drop into the detent (a catch hole) and does not allow pinion to disengage prematurely due to a false start or intermittent starting. The pinion continuously remain engaged with flywheel till the engine really gets started. At about 400 r.p.m., the lock pin comes out of the detent due to centrifugal force which exceeds the spring force. The pinion then disengages as usual in Bendix drives. The anti-drift pin is spring loaded and prevents the drifting of pinion to engage with ring gear accidently. The over running clutch saves the motor from damage in case when pinion remains engaged with flywheel, due to some defect, even after the full starting of the engine.

(ii) Over Running Clutch Drive

An over running clutch and the drive have been illustrated in Figs. 6.10 (a) and (b).

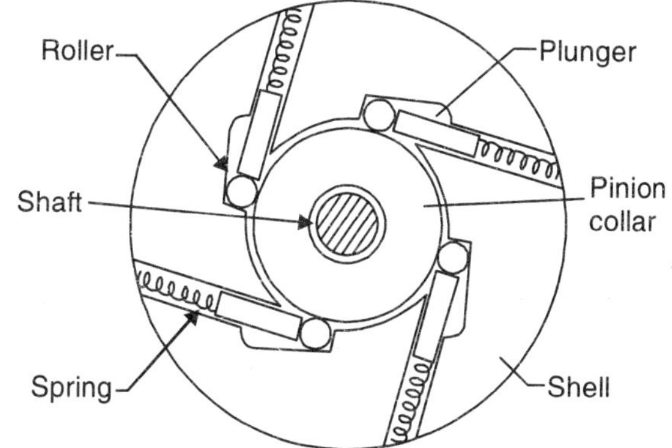

Fig. 6.10 (a) Over running clutch construction

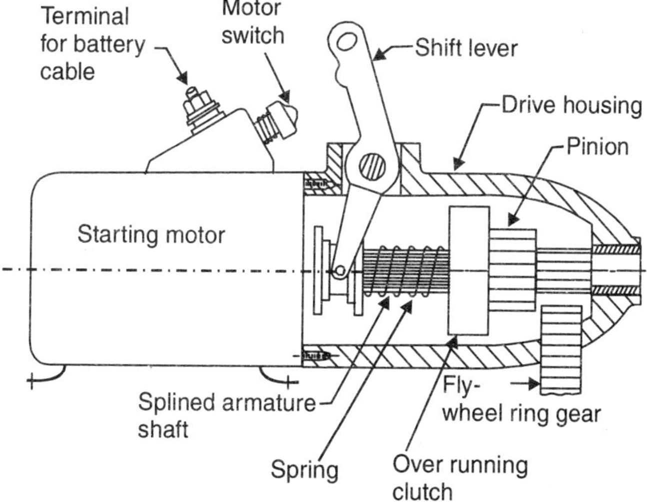

Fig. 6.10 (b) Over running clutch drive

In this drive, a over running clutch is placed between the splined armature shaft and the pinion. A shfit lever is attached to the starting pedal or a solenoid switch is commonly employed for its operation. The over running clutch has a pinion collar and a shell. When the shell rotates faster than pinion collar, it drives the collar. When speed of pinion collar increases more than the speed of shell, they are disengaged. This disengagement is caused by spring loaded rollers.

When the car driver presses the starting lever, the shift lever turns about its pivot and pushes the motor switch. It also moves the over running clutch and the pinion against the spring pressure, which causes the splined shaft to rotate with the motor and the pinion engages with ring gear to rotate it for engine starting.

When engine has started and picks up speed, the pinion rotates at much faster rate. Now the over running clutch disconnects the pinion from motor. When engine starts running at idling speed the driver releaves the pedal, thereby causing shift lever to come back to its original position and the motor-switch is switched off stopping the starter motor immediately.

D.C. GENERATOR OR DYNAMO

The function of generator in automobiles is to supply electrical energy for charging the battery and other electrical devices of the vehicle. Shunt-wound type d.c. generator or dynamo are widely used in autovehicles for producing electricity.

On modern cars, the electrical load has increased a lot due to induction of so many other accessories such as A/C, Music system, Heating system, Radio, T.V., Auto locking, Anti-theft devices, Sliding seats, Automatic transmission, Overdrives. Electric brakes, Power brakes, Mobile charging etc. The traditional d.c. generator is not able to meet these higher demands. So A.C. alternator is employed these days to meet the growing demands of electricity. The A.C. output of alternator is first rectified by in-built semi-conductor rectifier diodes and such the d.c. is obtained, which is utilised for charging the battery.

D.C. Generator or Dynamo has the following parts:

(i) *Armature*. It is the central part consisting of a central shaft, core, commutator and windings. The commutator is made in segments of copper which are insulated by mica strips.

(ii) *Field windings*. Generally, car dynamo is shunt-wound. There are two poles wound by field coils. These coils are connected to armature by a wire through slip rings. The other end of the armature is ground/earth. The current flows to the load as well as some current flows through the field windings. This is the way the electromagnets are produced in the dynamo.

(iii) *Frame or casing*. Magnet pole pieces are mounted inside the frame. The end covers of the frame support the Armature bearings, generally ball race type. The brush gear is fitted on the commutator.

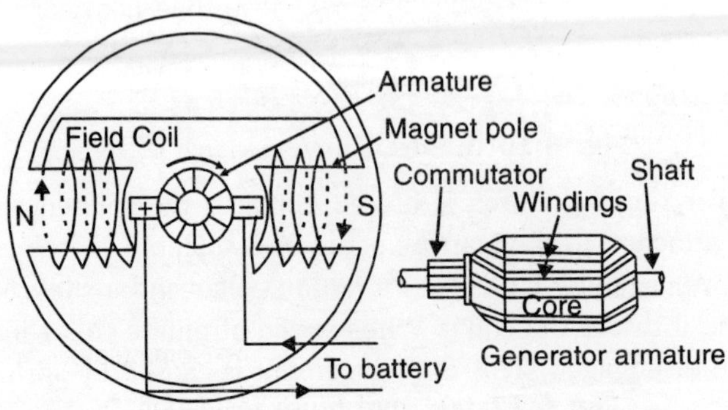

Fig. 6.11 Shunt wound dynamo and its armature

Carbon brushes kept in carbon holders bear on the commutators under a spring pressure. They collect the current from the armature. Two main brushes are provided at 180° apart for a 2-pole dynamo. In a 4-pole dynamo, four brushes are fitted with 90° angle in between the two consecutive brushes. The a.c. is induced in the stater as the field coil turns. This is then converted to d.c. by the rectifying diodes. A shunt wound dynamo is shown in Fig. 6.11. Diode allow flow of current only in one direction.

Output Regulation of Dynamo

When the speed of the dynamo increases, the supply of current to the shunt winding also increases, which increases the magnetic flux. The increase in flux results in higher output voltage of the dynamo. The output goes on increasing in proportion to the speed of dynamo/speed of engine. If this output is not controlled, the generator or dynamo may finally burn out. To avoid such type of damages, the output regulators are placed in the circuit.

1. Third brush generator output control

We know that the d.c. generator has a rising output as the speed increases. If the output is not controlled, it will further speed up the generator and a limit may reach when the armature parts will be damaged. Therefore, control of the generator output is a necessity. The control is done either by a third brush (now almost obsolete) or with external control.

In a generator, the armature containing a number of conductors is rotated within a magnetic field. As the armature rotates at high speeds distortion of field occurs and lines of magnetic force get tilted at traiting edges of magnets. This effect is termed as **Armature Reaction**.

In a third brush control system, there are three brushes in the generator. The main two brushes which carry the load current, are attached at the usual positions. The third brush leads the current to the field windings and is positioned at the place where the lines of force (Flux) become weaker at high speeds due to armature reaction. The current from third brush, through field windings, passes on to the main negative brush. Since the magnetic lines at higher speeds, decrease at the third brush due to which it gives low voltage to the field windings and consequently less voltage output is produced and such the speed is kept under control.

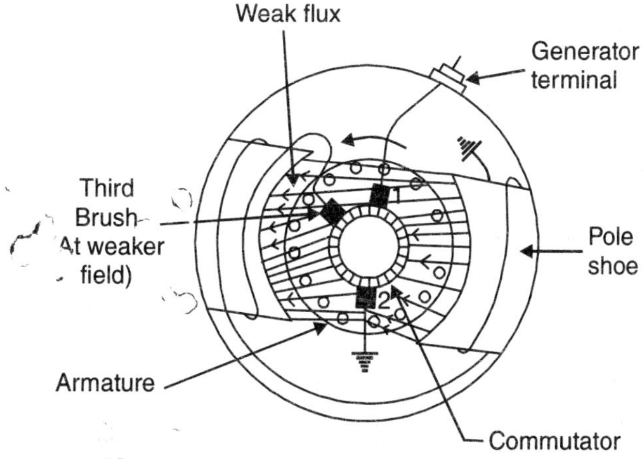

Fig. 6.12 (a) Third brush regulation

2. *External control of output*

In this system, a resistor is provided in the field winding (See Fig. 6.12(*b*)). The current flowing through the field coil is regulated with the help of this resistor and such the current output of dynamo is regulated within the prescribed limit. By increasing or decreasing the resistance inside the resistor unit, the regulation of current is achieved and the current in field coil is not allowed to exceed beyond a certain limit.

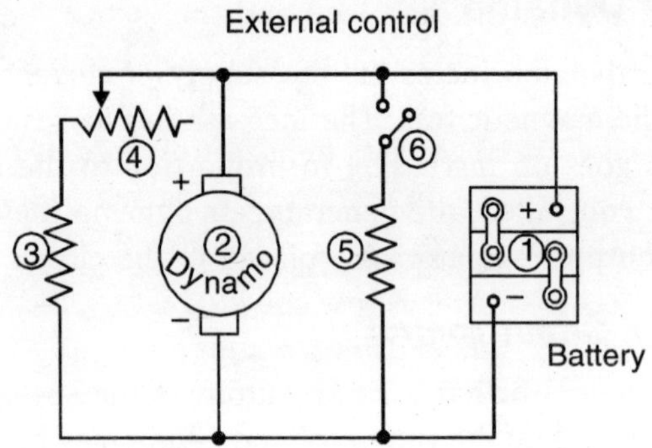

Fig. 6.12 (b) Output regulation of the dynamo (1) Battery; (2) Dynamo; (3) Field; (4) Variable resistor; (5) Load; (6) Switch

Cut-out Relay

During the running condition of the engine, the battery is charged by dynamo. When speed of dynamo is slow, there is a tendency of the battery to discharge the current to the dynamo. The dynamo then begins to run as a motor, which is totally undesirable. Therefore, it is must to employ a device that would permit the dynamo to charge the battery, but prevent the battery current from flowing back to dynamo at slow speeds. The device used for this purpose is called a 'cut-out relay'.

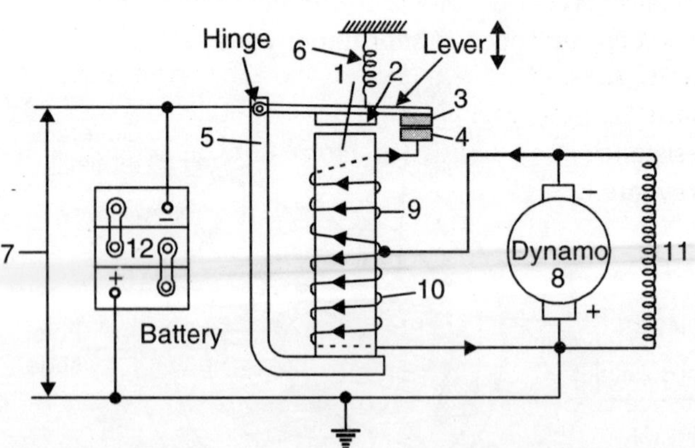

Fig. 6.13 Cut-out relay 1. Soft iron core; 2. Armature; 3, 4. Contact points; 5. L-shaped member; 6. Spring; 7. Load between the two leads, 8. Dynamo; 9. Current coil; 10. Voltage coil; 11. Field; 12. Battery

The Fig. 6.13 shows a cut-out relay. The dynamo is connected in parallel to a coil, called voltage coil (10). Another coil (9) is connected in series with dynamo and battery. Both coils (9) and (10) are wound round a soft iron core. An L-shaped member (5) is attached to the bottom of soft iron core at one end and the other end carries a hinge in which an iron lever is attached. The lever's free end is fitted with a contact point (3) which makes and breaks the circuit with a fixed contact point (4) which is attached to one terminal of the coil (9). The lever is held in open position of contact points by a spring (6).

The dynamo initially rotates at a slow speed and the voltage in coil (10) slowly builds up. When speed of dynamo reaches a particular value, the dynamo voltage also rises up to a specified limit. The same voltage is carried at coil (10). The coil (10) contains many turns of very fine insulated copper wire. The coil (10) magnetises the soft iron core. It pulls the armature (2) closer and the contact points (3) and (4) come in contact. The current from the dynamo starts flowing to the battery. The battery is thus charged.

At slow speed of dynamo, the voltage accross coil (10) is reduced where as battery, being fully charged, is at a higher voltage. The current flows through the points (4) and (3) and along the series coil (9) to the dynamo. This coil (9) is made up of a thick wire and has lesser number of turns. The current flows from battery to the dynamo in the opposite direction. The reverse flowing current passes through the series coil (9) which produces a magnetic effect in the coil (9) itself. Coil (9) is magnetic effect opposes the magnetic effect produced in voltage coil (10). These opposite nature magnetic effects, demagnetise the soft iron core. The core no more attracts the armature (2) and the armature (2) alongwith lever is pulled back by the spring (6) opening the contact between the two points (3) and (4). Thus, the flow of current from battery to dynamo is prevented. When engine is stopped, the battery current tries to flow to dynamo but the cut-out does not allow it to reach to dynamo.

Current Regulation

Current-regulator has a pair of contact points and a resistance. When the current of the dynamo goes beyond a particular limit, the contact points open and the resistance comes into action with the dynamo field current and the current output is reduced by the resistance. As soon as the current output is reduced, the contact points close and the current begins to flow immediately. The process of contact points' opening and closing goes on repeatedly many times in a second. By the combined action of resistance and opening and closing of contact points, the flow of heavy current into the dynamo is prevented.

Voltage Regulator

The voltage regulation is done by a shunt coil with the generator/dynamo and a resistor. There is also a pair of contact points. As the voltage of the dynamo crosses a certain limit, the contact points open and resistance comes into action simultaneously. Now the voltage output gets reduced and the contact point reclose. The value of the voltage then again starts rising and when reached at a certain limit, the contacts points re-open. This opening and closing of contacts points goes on many times (50 to 200) every second and the voltage value is more or less kept within a certain range. The contact points have to vibrate frequently for voltage regulation. The Voltage Regulator is shown in Fig. 6.14 (a).

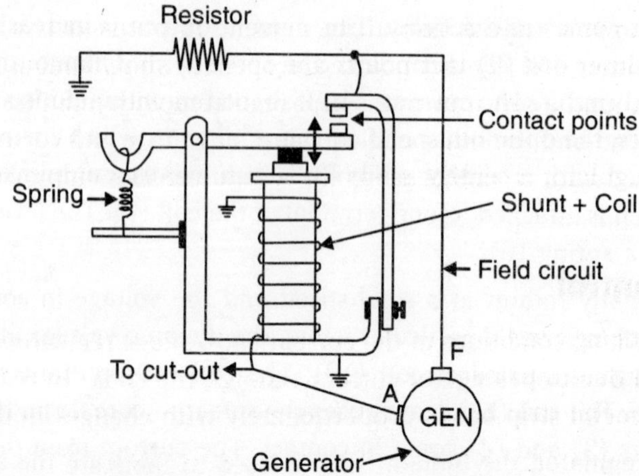

Fig. 6.14 (a) A voltage regulator

3-Unit Regulator

Various parts of a three unit cut-out relay, voltage and current regulator is shown in Fig. 6.14 (b) It consists of a cut-out relay (1), a current regulator (2), a voltage regulator (3), dynamo (9) and field coil (10). When the dynamo is started, the voltage starts building up and reached to a certain limit, the cut-out relay pulls the armature (4) towards the soft iron core of the cut-out relay. This makes a connection by closing two contact points C between the dynamo and battery. Cut-out relay prevents flow of battery current to dynamo.

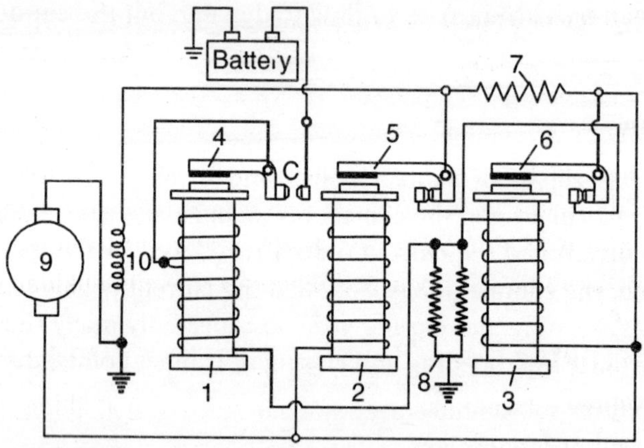

Fig. 6.14 (b) Three unit regulator; 1. Cut-out relay; 2. Current regulator; 3. Voltage regulator; 4–6. Armatures; 7. Point resistance; 8. Swamp resistance; 9. Dynamo; 10. Field coil

When the dynamo gains higher speed, the voltage regulator swifty comes into action. The armature (6) is pulled away from the contact points of the voltage regulator and the points open. Point resistance (7) is then introduced in the field coil, thereby reducing the voltage.

Sometimes, the voltage of the dynamo does not reach the operating value of the voltage regulator due to the discharge of battery. This happens when there is heavy electrical loading. Now the current regulator comes into action. If the current output is increasing, the current regulator pulls armature (5) closer and contact points are opened; simultaneously points resistance (7) is connected to the field circuit. The current is kept regulated within limits. By combining the voltage and current regulators, the optimum specified values of voltage and current outputs of dynamo are maintained. 3-unit regulator is widely used on the auto vehicles employing Dynamo for electricity supply.

Temperature Control

The ambient working conditions of the current and voltage regulators, the temperature of coils tends to rise and fall due to passage of current through the coils. To reduce this effect, a bimetal strip is used. This bimetal strip bends proportionately with changes in the temperature.

In the voltage regulator, the bimetal strip is used to fabricate the armature hinge spring.

In the cut-out relay, the bimetal strip is used to supplement the armature control spring.

The effect of the temperature-changes on control box settings is further minimised by a device called the DOUBLE SWAMP RESISTOR. This device is connected in series with the two shunt coils which enables the use of coils of lower resistance.

ALTERNATOR OR A.C. GENERATOR

Alternator is an alternating current (A.C.) generator for producing large quantities of current for operating quite a good number of devices fitted on the modern vehicles.

The modern cars are provided electrically operated windows, sliding front seats, air-conditioners, electrical equipments for automatic transmission, over-drives, anti-theft alarms, DVD players, computer facility, cigarete lighter, mobile phone charging etc. These vehicles require more electrical energy to operate these equipments. A dynamo is unable to supply heavy amount of electric power and it gets overheated at higher speeds. Sometimes the vehicles have to run at very slow speeds in congested areas or muddy roads, at that time, dynamo fails to charge the battery. Therefore, on A.C. generator or alternator is used in place of dynamo in luxury cars and even in non-luxury cars. Alternator continues to charge battery even at idling speed of the engine.

Advantages of Alternator

1. Alternator produces substantial current (about 20A) even at idling speed, whereas dynamo does not do well at low speeds.
2. For the same current output, the alternator is lighter in weight and smaller in size than dynamo.
3. No cut-out and current regulator unit is needed in alternator. The rectifier diodes do not allow reverse flow of current. The dynamo needs a cut-out unit to check reverse flow of current.
4. Its maintenance is easier because it does not employ commulator, slip ring brushes and brush gear.

Working of Alternator

Its principle of working may be understood by the following example:

An armature magnet is shown in Fig. 6.15 (i). The armature magnet is placed inside a coil of wire with magnet's north pole at top and south pole below it. The armature magnet is now given a half revolution; it will produce current in the coil wire in one direction.

Now the south pole comes at top and north pole reaches the bottom. See Fig. 6.15 (ii).

The armature magnet is further rotated by half revolution, the flow of current in the coil wire is reversed. Thus, we see that in one revolution of the armature magnet, the current flows in one direction in the first half revolution and then reverses in opposite direction in the next half revolution. Therefore, this type of current generation is called an Alternating Current and the machine is called Alternator. To change a.c. into d.c., a silicon rectifier is used. D.C. is required for charging the battery. The rectifier is a circuit consisting of many diodes which allow flow of current in one direction only from alternator to battery.

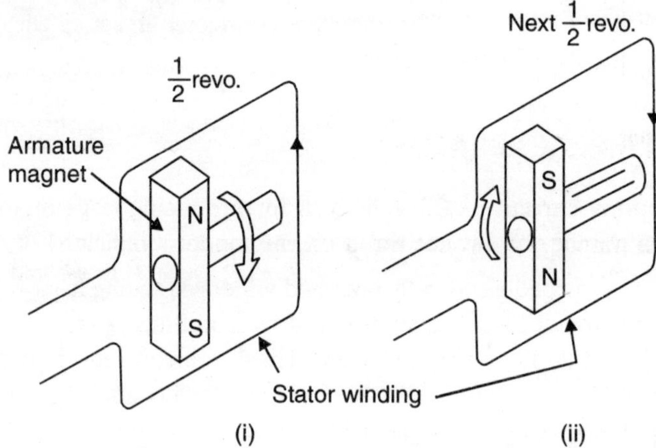

Fig. 6.15 Principle of production of alternating current

Construction of Alternator

Various parts of the alternator are shown in Figs. 6.16 (i) and (ii). The alternator has two main parts—a rotor and a stator. The rotor is called armature which is mounted on bearings and placed inside the stator. It is driven by fan belt. The armature slots contain one winding coil. The armature coil winding is connected to a separate slip ring. The current from the battery passes through the slip rings to this coil. This current passes through two small stationary carbon brushes. The slip rings make a running contact with the brushes. When the current passes through the armature coil, the coil gets magnetised and one of its ends becomes north pole while the other end becomes south pole. There may be 4 to 12 magnetic poles in a rotor depending upon the size of alternator.

The stator assembly consists of laminated iron core with windings connected in star (Y) shape. See Fig. 6.17(*i*). Whenever a magnet passes through each stator coil, current is generated. It more magnets pass through the each coil in a given time, the current generated is higher. The armature is a single magnet, but it acts like a set of many magnets. For this purpose the ends of the armature are formed into metal fingers as shown in Fig. 6.17(ii). Each finger becomes a magnet. These fingers of each end interlock but do not touch each other. Both poles of armature magnet passes each stator winding producing alternating current (a.c.) in it. This a.c. is changed into d.c. or direct current by using silicon rectifiers. The d.c. is required to charge the battery. These rectifiers containing several diodes are built into the alternator. The diodes permit flow of current in one direction only and provide a direct current.

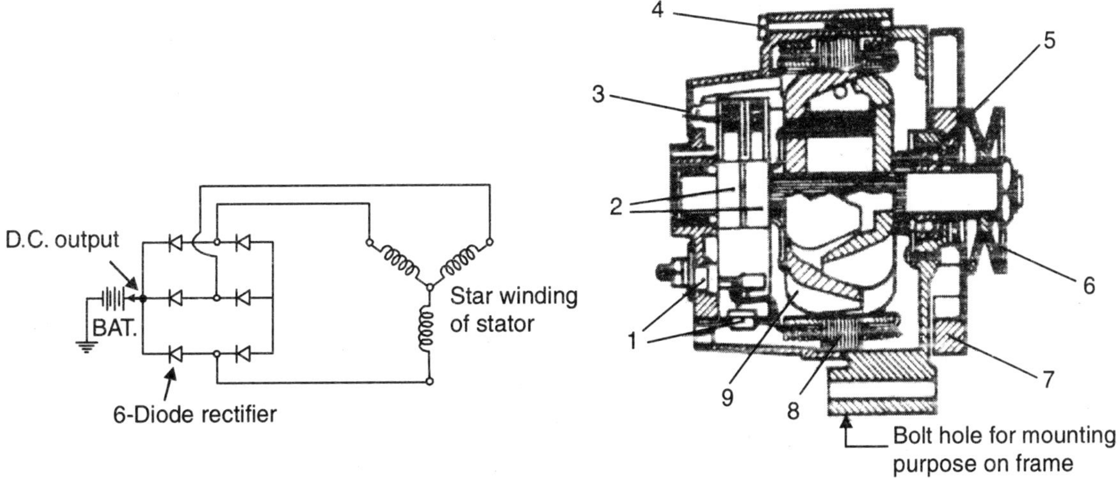

Fig. 6.16 (i) Wiring circuit of a Y-connected stator alternator having a 6-diode rectifier

Fig. 6.16 (ii) Alternator details: 1. Diodes; 2. Slip rings; 3. Brush and terminal assembly; 4. Through bolt; 5. Ball bearing; 6. Pulley; 7. Fan; 8. Stator assembly; 9. Rotor

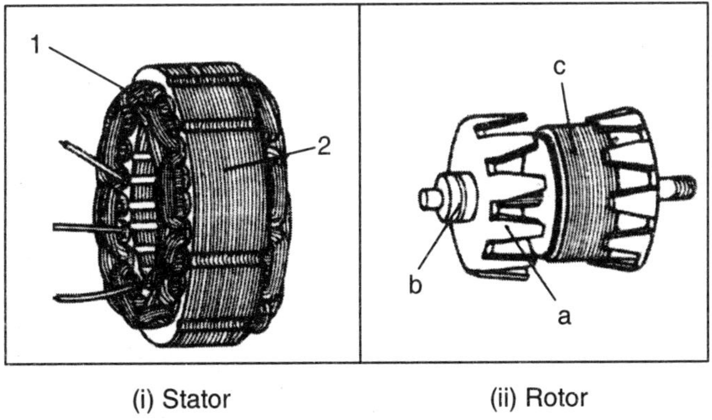

Fig. 6.17 (i) Stator and (ii) Rotor: 1. Magnets, 2. Windings, a. Pole pieces, b. Slip rings, c. Windings

The alternator controls its own current output. The rectifiers do not pass current in the reverse direction and do not act like a cut-out. Diodes in the rectifier do not permit the battery to discharge back through the stator of the alternator. Due to this reason, the alternator does not need a cut-out and current-regulator but needs only voltage regulator as shown in Fig. 6.14 (a). Fully Transistorised regulators are also used for controlling the alternator output.

Minor Adjustments/Repairs of Alternator

1. Engine is running but battery is not charged.
   (*a*) Belt may be loose—Adjust belt tension.
   (*b*) Brushes may be worn-out—Change the brushes.
   (*c*) Slip rings may have worn-out—Install new one.
   (*d*) Rectifiers not working—Install new rectifiers.
2. Alternator makes noise during running.
   (*a*) Mounting may be loose—Tighten and adjust the mounting.
   (*b*) Worn-out bearings—Replace them.
   (*c*) Interference between starter and fan wires—Correct the leads and insulate them properly.
3. Battery attained full charge but alternator continues to charge at high rate.
   (*a*) Stuck regulator contacts—Change contact assembly or install new regulator.
   (*b*) Voltage regulator set at a too high limit—Reset the regulator as per specification.
   (*c*) Poor regulator and ground connection—Correct and firmly make the ground connections.
4. Alternator gives no output.
   (*a*) Sticking Brushes—Clean or change the brushes.
   (*b*) Windings short-circuited or burnt—Rewinding is must.

HORN

It is a warning device for road users moving in the vicinity of the vehicle. All the automobile vehicles are fitted with horns. Two types of horns are commonly used in auto vehicles:

(*i*) Air-pressure Horn

(*ii*) Vibrating Tone Type Electric Horn.

(i) Air-Pressure Horn

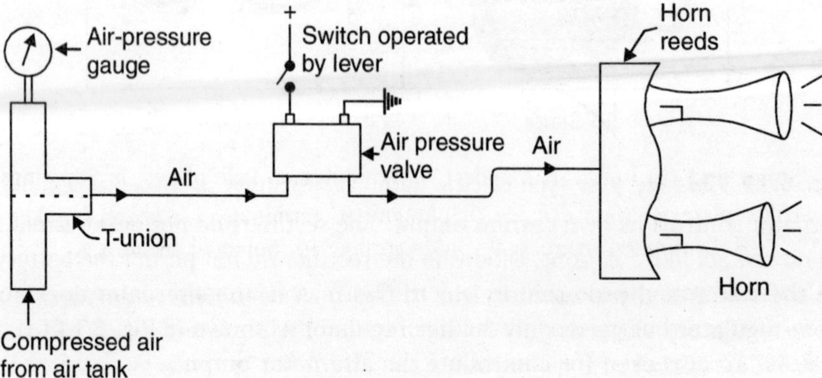

Fig. 6.18 Air pressure horn

Compressed air is used in air-pressure horns. The high pressure air is made to pass through reeds which produce alarming sound. These horns are fitted only on vehicles having air pressure assisted brakes. These vehicles have a tank for storing compressed air. A flexible pipe is taken from the air-tank to the horn through a lever fitted under the steering wheel. The lever presses an air pressure valve to lead the air to the horn reeds. It is shown in Fig. 6.18.

(ii) Vibrating Tone Type Electric Horn

A simple type of high frequency electric horn is shown in Fig. 6.19. Horns are employed to produce loud noise which will warn the traffic and other obstacles (animals and cyclists etc.) to clear the road for safe running for vehicle. It has a laminated magnet core and a insulated thin wire winding is done over it. The armature is attached to a central spindle. The spindle is supported at its one end by a guide spring and is supported at the other end by a diaphragm. The system also contains a contact breaker, and is connected in series with the solenoid (a core inside a coil). When the horn is switched 'ON', the current from battery flows through the contact breaker and solenoid. The current circuit is completed through an earthed terminal. When the circuit is completed, the laminated core gets magnetised. The magnet attracts the armature towards it. There is provided a protruding plate on the armature (See Fig. 6.19). This plate separates the contact breaker points and consequently the magnetic flux is collapsed and core is demagnetised. The armature is then pulled back to its normal position by spring. This cycle of operations is repeated frequently. Due to this high frequency vibrations, the vibrations developed in armature are transferred to the diaphragm which also vibrates with the same frequency and produces high notes of sound in the air column. It is also known as wind tone horn.

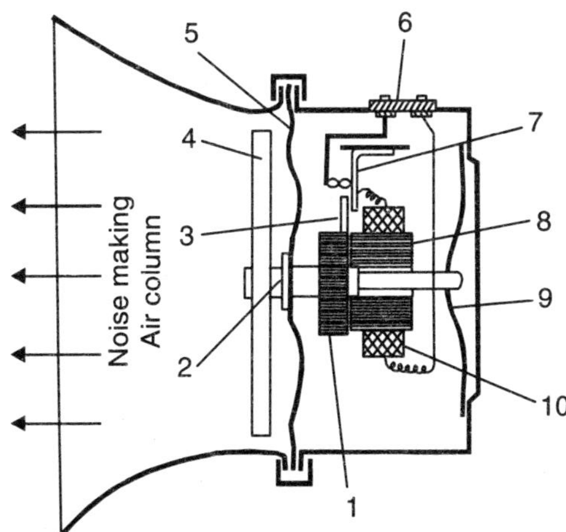

Fig. 6.19 Vibrating tone type electric horn: 1. Armature; 2. Armature spindle; 3. Striker plate; 4. Tone disc; 5. Diaphragm; 6. Terminals; 7. Contact breaker; 8. Laminated magnet; 9. Guide spring; 10. Solenoid winding

There are two notes of sound produced by electric horn. A deep note is produced due to bigger amplitude or low frequency vibrations of the flexible diaphragm and the high notes of sound are produced by higher frequency vibrations of a tone disc fitted at the outer end of the armature spindle.

The Fig. 6.20 shows a commonly used horn circuit in an exploded view. It contains two horns, each horn is fitted with flexible diaphragm which vibrates causing noise in air column.

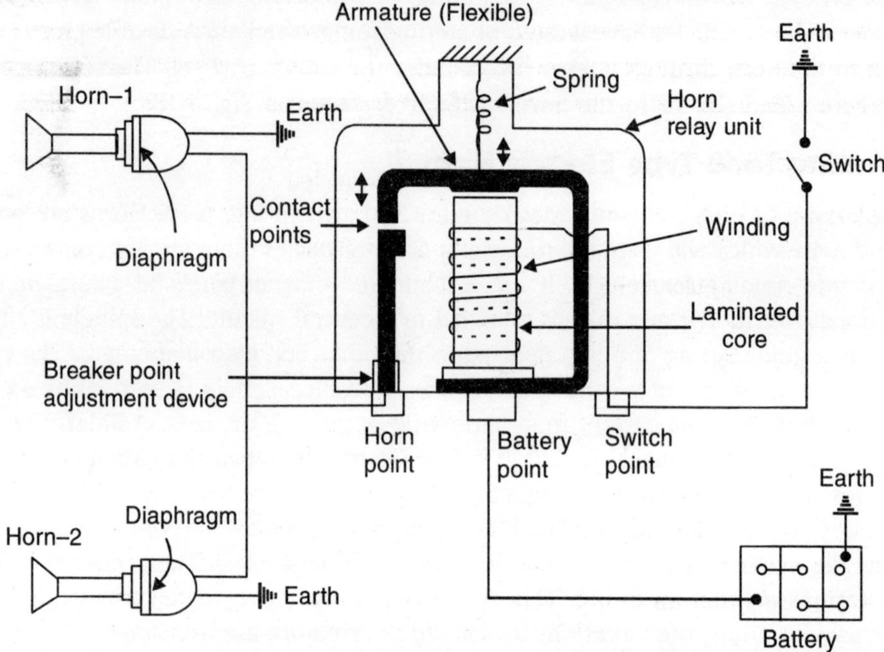

Fig. 6.20 Circuit diagram of an electric horn

When horn switched is pressed, the current from battery flows due to the completion of the circuit. The horn's laminated iron core gets magnetised and the armature is pulled down which closes the contact points. As soon as the contact point close, the diaphragm circuit is also completed and diaphragrm is also pulled back. As the points close, the current will not pass through core winding but it will pass through the armature shown by thicker lines as shown in Fig. 6.20 in the Horn Relay Unit. As a result of which the core is demagnetised and the armature lifts up, points open and horn diaphragm also return to its original position.

The circuit make and brake process goes on happening continuously at a very high speed. The diaphragms reciprocate at a very high frequency and generate vibrations in the air column which produces sound.

The output of sound depends on the correct setting of breaker points and the supply of proper voltage by the battery. A wrong setting of breaker points and a low voltage or discharged (partially) battery result in poor sound of the horn.

WIND-SCREEN WIPER

A set of wind-screen or windshield wipers is essential on every four wheeler vehicle. The wipers help in cleaning the front glass screen during rains, fog and dew. It also remove dirt and dust from the wind-screen, but wipers should not be used dry, otherwise, scratches will develop on the wind-screen. If wipers are being used to clean dust from the wind-screen, the screen should first be sprayed with liquid detergent shower; this device is provided on almost all cars. Wipers are also being provided at back glass sheet or tail gate screen on many cars.

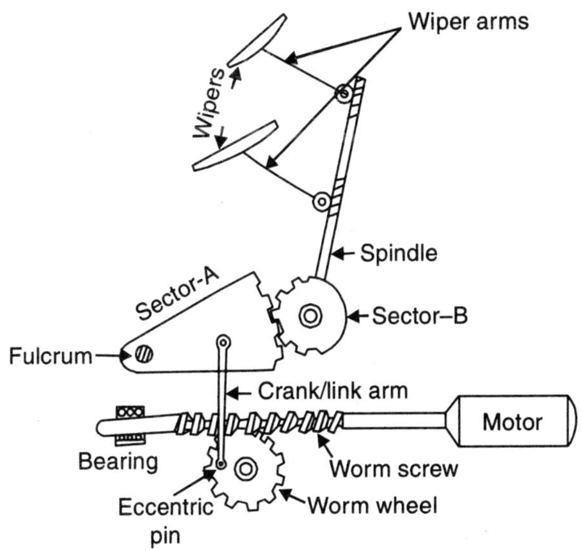

Fig. 6.21 Wind screen wiper drive mechanism

Majority of vehicles have a pair of wipers driven by a d.c. shunt-wound motor. Figure 6.21 shows the various parts of an electricity operated wind-shield wiper.

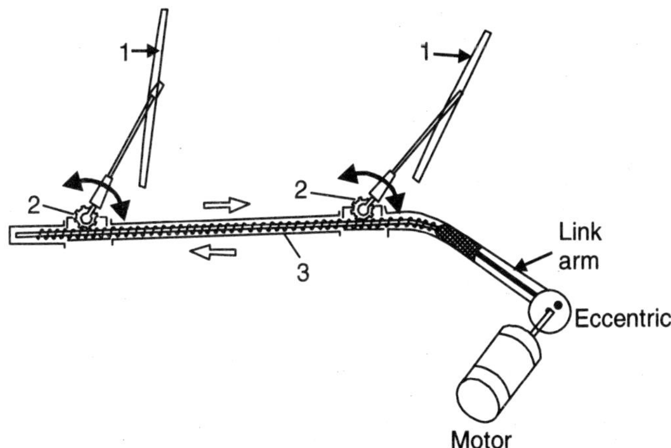

Fig. 6.22 Flexible rack driven wipers: 1. Wiper blades; 2. Pinion; 3. Flexible rack

A worm screw is attached to the shaft of the wiper motor. The worm rotates a worm wheel resulting in speed reduction. A crank lever or link arm is attached to worm wheel eccentrically at one end and the other end of link arm is attached to a sector-'A' which has fulcrum at its centre of circle. The sector-'A' drives another sector or pinion B. This sector B revolves clockwise when sector A moves upwards about fulcrum and it moves anti-clockwise when sector-'A' moves downwards under the action of reciprocating end of the crank arm attached with the sector A. A spindle is attached at the centre of the sector B. The wiper blades are mounted on wiper arms which are rotated by the spindle (in some designs, spindle is replaced by a flexible rack see Fig. 6.22). The wiper blades move with the spindle to wipe and clean the windscreen as and when desired by the car driver.

HEAD LIGHTS

The head lights are must for every vehicle. It provides light to see the road for driving the vehicle. The head lamp is made in the shape of a parabola. A parabolic reflector is placed along the parabola from inside. The source of light (Bulb) is placed at the focus 'F' of the parabola which reflects the light rays in such a way that they travel parallel to the axis.

Double Filament Lamp

This is most widely used head lamp. In this the upper beam filament is positioned "at the focal point" of reflector which throws light rays for long distance.

The lower beam filament is placed 'in front of the focal point' but it is shielded at the bottom by a small hood due to which light rays are emitted only towards the top of reflector. They are reflected towards nearer distance ahead of vehicle. It prevents glare to the driver coming from opposite direction. A double filament head lamp is shown in Fig. 6.23.

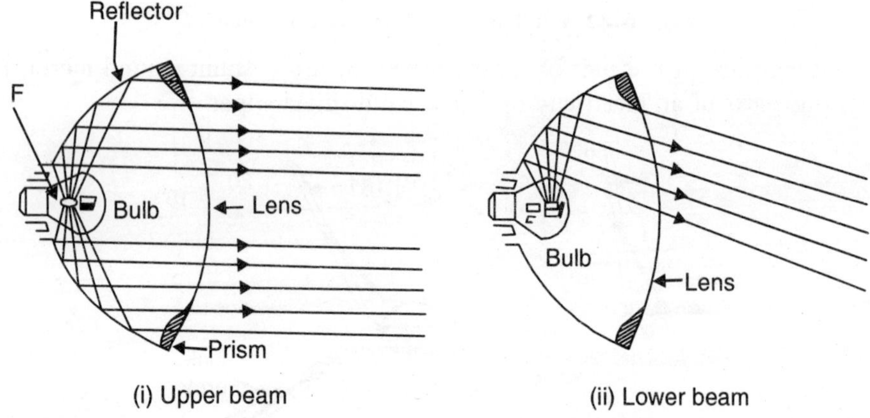

Fig. 6.23 Double filament head lamp

Sealed Beam Head Lights

The filament, reflector and lens cover are made as one complete sealed unit. If there occurs any failure in the filament, a new sealed beam has to be used which is the only disadvantage of sealed beam lights, otherwise it is durable and its focus never disturbs.

INSTRUMENTS AND CONTROLS ON DASH BOARD

The Fig. 6.24 and Fig. 6.25 shows the various instruments and controls for some devices fitted at the Dash Board of a car. Driver can easily see them and operate the instruments during driving. He can watch his speed, temperature of cooling water, fuel level and many other instructions/ warnings displayed on board. Figures 6.24 and 6.25 show the instruments and control in details.

Instruments and Controls on Dash Board

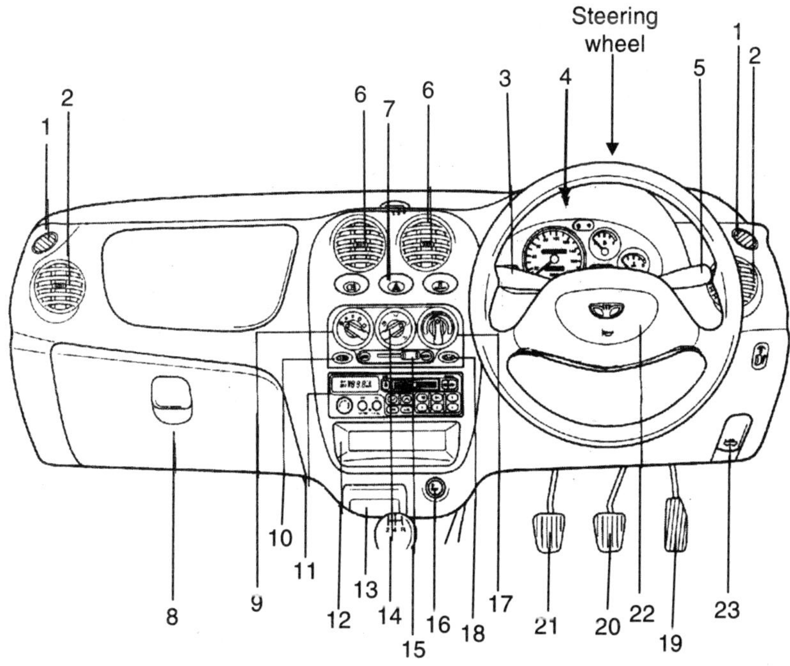

1.	Front door window defroster vents	12.	Deposit box
2.	Side ventilation vents	13.	Ashtray
3.	Light switch, turn signal switch	14.	Air distribution switch
4.	Instrument cluster (Details in Fig. 6.25)	15.	Air intake lever
5.	Windscreen wiper and washer switch, Tailgate window wiper and washer switch	16.	Cigarette lighter
		17.	Temperature control switch
6.	Center ventilation vents	18.	Air conditioning switch
7.	Hazard warning flasher switch	19.	Accelerator pedal
8.	Glove box	20.	Brake pedal
9.	Fan control switch	21.	Clutch pedal
10.	Rear window demister switch	22.	Horn button
11.	Radio (AM/FM) and cassette player	23.	Bonnet release handle

Fig. 6.24 Instrument and controls on dash board (Courtesy: Daewoo Motors)

Instruments Fitted in front of steering wheel on dash board or Instrument cluster shown at point 4 in Fig. 6.24

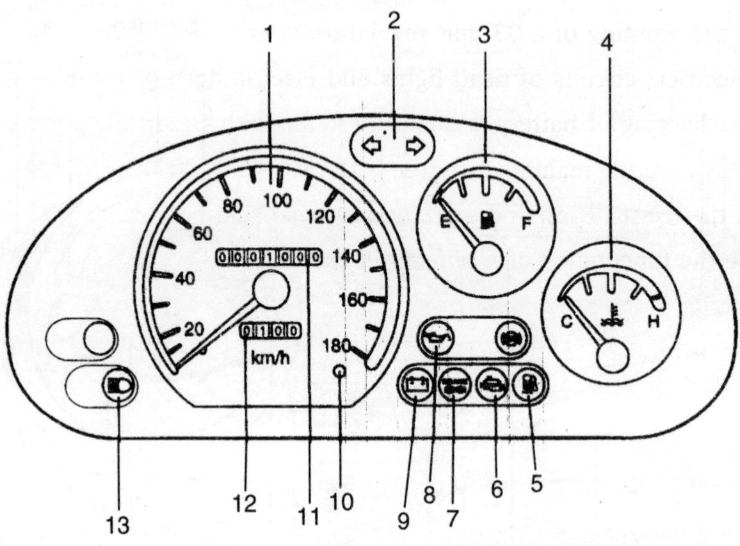

1. Speedometer
2. Turn signal/hazard warning flashes indicator
3. Fuel gauge
4. Temperature gauge
5. Low fuel level warning light
6. Engine control indicator (Service engine soon light)
7. Brake system warning light.
8. Engine oil pressure warning light
9. Alternator warning light
10. Reset button for trip odometer
11. Odometer
12. Trip odometer
13. High beam indicator

Fig. 6.25 Instruments fitted at instrument cluster on Dash Board (Courtesy : Daewoo Motors)

QUESTIONNAIRE

1. Write short note on the necessity of a dynamo in a auto vehicle.
2. Describe the working of a dynamo (2-Pole type) with the help of a neat sketch.
3. Why output of dynamo is regulated?
4. Describe a cut-out relay with the help of a diagram.
5. Explain the working of a voltage-regulator.
6. How third brush control of generator output is done?
7. Draw a neat sketch of voltage regulator, name its parts and describe its functioning.

8. How a shunt wound dynamo works?
9. Explain the functioning of an alternator with diagram.
10. What are the advantages of an alternator over the d.c. generator?
11. Explain the working of a 03 unit regulation of generator output.
12. Draw electrical circuits of head lights and electric horn of a car.
13. (*i*) How charging of battery is done? Describe with circuit diagram.

    (*ii*) Briefly describe maintenance free batteries.
14. Explain the construction and working of a lead-acid battery.
15. Describe the functioning of a self-starter motor.
16. (*i*) Describe a Bendix drive.

    (*ii*) What is an over-running clutch?
17. Describe an over running clutch drive.
18. (*i*) What are purposes of an electrical system of a car?

    (*ii*) Describe: (*a*) Wiring colour code; (*b*) Wiring harness.
19. (*i*) How a battery is rated?

    (*iii*) When battery electrolyte is top up or level-up?
20. Describe briefly the following:

    (*a*) Bendix drive

    (*b*) Over-running clutch.
21. Draw a simple wiring diagram of a car.
22. Describe the construction and working of a double filament head light of a car.
23. Describe working of the following:

    (*a*) Windshield wiper

    (*b*) Air-pressure horn.
24. How many electrical systems are employed in a car?
25. Describe with neat sketch the working of a Folo-Thru drive of starting.

CHAPTER 7

Principles of Design of Main Components of Power Unit

An automobile vehicle gets its propelling power from the engine fitted in it. The engine is known as the power unit of the vehicle. The main parts of an automobile engine, *i.e.*, internal combustion engine, have already been discussed in chapter number, 1. In this chapter, we shall discuss some considerations for design of main components.

1. Cylinder Block: It is the main block of the engine which houses almost all the essential and moving parts of the engine. It forms the basic structure of the engine. Blocks are made from grey cast iron (more heavy) or Nickel/Chromium alloyed cast iron (rough and tough) or Aluminium alloy (light but lesser toughness).

The main parts housed or fitted in it are:

1. Engine cylinders and liners, cylinder head.
2. Piston and connecting rod assembly.
3. Crankshaft and cam shaft.
4. Main bearings and cam shaft bearings.
5. Water jacket and openings for valves and lubricant, bearings.
6. Oil pan or oil sump.
7. Water pump, oil pump, pulleys etc., fuel filters, fuel pump etc.
8. Timing gears, valve lifting mechanism.
9. All other necessary systems—Fuel injection and fuel ignition systems etc.
10. Flywheel, exhaust and inlet manifolds etc.

The block is designed in such a way that it should be strong enough to contain all above parts as per their requirements. It is cast in one piece from grey iron alloyed with Nickel or Chromium. Some engine blocks are made of aluminium alloys in which cast iron or steel cylinder sleeves (liners) are inserted.

Most of the engine blocks are cast in one piece for in-line and V-shape cylinders. In radial or aero engines, there are different barrels for cylinders which are held together within the crankcase.

LINERS OR CYLINDER SLEEVES

In old engines, liners or cylinder sleeves are fitted in cylinder bores to obtain standard size bore for over size pistons after over hauling. Liners are also provided as original component on some engines. The liners provide a wear resistant surface for the cylinder bores of engine blocks. Liners are made from chromium plated mild steel tubes. Liners are of two types:

(*i*) *Dry liner.* They form metal-to-metal contact with the engine block all over the outer surface.

(*ii*) *Wet liner.* These liners are in touch with cooling water of the jacket at the outside surface. They are used in diesel engines to facilitate fast cooling. When liner is worn-out, generally it takes oval shape, then it is replaced by new one or rebored if tolerances permit.

PISTON

Piston of an I.C. engine is a cylindrical member open at lower side. It fits into the engine cylinder. It is connected to crankshaft through connecting rod. The upper end of connecting rod is attached to piston by inserting a gudgeon pin or piston pin. The lower end (Big end) is connected to crankshaft at crank pin. The high pressure created by combustion of fuel and air at the top (crown) of the piston moves the piston rapidly towards B.D.C. This reciprocating motion is transformed in to rotary motion through connecting rod and crank. The inertia of flywheel rotation helps to move piston from B.D.C. to T.D.C. Flywheel stores the energy obtained during power stroke.

Design Considerations of Piston

It should be able to sustain hammering effect of burning and expanding gases, fluctuating side thrust loads and high temperature of gases. For the above reasons, the piston must be capable of bearing the high temperature and pressure and impact loads, as well as, it should also be light in weight. A light weight piston reduces energy losses due to inertia, and inertia loads on bearings due to change in motion. Pistons are made of Cast Iron and Aluminium alloys. All alloy pistons are light in weight and have higher thermal conductivity—therefore widely used. The expansion of the piston skirt due to high temperature is controlled by making horizontal and/or vertical slots. On top side outer surface, at least two ring grooves are provided for inserting compression rings and towards bottom side-one groove for oil ring.

(*i*) The aluminium alloy piston are reinforced under the head (top) by 4 to 6 ribs, each of thickness $0.3t$ to $0.5t$ mm where t is the thickness of piston head in mm. which varies from 0.13 D to 0.15 D in 4-stroke engines and 0.18 D to 0.20 D in 2-stroke engines, where D = External diameter of piston.

(*ii*) Thickness of piston wall under rings = thickness of head (t).

(*iii*) The thickness of lower side skirt = $0.3t$.

(*iv*) Distance of first compression ring from piston crown = 0.15 D to 0.30 D for low speed engine and 0.10 D to 0.20 D for high speed engine where D = diameter of the piston.

(v) Width of the lands between the grooves = 0.75 t_r.

Where t_r = axial thickness of piston ring, mm.

Numerical Problem 1. *A 4-cylinder 4-stroke diesel engine is required to develop 14.7 kW at 1000 r.p.m. The mean effective pressure should be 5.5 bar. Calculate the bore and stroke of the engine. Assume length of piston stroke as 1.5 times the bore.*

Solution. Number of cylinder, $n = 4$

Indicated power, $IP = 14.7$ kW

Speed, $N = 1000$ r.p.m.

Length of Stroke $L = 1.5$ D, where D = bore diameter or inside diameter of engine cylinder.

Since indicated power,

$$IP = \frac{n.P_{mep.}L.A.N.\,k \times 10}{6}$$

$$A = \text{Area of Piston/Cylinder} = \frac{\pi}{4}D^2$$

k = Number of power strokes per revolution of crankshaft

= 1 for 2-stroke cycle engine

$= \frac{1}{2}$ for 4-stroke cycle engine

and $P_{mep} = 5.5$ bar (1 bar = 10^5 N/m^2)

Now putting the values:

$$14.7 = \frac{4 \times 5.5 \times 1.5D \times \frac{\pi}{4} D^2 \times 1000 \times \frac{1}{2} \times 10}{6} \quad (\pi = 3.14)$$

or

$$D^3 = \frac{14.7 \times 6 \times 4 \times 2}{4 \times 5.5 \times 1.5 \times 3.14 \times 1000 \times 10}$$

$$= 0.0006809$$

or

$$D = 0.08798 \text{ m}$$

$$= 87.98 \text{ mm} \simeq 88.00 \text{ mm}$$

$$L = 1.5 \text{ D}$$

$$= 1.5 \times 88 = 132.0 \text{ mm}$$ **Ans.**

Numerical Problem 2. *A 4-cylinder, 4-stroke, S.I. engine develops a maximum brake torque of 160 N-m at 3000 r.p.m. Calculate the engine displacement, bore and stroke. The brake mean effective pressure at the maximum engine torque point is 960 kPa. Assume bore = stroke. Also determine radius of crank. (GATE – 1997)*

Solution. Data given: $n = 4$, $k = \frac{1}{2}$ (for 4-stroke cycle), $T_b = 160$ N-m r.p.m., N = 3000, $P_{bmep} = 960$ k.Pa, D = L

$$P_{bmep} = 960 \text{ k.Pa} = 960 \times 1000 \text{ N/m}^2 = 9.6 \text{ bar } (1\text{Pa} = 1 \text{ N/m}^2)$$

$$\text{Brake Power} = \frac{2\pi N T_b}{60 \times 1000} \text{ k.w.} = \frac{P_{bmep} \text{ L.A.N.k.10}}{6} \text{ k.w.}$$

or
$$\frac{2\pi \times 3000 \times 160}{60 \times 1000} = \frac{9.6 \times D \times \frac{\pi}{4} D^2 \times 3000 \times \frac{1}{2} \times 10}{6}$$

or
$$\frac{2 \times 160}{60 \times 1000} = \frac{9.6 \times D^3 \times 10}{4 \times 6 \times 2} = \frac{96 \text{ D}^3}{48} = 2 \text{ D}^3$$

or
$$\frac{16}{3000} = 2 \text{ D}^3$$

or
$$D^3 = \frac{16}{6000} = 0.0026666$$

or Diameter/Bore,
$$D = [0.0026666]^{\frac{1}{3}}$$
$$= 0.1387 \text{ meter}$$
$$= 138.70 \text{ mm.}$$

Length of stroke,
$$L = D = 138.70 \text{ mm.}$$

Displacement or swept volume
$$= \frac{\pi}{4} D^2.L.$$
$$= \frac{22}{7 \times 4} \times D^3$$
$$= \frac{22}{7 \times 4} \times (0.0026666)$$
$$= 0.7857 \times 0.0026666$$
$$= 0.0020951 \text{ m}^3$$

Crank radius,
$$r = \frac{L}{2} = \frac{138.70}{2} = 69.35 \text{ mm. } \textbf{Ans.}$$

Design of Compression Rings of Piston

The compression ring's width has been reduced considerably from 3.0 mm to 1.5 mm in present days cars. The reduction in ring width provides better resistance to ring flutter and ring scuffing. By using a thin ring width, (*a*) the piston height also shortens, which means, (*b*) overall size of the engine is also reduced and (*c*) problem of ring inertia is also diminished. But, reduction of ring width without a suitable thickness/width ratio, may render its unstable in the ring groove and machining of narrow ring grooves in the piston is also very difficult and time-consuming and require special tools. Most of the rings are made of chrome-plated cast iron for bearing high pressures and temperatures.

Refer to Fig. 1.4 (b) in the chapter 1, the first compression ring's cross-section is generally made plain rectangular but second compression ring is tappered towards cylinder wall because the pressure on second ring is less.

Some emperical dimensions are taken as below:

Radial thickness, t_r = 0.05 D to 0.1 D mm

Width of ring, W = 0.8 t_r mm

Where D = Diameter of cylinder bore in mm.

Ring-groove depth clearance is taken 0.7 mm to 0.95 mm depending upon the thickness of rings and side clearance varies from 0.1 to 0.20 mm and ring gap between two ends (after fitting) is about 1 to 1.5 mm.

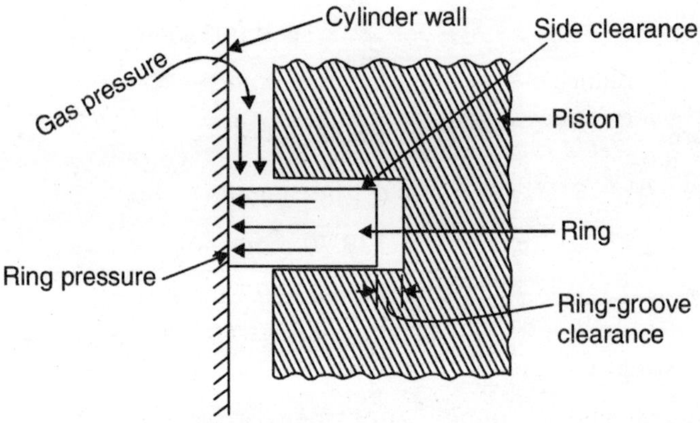

Fig. 7.1 Forces acting on ring (compression)

Connecting Rod

Cylinder, piston, connecting rod and the crank form a four-bar mechanism. The function of connecting rod is to transform the reciprocating motion of the piston into rotary motion of the crankshaft with the help of a crank.

The connecting rod is subjected mainly to axial stresses and bending stresses. The reciprocating motion of piston, gas pressure on it and inertia forces produce axial stress in the connecting rod. The centrifugal forces produce bending stresses within the plane of rotation.

To reduce these forces, the cross-section of connecting rod is made in I-section which reduces the weight and area of cross-section for minimising the stresses.

The connecting rods are formed by drop forging process or by casting. The material used are Alloy steels and Al-alloy or Duralumin.

Design considerations for connecting rod and piston-pin

To reduce the weight and height of the engine, it is essential to reduce the length of connecting rod. Following stresses act on the connecting rod:

1. Axial stresses due to gas-pressure on piston.
2. Inertia force of reciprocating parts.
3. Bending stresses due to centrifugal forces.

To have a stronger and rigid connecting rod with lesser weight, the rod is made in I-section which easily takes on all the above mentioned forces.

The small end of connecting rod holds the piston with the help of a piston pin. For keeping light weight, piston pin is made in tubular form. The surface of pin is heat treated by carburising at about 900°C and then hardened and tempered. It is given a very fine finish of about 0.10 μm. The clearance between pin and its bushes is kept 7.5 μm to minimise noise. The big end of connecting rod fits at crank pin. The big-end is always split. A hole is also drilled from big end to small end for forcing lubricating oil, from big end, to lubricate piston pin and its bushes.

Flywheel Effect

A large flywheel will reduce the speed fluctuations to a minimum but its response to acceleration or deceleration will be very slow. In automobiles, the response to pick-up must be quick, therefore, there is a limitation to the size of flywheel. That is the reason the automobiles use multi-cylinder engines of small sizes which give a smooth torque ratio which reduces cyclic fluctuations and hence a small size flywheel becomes sufficient to act as energy reservoir.

A single cylinder engine's torque variation with crankshaft rotation is shown in Fig. 7.1 (a). The torque ratio $\frac{AB}{AC}$ is quite high for single cylinder engine. It will become more smooth for multi-cylinder engines and give smooth running of engine with less vibrations.

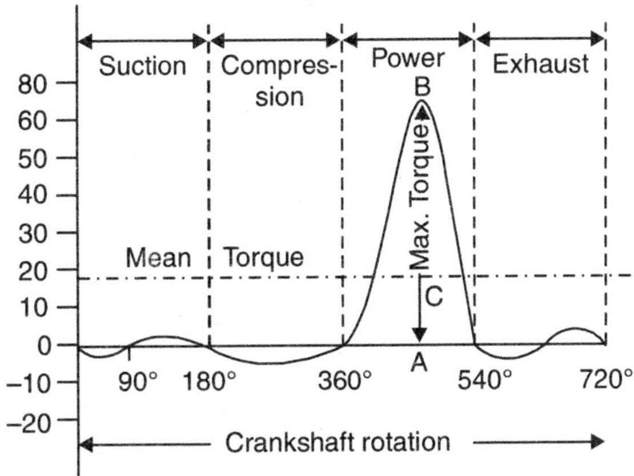

Fig. 7.1(a) Torque variation

CAM SHAFT DRIVE

The cam shaft in 4-stroke engines rotates at half the speed of crankshaft. The following three mechanism are used to drive the cam shaft by taking power from crankshaft:

(*i*) Timing gears—used on cars.

(*ii*) Sprocket and chain system—used on motor cycles and some scooters.

(*iii*) Toothed timing belt mechanism—used on some cars.

(i) Timing Gears

In this drive, a small gear (suppose of pitch diameter '*d*') is mounted on the crankshaft and the other bigger diameter gear (2*d*) is mounted on the cam shaft. This size of gears makes the cam shaft to rotate at half the speed of crankshaft (N). The cam shaft is rotated at half speed (N/2) because the valves have to open and close once in every two revolutions of the crankshaft.

The timing gears are shown in Fig. 7.1 (b). The cam shaft gear has twice the number of teeth than crankshaft gear.

Both the timing gears are such designed that the sum of their diameters ($d+2d = 3d$) equals the centre to centre distance between the axes of crankshaft and cam shaft. The both gears are called timing gears because they are provided with timing marks—the two marks are matched together in such a manner that the opening and closing of valves takes place at the accurate and correct timings of the working cycle of the engine. After repair and overhauling of engine, the two gears are re-assembled by keeping timing marks in line as shown in the Fig. 7.1 (b).

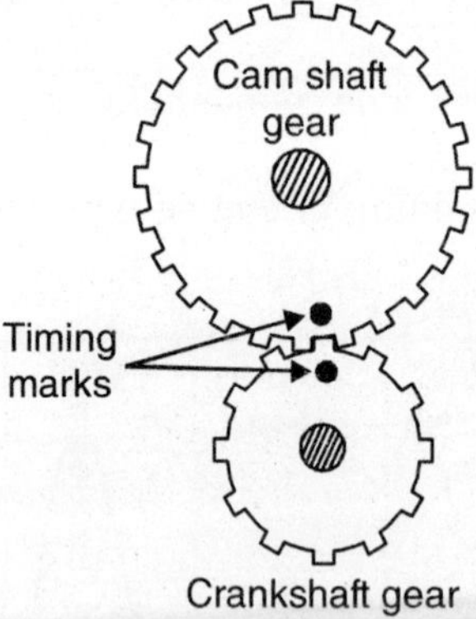

Fig. 7.1 (b) Timing gears

Timing gears are made of cast iron, steel, aluminium alloy or some special plastics. These days bakelite gears are becoming popular as they are light in weight, make no noise and lubrication is also not required.

Numerical Problem 3. *Design a set of timing gears for transferring power from crankshaft to the cam shaft. The cam shaft must rotate at half the speed of crankshaft. The axis to axis distance between the two shafts is 36 cms.*

Solution. Distance between the axes of the two shafts = 36 cm

If d = pitch diameter of crankshaft gear

then diameter of the cam shaft gear should be '$2d$' to rotate it at half speed.

$$\therefore \quad d + 2d = 36 \text{ cm}$$

or $$3d = 36$$

or $$d = \frac{36}{3} = 12 \text{ cm} = \text{pitch diameter of crankshaft gear.}$$

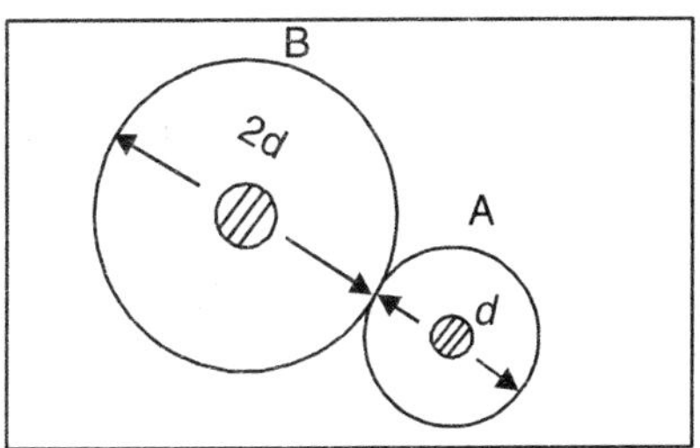

Fig. 7.2

$\therefore$ Pitch diameter of cam shaft gear $= 2d$

$$= 2 \times 12 = 24 \text{ cm.}$$

Similar and suitable size and number of teeth may be selected for both the gears.

Gear ratio or velocity ratio $= \frac{N_B}{N_A} \frac{T_A}{T_B}$ (Fig. 7.2)

Where N = speed in r.p.m. T = No. of teeth

If speed of crankshaft is N_A r.p.m.

Then speed of cam shaft = $N_B = \frac{1}{2} N_A$

$$\therefore \quad \frac{N_B}{N_A} = \frac{\frac{N_A}{2}}{N_A} = \frac{T_A}{T_B}$$

or $$\frac{1}{2} = \frac{T_A}{T_B}$$

or $$T_B = 2\,T_A$$ **Ans.**

PERFORMANCE OF AUTOMOBILE ENGINES

Engine performance is an indication of the degree of success with which it converts the chemical energy contained in the fuel into the useful mechanical work. The following are the basic parameters of knowing about the engine performance:

1. Power developed by combustion of fuel (I.P.).
2. Power output or power available at the crankshaft for conversion into useful work (B.P.).
3. Mean effective pressure (m.e.p.)
4. Maximum value of Torque at the maximum speed.
5. Thermal Efficiency, Mechanical Efficiency and Volumetric Efficiency.
6. Specific fuel consumption—Fuel consumed per brake kw per hour $= \frac{m_f}{\text{B.P.}}$.
7. Exhaust emissions etc.
8. Power loss in overcoming friction of the moving parts of the engine (F.P.).

1. **Indicated Power (I.P.):** The power developed by combustion of fuel in the engine cylinder is called indicated power.

$$\text{I.P.} = \frac{n.\,P_{mepi}\,.\,L.A.N.k.\,10}{6}\ \text{kw}$$

where n = Number of cylinders in the engine.

P_{mepi} = Indicated mean effective pressure in bars. It is the hypothetical pressure which is assumed to be acting uniformly on the piston throughout the power stroke of the working cycle.

L = Length of piston stroke in m.

A = Area of piston head in $m^2 = \frac{\pi}{4} D^2$ where D is the diameter of piston in meter.

k = $\frac{1}{2}$ for 4-stroke cycle and 1 for 2-stroke cycle.

= Number of power stroke per revolution of crank.

2. **Brake Power (B.P.):** The power available at the output shaft or crankshaft is called brake power.

$$\text{B.P.} = \frac{2\pi NT}{60 \times 1000}\ \text{kw}$$

where N = Speed of crankshaft in r.p.m.

T = Torque or turning effort at crankshaft in N-m

Frictional Power (F.P.): The difference between I.P. and B.P. is called F.P. It is the power lost in frictions.

$$\text{F.P.} = \text{I.P.} - \text{B.P.}$$

Mechanical Efficiency: The ratio of B.P. to I.P. is called mechanical efficiency (η_{mech}).

$$\eta_{mech} = \frac{\text{B.P.}}{\text{I.P.}}$$

Specific Output: It is the brake output per unit of piston displacement.

$$\text{specific output} = \frac{\text{B.P.}}{\text{A.L.}} = \text{C.P}_{\text{mep}_b}.\text{N.}$$

where C = a constant

P_{mep_b} = brake m.e.p., in bars

N – R.P.M. (Revolutions per minute)

A – Area of piston, m^2

L – Stroke length, m

Torque

The torque or turning moment is a rotary or twisting effect that an engine applies to the crankshaft through piston, connecting rod and crank. It is taken as the product of pushing force 'F', exerted by expansion of gases after ignition through piston and connecting rod, and its perpendicular distance '*r*', the crank length or radius at the central axis of the crankshaft (Fig. 7.3).

If T = Torque in N-m or kgf-m on crankshaft

then $T = F \times r$ N-m

if l = length of stroke = $2r$ = 2 × crank radius

or $r = l/2$

$\therefore$ $$T = \frac{F.l}{2}$$

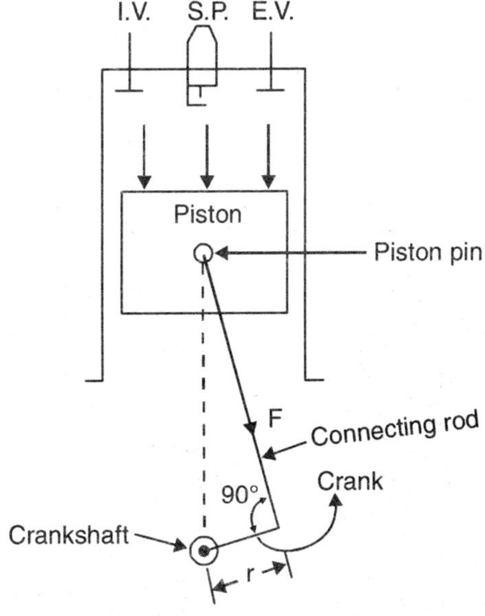

Fig. 7.3

which shows that torque will be higher for engines having larger stroke length (l) or crank radius or its length (r) as $l = 2r$.

POWER AND TORQUE CHARACTERISTICS

The torque developed by engine varies with speed. At lower speeds, the air-fuel mixture gets sufficient time to enter into the cylinder, resulting in higher volumetric efficiency. This gives higher combustion pressure during power stroke and consequently, greater torque is available at the crankshaft and it acquires a maximum value at an intermediate speed.

But at higher speeds, the mixture gets very less time to enter into the cylinder during suction stroke and therefore volumetric efficiency drops. As lesser amount of air-fuel mixture burns during power stroke, the combustion pressure does not reaches at its high value and hence lesser torque is developed. The variation of power and torque with speed has been shown in Fig. 7.4.

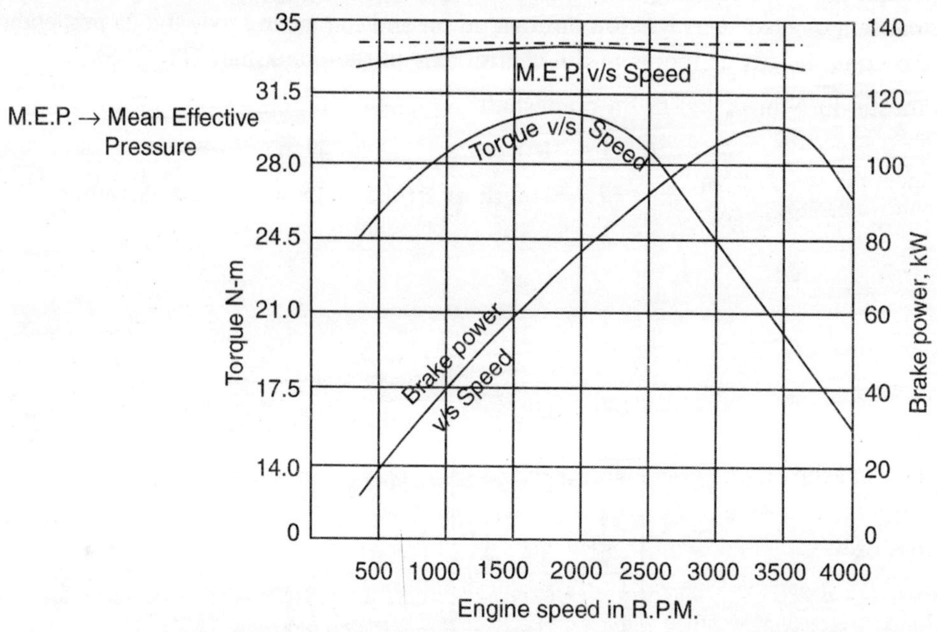

Fig. 7.4 Variation of torque, power and M.E.P. with engine speed

The mean effective pressure remains more or less constant at all speeds but slightly falls at low speeds due to leakages past valves and rings, it also falls at higher speeds due to lesser volumetric efficiency. Higher the mean effective pressure, more power will be developed by the engine for a given displacement. The mean effective pressure is also known as brake m.e.p. if it is calculated from brake power.

If the friction loss is negligible,

If $n = 1$ for single cylinder engine and $k = \dfrac{1}{2}$ for 4-stroke engine then, B.P. = I.P. .... (Neglecting friction losser)

$$\frac{2\pi NT}{60 \times 1000} = \frac{n.P_{mep}.\,L.A.N.k.10}{6}$$

or Torque,T = (A constant) × P_{mep}

or Torque ∝ m.e.p. .... (Torque is directly proportional to m.e.p.)

Comparison between Single Cylinder and Multi-cylinder Engines

Single Cylinder Engine	Multi-Cylinder Engine
1. Greater unbalance torque at the crankshaft, thereby, jerky motion and lot of vibrations.	1. Balanced torque characteristics thereby smooth running.
2. The reciprocating mass (piston and connecting rod) generate high inertia forces which impose a higher limit to the size of piston and cylinder and which limits the production of power.	2. The total engine displacement is divided into multi-cylinders which produce less inertia forces and higher power output is obtained.
3. Inertia forces increase as square of speed— which limits the speed. It will be dangerous to run a single cylinder engine at high speed.	3. The multi-cylinder engines do not produce much inertia forces, therefore, they run at higher speeds.

TRACTIVE EFFORT

The force available at the contact line between road and the driving wheels of the vehicle is called Tractive Effort.

The drive from the engine crankshaft is fed to gear box and to propeller shaft and final drive at the differential at rear axle or at front axle. The speed is decreased at gear box and final drive to increase the torque.

If T_d = Torque at driving wheels axle shaft

= [Gear box ratio × drive axle or final drive ratio] × overall transmission efficiency × engine torque

$$= G.\eta_o.T. \qquad ...(1)$$

Where G is called overall gear ratio = (gear box ratio × final drive ratio)

$$\text{Engine torque, } T_E = \frac{B.P. \times 60 \times 1000}{2\pi\, N} \text{ N-m} \qquad ...(2)$$

If r_w = radius of driving wheel in m (Fig. 7.5)

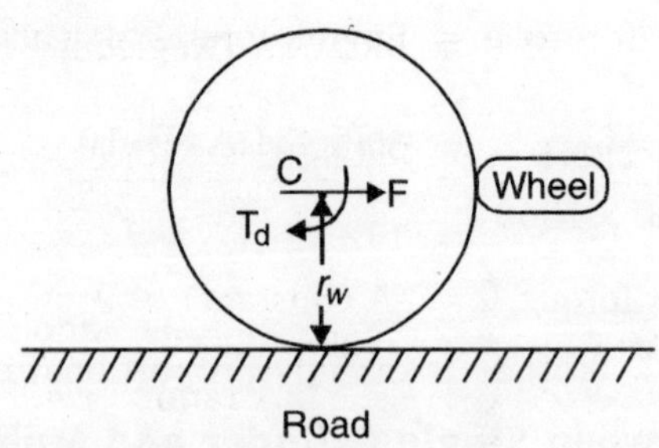

Fig. 7.5 Driving wheel

The tractive effort, $F = \dfrac{T_d}{r_w}$

$$= \frac{G.\,\eta_o\,.\,T_E}{r_w} = \frac{G.\eta_o\,.B.P.\times 60{,}000}{2\pi r_w\,.N} \qquad ...(3)$$

Vehicle speed in m/min., $V = \dfrac{2\pi r_w \times N}{G}$ m/min. r_w = radius of road wheel in m

N = r.p.m. of engine.

If V is given in kms per hour

then, $$\frac{V \times 1000}{60} = \frac{2\pi\, r_w\, N}{G} \qquad ...(4)$$

or $$\frac{V}{N} = \frac{1000 \times G}{2\pi\, r_w \times 60} = 2.65 \left(\frac{G}{r_w}\right) \text{ where V is in kms/hr.} \qquad ...(5)$$

Overall gear ratio, $$G = \frac{2\pi\, r_w\, N}{V} \text{ where V is in m/min.} \qquad ...(6)$$

Numerical Problem 4. *An engine develops 10 kgf-m torque at 2500 r.p.m. If gear box ratio is 2.1 : 1 and final drive reduction is 5 : 1. Determine the tractive effort at rear driving wheels and the speed of propeller shaft. The radius of wheel may be taken as 51 cms.*

Solution. (*i*) Speed of propeller shaft $= \dfrac{2500}{2.1} = 1190.5$ r.p.m.

(*ii*) Torque at rear wheel centre, $T_d = 10 \times 2.1 \times 5 = 105$ kgf-m

∴ Tractive effort, $F = \dfrac{T_d}{r_w}$ r_w radius of wheel in m = 0.51 m

$$= \frac{105}{0.51}$$

$= 205.8$ kgf. **Ans.**

Numerical Problem 5. *A car engine develops 60 kg-m torque at 4000 r.p.m. If the car is running in second gear with 4 : 1 as gear ratio, find the torque and speed of propeller shaft if the mechanical efficiency of the drive is 80%.*

Solution. Propeller shaft torque = Engine torque × gear ratio × mechanical efficiency.

$$= 60 \times 4 \times \frac{80}{100}$$

$$= 192 \text{ kg-m}$$

Speed of propeller shaft $= \dfrac{\text{Engine speed}}{\text{gear box ratio}} = \dfrac{4000}{4}$

$$= 1000 \text{ r.p.m.}$$ **Ans.**

The power developed by engine is lost in driving lot of moving parts-such as piston, connecting rod, crankshaft, and flywheel. From crankshaft the power is further lost in driving clutch, gearbox, universal joints, propeller shaft and final drive, that is the reason that mechanical efficiency is always less than 1 or less than 100%.

FACTORS AFFECTING VEHICLE PERFORMANCE

Following factors are responsible for good or poor performance of the power unit of the vehicle:

1. Effect of compression ratio: Increased compression ratio $\left(\dfrac{\text{swept volume}}{\text{clearance volume}}\right)$ has the following advantages:

(*i*) Enhanced power output.

(*ii*) Higher speed.

(*iii*) Increased air standard efficiency (Fig. 7.6)

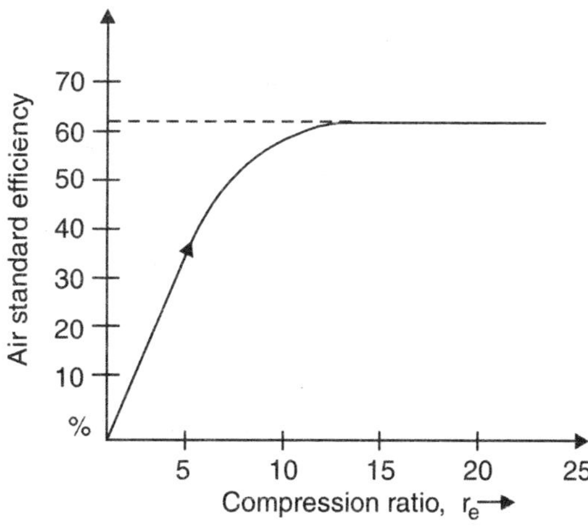

Fig. 7.6 Variation of air standard efficiency with compression ratio

(*iv*) Greater economy.

(*v*) Higher load carrying capacity.

Increased compression ratio has some disadvantages also, which are:

(*i*) Size and weight of the engine has to be increased.

(*ii*) Problems of detonation and knocking arise.

(*iii*) Carbon deposits inside cylinder are increased.

(*iv*) There is incomplete fuel combustion at slow speeds.

(*v*) Maintenance is increased.

2. Power–weight ratio of the vehicle: Power/weight ratio is an important parameter in working out the performance of the auto vehicle. A high power/weight ratio is desirable. It gives following benefits:

(*a*) Attain higher speeds.

(*b*) Provides better hill climbing ability or gradeability.

(*c*) Quick pick-up (acceleration).

(*d*) More hauling (load carrying) capacity.

(*e*) Faster rate of climbing up hill.

Power/weight ratio in racing cars is highest, thereby, they give superb performance.

(*f*) Fuel consumption is also low.

(*g*) More seating and luggage space in comparison with the same fuel consuming cars.

Weight/Power (W/P) ratio for some vehicles

Car	*W/P (kg. per B.P.)*
Matiz	15.38
Santro	14.10
Maruti 800	16.20
Zen Vx	15.10
Indica (Petrol)	16.30
Esteem	12.86
Accent	10.70

Engine Power Output may be increased by taking the following steps:

(*a*) By increasing compression ratio.

(*b*) Using high octane value fuels.

(*c*) By improving design of engine components.

(*d*) By reducing weight of components by using lighter but strong materials.

(*e*) By proper balancing (static as well as dynamic) of the engine.

(*f*) By better lubrication of moving parts.

(*g*) By atomising the fuel before combustion.

(*h*) By preventing leakages of high pressure gases past the piston.

(*i*) By minimising the unburnt fuel losses through exhaust gases.

(*j*) By complete combustion of the fuel.

(*k*) By proper governing and tuning of the engine.

(*l*) By increasing input energy (using high grade fuels).

3. Pick-up or Acceleration. A good vehicle must take negligible time to raise its speed as desired by the driver. Good pick-up means a fast acceleration of the vehicle.

The Fig. 7.7 shows a variation curve between velocity 'V' and time 't'.

At V = 0, t = 0

If Velocity V_p, reaches in time t_p and V_Q velocity reaches in time t_q. V_p to V_Q is a constant or uniform velocity during the normal running of the vehicle.

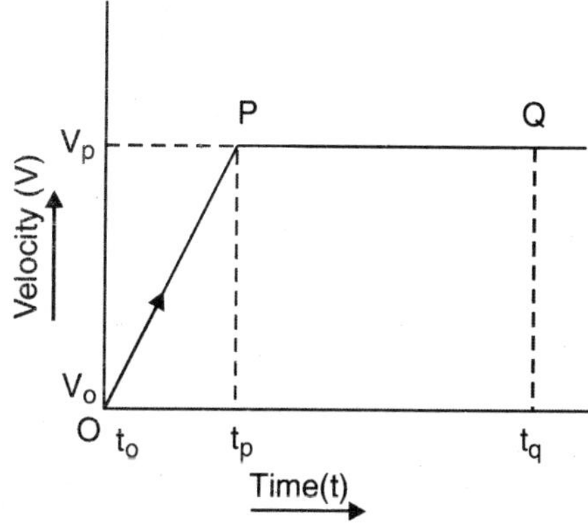

Fig. 7.7 Velocity-time variation

The vehicle picks up velocity during O to P.

at O, V_o = 0, and t_o = 0

If f = pick-up or accelerator in m/sec/sec or m/sec^2

$$= \frac{V_p - V_o}{t_p - t_o} = \frac{V_p}{t_p} = \text{m/sec}^2 \qquad ...(1)$$

Pick-up times for some vehicles:

Name of the Car/Bike	Speed range, Kmph	Pick-up time in seconds
Hero Honda CBZ Bike	0–60	5.0
Bajaj Chetak Scooter	0–50	8.0
Maruti and Zen. Car	0–60	7 to 12.0
Fiat Uno Car	0–100	9 to 18
Old Fiat Car	0–60	12.0

Drawbar Horsepower (D.H.P.): This is the actual power available which is utilised to propel the vehicle against all the resistances.

Dhp = Bhp – Resistances to motion of vehicle.

4. Gradeability of a vehicle: The gradeability may be defined as the hill climbing ability of a vehicle. The gradeability is expressed in degrees or in percentage as explained below:

Grade or gradeability, in percentage, refers to the ratio of the amount of vertical movement to the amount of horizontal movement of the vehicle multiplied by 100.

Suppose the vehicle shown in Fig. 7.8 climbs up a hill ABC. If it travels up from A to B, the vertical distance covered will be BC and horizontal distance will be AC.

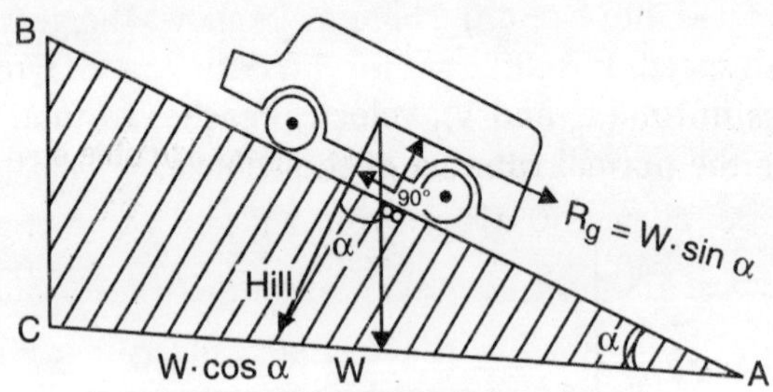

Fig. 7.8 Vehicle climbing a hill

$$\therefore \quad \text{\% Gradeability} = \frac{BC \times 100}{AC} = \tan\alpha \times 100 \qquad \text{...(2)}$$

$$AB = \text{Actual distance covered by vehicle}$$

$$BC = AB \cdot \sin\alpha$$

$$AC = AB \cdot \cos\alpha$$

If
$$\alpha = 26.1° \text{ then } \tan\alpha = 0.49$$

$$\text{the \% Gradeability} = 0.49 \times 100 = 49\%$$

The grade force is the component of the vehicle weight along the grade, it is also called Road Gradient Resistance, R_g.

$$R_g = W \cdot \sin\alpha = m.g.\sin\alpha \qquad \text{...(3)}$$

$$= \text{Resistance due to road gradient.}$$

where m = mass of the vehicle.

Gradeability is higher in vehicles having powerful engines and a well-balanced aerodynamic body. The gradeability is always maximum in first gear and decreases in second, third and fourth gears. Maruti 800 car has 14.3° or 25.2% gradeability whereas a Bajaj motorcycle also has almost same gradeability about 25% or 14° slope.

RESISTANCES TO THE MOTION OF A VEHICLE OR RETARDING FORCES

1. **Braking force:** It is applied by brakes, as and when required by driver. It does not work constantly all the time.
2. **Road Resistances:** It includes the following resistances:

(*a*) ***Rolling Resistance:*** It is the force required to maintain a constant speed on a levelled road. It's average value is assumed as 15.0 kgf/tonne. It varies from 4.5 to 27.0 kgf/tonne weight of vehicle on pacca roads. In loose sandy road, it is high upto 252.0 kgf/tonne. The rolling resistance of the tyres of wheels is proportional to the tyre load. The rolling resistance helps to slow down the vehicle. It remains constant upto a certain speed but increases fastly with speed. The rolling resistance is expressed as a coefficient equivalent to the drag per thousand kilogram of tyre load which can easily be converted into tyre drag. Figure 7.9 shows the general tendency of the rolling resistance with speed. It is different for different type of tyre treads and types of tyres. The rolling resistance increases fuel consumption. Roughly, 10% of rolling resistance reduction will reduce 2% fuel consumption. The average value of rolling resistance co-efficient may be taken as 0.015 or 15.0 kgf per tonne of vehicle weight.

General Motors of USA has suggested an emperical formula for it.

$$R_r = (0.0112 + 0.00006\ V)\ m.g. \quad ...(1)$$

where R_r = Rolling Resistance in Newton

V = Velocity of vehicle in kmph.

m = Mass of vehicle in kg.

g = Acceleration due to gravity

R_r depends on mass of vehicles, quality and material of road, material and inflation pressure of tyres and tyre tread pattern.

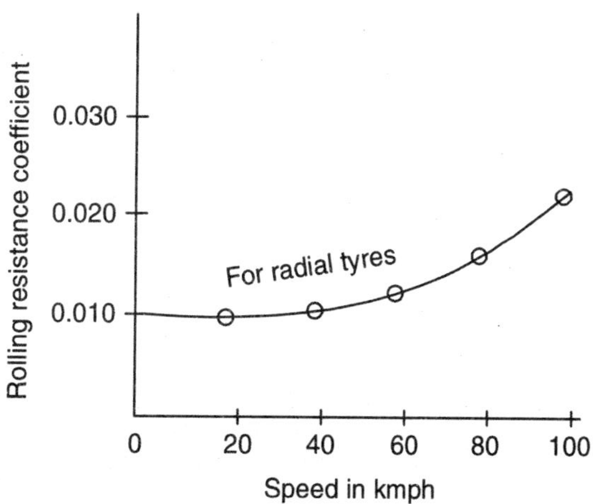

Fig. 7.9 Rolling resistance-speed curve

Sufficient amount of energy of the vehicle is wasted in deformation of road and tyres and damping of impacts of road irregularities. The rolling resistance, therefore, depends on velocity of vehicle, mass of vehicle, material of road surface, quality of road, material, tread and type of tyres and inflation pressure in the tyres.

Rolling resistance, for general purpose, may also calculated by another formula,

$$R_r = C_r \cdot mg \quad ...(2)$$

where C_r = Rolling resistance coefficient (C_r = 0.001 for well finished roads and 0.18 for ordinary roads)

m = mass of vehicle in kg

The maximum value of R_r will be μ.W.

where μ = Coefficient of friction between tyre and road

W = Weight of vehicle in kg

(b) ***Frictional Resistance or Transmission Friction Losses:*** Frictional resistance includes resistance due to transmission components (generally in Ist gear 25%, in second gear 20%, in third gear 15%, in top gear 10%) and resistance due to churning of oil in gear box and rear axle drive system or differential box and the resistance between road and tyres (first gear 15–17%, second gear 12–14%, third gear 9–10%, top gear 6–7%). The vehicle fitted with overdrive and/or with automatic gear box has lesser resistance. The resisting force, R_f is calculated by the following emperical formula,

$$R_f = 132.5 + 50.5\text{ m} \quad \text{Newtons.} \qquad \text{...(3)}$$

where m = mass of vehicle in kg.

If transmission efficiency (η_t) is given, then this resistance will not be separately calculated, as the η_t will be considered in determining the horsepower.

(c) ***Road Gradient Resistance:*** The forces due to road gradient G = tan θ depend on mass of vehicle (m) and the slope of the road (θ).

R_g = Resistance due to road gradient (Fig. 7.10)

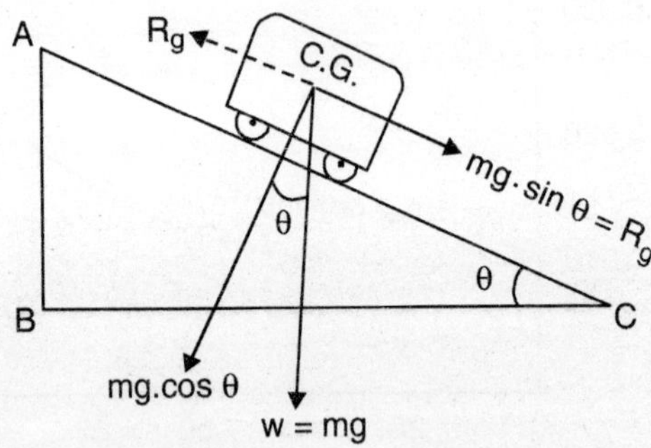

Fig. 7.10 Road gradient forces

$$R_g = \text{mg. sin } \theta \qquad \text{...(4)}$$

If θ = 0, *i.e.*, the road is perfectly levelled or horizontal,

then $$R_g = 0 \quad \text{as sin } 0^\circ = 0 \qquad \text{...(5)}$$

3. **Air Resistance:** It is the aerodynamic drag which is offered by wind passing over the vehicle. The Air resistance is quite high at high speeds. It is given by the formula,

$$R_a = \text{Air resistance} = \rho\,.C_a.A.V^2 \text{ Newtons} \qquad \text{...(6)}$$

where ρ = Density of air (about 1.3)

C_a = Coefficient of air resistance or aerodynamic coefficient

A = Projected frontal area of the vehicle in m^2 (Fig.7.12)

V = Velocity of the vehicle in kmph.

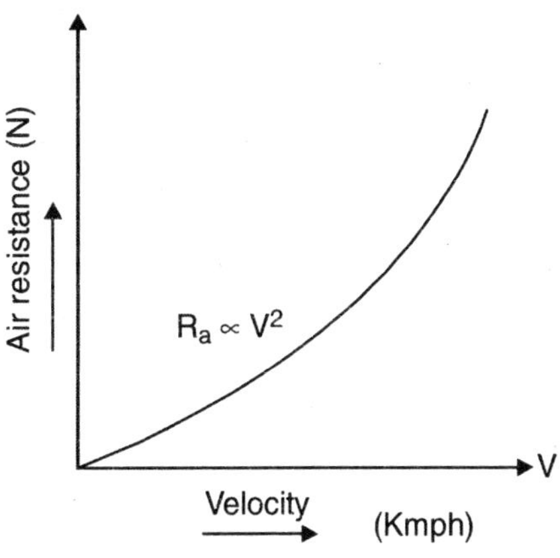

Fig. 7.11 Variation of air resistance with velocity

The value of C_a varies from 0.0230 to 0.045 N^2/m^2-h/km^2.

The air resistance increases as square of the velocity, the (refer to Fig. 7.11) values of ρ, C_a and 'A' may be taken as constant

$$R_a = \text{A constant} \times V^2$$

or

$$R_a \propto V^2.$$

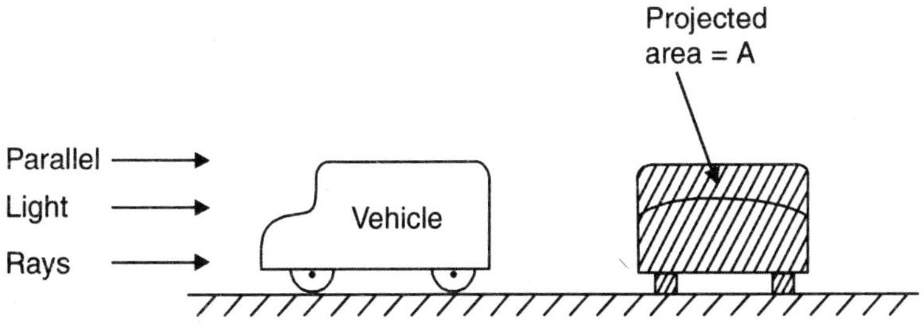

Fig. 7.12 Projected area

The value of coefficient of air resistance or aerodynamic drag for few vehicles is given below:

Ordinary vehicles ... 0.0314 (N^2/m^2-hr/km^2)

Luxury cars ... 0.0268

Trucks and buses ... 0.0450

4. Tractive resistance and propelling power: Tractive resistance, R_T, is the sum of all the resistances acting at the driving axle of the vehicle,

$$R_T = R_r + R_f + R_g + R_a + R_{acc}. \quad ...(7)$$

where R_r = Rolling resistance

R_f = Frictional (Transmission) resistance

R_g = Road gradient resistance

R_a = Air or Aerodynamic resistance or drag

R_{acc} = Acceleration resistance = m.f. It is experienced only when the vehicle is accelerated.

If W_{ax} = Work required at Axle to propel the vehicle

$$= R_T \times V \quad \text{N–km/hr}$$

$$= \frac{R_T \times V \times 1000}{60 \times 60} \text{ N-m/sec or watts}$$

$$= \frac{R_T \times V \times 1000}{3600 \times 1000} \text{ kW}$$

$$= \frac{R_T \times V}{3600} \text{ kW} \qquad \text{...(8)}$$

If η_t = transmission efficiency (about 85%)

Then actual horsepower required to propel the vehicle.

$$\text{Actual H.P. to propel the vehicle} = \frac{R_T \times V}{3600 \times \eta_t} \text{ kW.} \qquad \text{...(9)}$$

Numerical Problem 6. *A truck is weighing 30 kN and has its frontal projections area as 3.5 m^2 and running at its peak speed 80 kmph. Calculate the power required to propel it. Assume mechanical efficiency of the engine as 80% and transmissions efficiency 85%, C_r = 0.02 and C_a = 0.035 N^2/m^2-hr/km^2 and ρ = 1.*

Solution. At peak speed, $R_{acc} = 0$

At levelled road, $R_g = 0$

$$R_r = C_r.\ m = 0.02 \times (30 \times 1000) = 600 \text{ N}$$

$$R_a = \rho.\ C_a.\ AV^2$$

$$= 1 \times 0.035 \times 3.5 \times (80 \times 80)$$

$$= 728 \text{ N}$$

Since η_t is given, therefore, there is no need to calculate R_f. η_t will be taken into account in H.P.

$$\therefore \quad R_T = R_r + R_a + R_g + R_{acc} \text{ (}R_g \text{ and } R_{acc} \text{ being zero)}$$

$$= 600 + 728 + 0 + 0$$

$$= 1328 \text{ N}$$

Assuming the resistance to propulsion 10% more, *i.e.*, 100 + 10 = 110 per 100 value.

then propulsive $$R_T = \frac{110}{100} \times 1328 = 1460.8 \text{ N}$$

$$\text{H.P. to drive the vehicle} = \frac{R_T \times V}{3600 \times \eta_m \times \eta_t}$$

$$= \frac{1460.8 \times 80}{3600 \times 0.8 \times 0.85}$$

$$= 47.74 \text{ kW.}$$

Numerical Problem 7. *The gross vehicle weight (GVW) of a passenger car is 1340 kg including weights of passengers and their luggage. If its front area is 2.5 m*^{2} *and car run at a maximum speed of 100 km per hour on a levelled road, calculate the maximum inclination it can climb at 30 kmph in top gear. Take* $C_r = 0.03$, $C_a = 0.032$ *density of air,* $\rho = 1.3$.

Solution. $$\text{GVW} = 1340 \text{ kg} = \text{m.g. N}$$

At peak speed, $R_{acc} = 0$, and at levelled road $R_g = 0$

$$\therefore \quad R_T = C_r.\text{mg (max. value of } R_r) + R_a \text{ (Assuming } R_f = 0)$$

$$= 0.03 \times 1340 + 1.3 \times 0.032 \times 2.5 \times (100)^2$$

$$= 40.2 + 1040.0$$

$$= 1080.2 \text{ kg at 100 kmph} \qquad ...(i)$$

At a speed of 30 kmph, $$R_{acc} = 0$$

$$R_g = W \cdot \sin\theta \text{ .... If slope is } \theta.$$

$$= 1340 \times \sin\theta$$

$$\text{Maximum } R_r = C_r.\ W = 0.03 \times 1340$$

$$= 40.2 \text{ kg.}$$

$$R_a = \rho\, C_a.A.\ V^2$$

$$= 1.3 \times .032 \times 2.5 \times (30 \times 30) \text{ ..... as V = 30 kmph}$$

$$= 93.6 \text{ kg.}$$

$$\therefore \quad R_T = R_r + R_a + R_g + R_{acc} \qquad \text{(Assuming } R_f = 0)$$

$$= 40.2 + 93.6 + 1340 \sin\theta + 0$$

$$= (133.8 + 1340 \sin\theta) \text{ at 30 kmph} \qquad ...(ii)$$

Equating R_T at 100 kmph with R_T at 30 kmph—

$$1080.2 = 133.8 + 1340 \sin\theta$$

$$\sin\theta = \frac{1080.2 - 133.8}{1340} = 0.63$$

$$\theta = 39° \text{ approx.} \qquad \textbf{Ans.}$$

MORSE TEST FOR MULTI CYLINDER I.C. ENGINES

This test is done on multi-cylinder I.C. engines for determining indicated horsepower or indicated power developed by each cylinder. This test also helps in diagnosing the faulty cylinder of the engine, the cylinder producing less I.P. will be faulty. Frictional power can also be found out by substracting brake power from indicated power procedure. Under the load W, the engine is run at its rated speed at about 3000 r.p.m., and the torque is measured by Rope Brake Dynamometer (Fig. 7.13) by noting the values of W, S and N.

$$\text{The work done} = \text{Torque} \times \text{angle turned per revolution } (2\pi)$$

$$= (W - S)\,\frac{(D + d)}{2} \times 2\,\pi \qquad \text{...(1)}$$

where
W = Weight placed in the hanger of rope, N
S = Reading of spring balance, N
D = Diameter of brake drum, m
d = Diameter of rope, m
N = R.P.M. of the engine.
(D + d) = Effective diameter of drum, m

$$\text{Work done per second} = \frac{(W - S)\,\pi\,(D + d) \times N}{60} \text{ N-m/watts}$$

$$\therefore \quad \text{Brake power, B.P.} = \frac{(W - S)\,\pi\,(D + d) \times N}{60 \times 1000} \text{ kW}$$

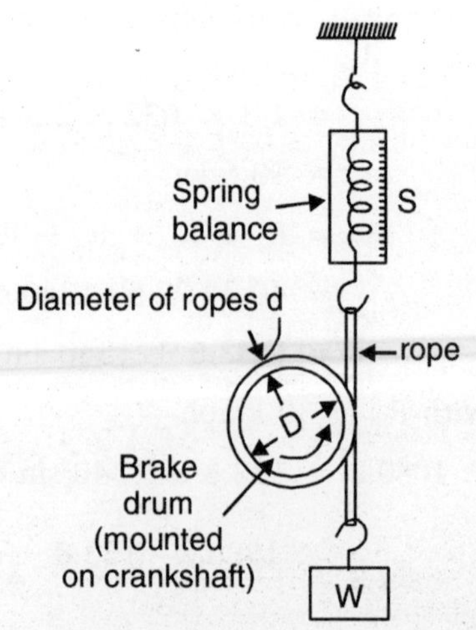

Fig. 7.13 Rope brake dynamometer

or $$\text{B.P.} = \frac{2\pi NT}{60 \times 1000} \text{ kW} \quad \text{...(2)}$$

where T = Torque, N-m = $\{(W - S) \times 2\pi R\}$ where $R = (D + d)/2$

After measuring N. T and R on 4 cylinders running, first cylinder is cut out by disconnecting the injector of first cylinder in case of diesel engine or spark plug in case of S.I. engine. The speed of engine will fall, restore it to the rated one by reducing the load. Now again note the readings of spring balance and the load in hanger and calculate the B.P. by the above formula. Now connect the cylinder number 1 again to the injector or spark plug connection and then repeat the procedure by disconnecting cylinder number 2, and then 3rd and lastly 4th cylinder. Only one cylinder will be cut at a time.

Suppose the engine has 4 cylinders and

B_1 = B.P. when 1st cylinder is cut.

B_2 = B.P. when 2nd cylinder is cut.

B_3 = B.P. when 3rd cylinder is cut.

B_4 = B.P. when 4th cylinder is cut.

If B = B.P. of the engine with all 4 cylinders in action

= (I – F)

where I = I.P. (Indicated Power of the engine with all 4 cylinders)

and F = Total Frictional power lost in overcoming the friction offered by all the moving parts of the engine. It will be same with any number of cylinders in running.

If I_1 = I.P. of the 1st cylinder and F_1 is its power lost in friction.

I_2 = I.P. of the 2nd cylinder and F_2 is its power lost in friction.

I_3 = I.P. of the 3rd cylinder and F_3 is its power lost in friction.

I_4 = I.P. of the 4th cylinder and F_4 is its power lost in friction.

Then
$$B = (I_1 - F_1) + (I_2 - F_2) + (I_3 - F_3) + (I_4 - F_4)$$
$$= (I_1 + I_2 + I_3 + I_4) - (F_1 + F_2 + F_3 + F_4) \quad \text{...(3)}$$
$$= (I - F)$$

When Ist cylinder is cut then I_1 is lost and actual brake power:

$$\text{B.P.} = B_1 = (0 - F_1) + (I_2 - F_2) + (I_3 - F_3) + (I_4 - F_4)$$
$$= (I_2 + I_3 + I_4) - (F_1 + F_2 + F_3 + F_4) \quad \text{...(4)}$$

Though, the cylinder, which is cut, does not develop any power but its parts remain moving and offer friction. Substracting equation (4) from equation (3), we get

$$B - B_1, = I_1,$$

Similarly, we shall get the values for other cylinders:

$$B - B_2 = I_2$$

$$B - B_3 = I_3$$

$$B - B_4 = I_4$$

Then total I.P. of the engine, $I = I_1 + I_2 + I_3 + I_4$ ...(5)

Numerical Problem 8. A morse test conducted on a 4-stroke, 4-cylinder petrol engine and following data were noted:

B.P. with all 4-Cylinders = 14.5 kW

B.P. with 1st Cylinder cut-off = 10.0 kW

B.P. with second cylinder cut-off = 10.2 kW

B.P. with third cylinder cut-off = 10.1 kW

B.P. With 4th cylinder cut-off = 10.3 kW

Fuel consumption per minute = 0.09 kg

Calculate (*i*) Mechanical efficiency (*ii*) Brake thermal efficiency (*iii*) Relative efficiency.

Specifications of the engine are as under:

Bore = 80 mm.

Stroke length = 120 mm.

Clearance volume = 8×10^4 mm^3

Calorific value of fuel = 44000 kJ/kg

Solution.

$$\begin{aligned} I &= I_1 + I_2 + I_3 + I_4 \\ &= (B - B_1) + (B - B_2) + (B - B_3) + (B - B_4) \\ &= 4\,B - (B_1 + B_2 + B_3 + B_4) \\ &= 4 \times 14.5 - (10.0 + 10.2 + 10.1 + 10.3) \\ &= 58.0 - 40.6 \\ &= 17.4 \text{ kW} \end{aligned}$$

Energy obtained by fuel combustion per second

$$= \frac{0.09 \times 44000}{60} = 66 \text{ kW}$$

(*i*) Mechanical Efficiency $= \dfrac{\text{B.P.}}{\text{I.P.}} = \dfrac{14.5}{17.4} \times 100 = 83.33\%$ ...(*i*)

(*ii*) Indicated thermal efficiency $= \dfrac{\text{I.P.}}{\text{Fuel energy/sec.}} = \dfrac{17.4}{66} \times 100 = 26.36\%$ ...(*ii*) (*iii*) Brake thermal efficiency $= \dfrac{\text{B.P.}}{\text{Fuel energy/sec}} = \dfrac{14.5}{66} \times 100 = 21.97\%$ ...(*iii*)

(*iv*) Air standard efficiency $= 1 - \dfrac{1}{r_c^{\gamma-1}}$

Where r_c = Compression Ratio

$$\gamma = \frac{C_p}{C_v} = 1.40 \text{ for air}$$

$$r_c = \frac{V_c + V_s}{V_c}$$

where, V_c = Clearance volume

V_s = Stroke/swept volume

$$= \frac{\pi}{4} D^2 \times L$$

$$= \frac{8 \times 10^4 + \frac{\pi}{4} \times (80)^2 \times 120}{8 \times 10^4}$$

$$= 8.54$$

$\therefore$ Air standard efficiency $= 1 - \dfrac{1}{(8.54)^{(1.40-1)}}$

$$= 1 - \frac{1}{(8.54)^{0.40}}$$

$$= 0.5760$$

$$= 57.60\,\%$$ ...(*iv*)

(*v*) Relative Efficiency $= \dfrac{\text{Indicated thermal efficiency}}{\text{Air standard efficiency}}$

$$= \frac{0.2636}{0.5760}$$

$$= 0.4576$$

$$= 45.76\%$$

Heat Balance Sheet

If a heat balance sheet is to be drawn for the engine, the following measurements are also required to be observed:

1. Speed of the engine. It is measured either by electrical counter or by mechanical techometer.
2. Fuel consumption during the trial period is measured by passing the fuel through a graduated glass tube or by a fuel flow meter.
3. Mass of cooling water flown during the trial is measured by water flow meter.
4. Temperatures of inlet air, exhaust gases, cooling water etc. are measured by digital thermometers using sensors at various points.
5. The air consumption is determined by a air flow meter or by air-box method.

The Air Box Method

It is a simple device for measuring mass of air consumed during the trial period. It is a square shape steel sheet box having its capacity about 600 times the swept volume of the engine cylinders. It is used for obtaining a steady flow of air. The box contains a small orifice of known diameter and coefficient of discharge (C_d). At its opposite side, outlet pipe is provided which leads the air to engine through an air-cleaner. On the same side, a water manometer is provided to measure the pressure difference causing the air to flow through the orifice as shown in Fig. 7.14.

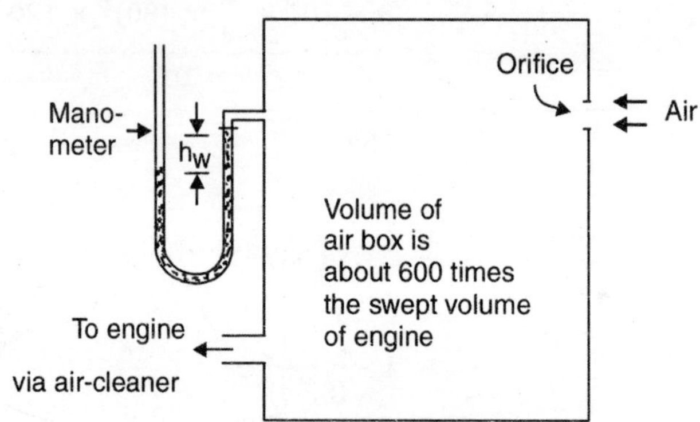

Fig. 7.14 Air-box

If d = Diameter of orifice in cms.

$$A = \text{Area of orifice} = \frac{\pi d^2}{4(100)^2} \text{ m}^2$$

C_d = Coefficient of discharge for orifice

h_w = Head of water in manometer causing the flow, in cms.

ρ_w = Density of water, in kg/m^3 = 1000 kg/m^3

ρ_a = Density of air at atmospheric pressure in kg/m^3

If $\quad H$ = Head of air in m

$$= H.\rho_a = \frac{h_w}{100} \times \rho_w$$

$$\therefore \quad H = \frac{h_w \times \rho_w}{100\,.\,\rho_a}$$

$$= \frac{h_w \times 1000}{100\,.\,\rho_a}$$

$$= \frac{10\,h_w}{\rho_a} \text{ m of air}$$

If $\quad V_a$ = Velocity of air passing through orifice in m/s

$$= \sqrt{2\,gH} = \sqrt{2g.\frac{10\,h_w}{\rho_a}} \text{ m/s} \qquad ...(6)$$

and $\quad v_a$ = Volume of air passing through orifice in m^3/sec.

$$= A.V_a.C_d$$

$$= C_d.\ A\sqrt{2g.\frac{10\,h_w}{\rho_a}} = C_d.\ A\sqrt{2 \times 9.81 \times 10.\frac{h_w}{\rho_a}}$$

$$= 14.07\ C_d.A\sqrt{\frac{h_w}{\rho_a}}\ m^3/\text{sec.}$$

$$= 844.2\ C_d.A\ \sqrt{h_w/\rho_a}\ m^3/\text{min.}$$

M_{air} = Mass of air passing through the orifice, kg per minute

$$= V_a.\ \rho_a = 844.2 \times C_d \times A\ \sqrt{h_w/\rho_a} \times \rho_a$$

$$= 844.2 \times C_d \times \frac{\pi\,d^2}{4\,(100)^2}\sqrt{h_w.\rho_a}$$

or

$$M_{air} = 0.0663\ C_d.\ d^2\ \sqrt{h_w.\rho_a}\ \text{kg/min.} \qquad ...(7)$$

where $\quad d$ is in cms

h_w is in cms

ρ_a is in kg/m^3 $= \dfrac{P}{RT}$

where $\quad P$ = Air pressure in bar = 1.013 bar = 1.013×10^5 N/m^2

T = Temperature (Absolute) in kelvin.

R = 287 J/kg.K for air.

Heat Balance Sheet Data

(*i*) Heat supplied by fuel = $m_f \times C$ kJ/min ...(8)

where m_f = Mass of fuel consumed in kg/min.

C = Lower calorific value of fuel in kJ/kg.

(*ii*) Heat absorbed in I.P. = $\dfrac{n.100\, p_{m}.\ \text{L.A.N.k}}{60}$ kW (1kW = 1kJ/sec.)

Heat absorbed in I.P./min = $n.100\ p_m$ L.A.N.k. kJ/min. ...(9)

where n = no. of cylinders

k = $\frac{1}{2}$ for 4-stroke cycle, power stroke/revo.

= 1 for 2-stroke cycle.

N = r.p.m.

p_m = m.e.p. in bars

L = Length of stroke, m

A = Area of piston/cylinder, m^2

(*iii*) Heat carried away by cooling water

$$= m_w.\ C_w.\ (t_1 - t_2) \text{ kJ/min.}$$

where m_w = Mass of water supplied for cooling in kg/min.

C_w = Specific heat of water = 4.2 kJ/kg.K.

t_1 = Inlet temperature of water, °C (into water jacket)

t_2 = Outlet temperature of water, °C (from water jacket)

(*iv*) Heat carried away by exhaust gases

If m_f = Mass of fuel supplied/min.

and M_{air} = Mass of air supplied/min = $0.0663\ C_d.d^2\ \sqrt{h_w \cdot \rho_a}$ in kg/min.

M_{aex} = Mass of exhaust gases, kg/min = $m_f + M_{air}$

If t_e = Temperature of exhaust gas °C.

t_a = Atmospheric air temperature, °C

and C_g = Specific heat of exhaust gases; 1kJ/kg.K

∴ Heat carried away by exhaust gases = $M_{ex}.\ C_g\ (t_e - t_a)$ kJ/min

(*v*) Heat unaccounted—It is the heat lost due to radiation and leakage etc. into the atmosphere and the heat lost in raising the temperature of engine parts. It is obtained by substracting total of (*i*) heat absorbed *w* I.P., (*ii*) Heat carried away by cooling water (*iii*) heat carried away by exhaust gases from the heat produced by burning of fuel.

i.e., heat unaccounted = $m_f.\ C - n.100\ p_m$ L.A.N.k. + $m_w.\ C_w\ (t_1 - t_2) + M_{ex}.\ C_g\ (t_e - t_a)$ kJ/min.

Heat Balance Sheet

S.No.	Particulars	kJ	%
1.	Total heat produced by fuel	H	100
2.	Heat absorbed in I.P. (or B.P.)	—	—
3.	Heat carried away by cooling water	—	—
4.	Heat carried away by exhaust gases	—	—
5.	Heat unaccounted for (By difference) —	—	
	Total	H	100

Numerical Problems on heat balance sheet are the part of the syllabus of Applied thermodynamics, which B.Tech (Mech.) students have already learnt in their previous semesters.

QUESTIONNAIRE

1. Describe the various types of retarding forces encountered by a vehicle when in motion.
2. What is drawbar horsepower? How it differs from frictional horsepower?
3. What do you understand by Air Resistance? How it varies with velocity?
4. What is gradeability of a vehicle? What factors affect it?
5. What is propelling power? What is its relation with tractive resistance?
6. What are design criteria for the following engine parts:
   (*a*) Engine block
   (*b*) Connecting rod
   (*c*) Piston.
7. What are timing gears? Explain their need and working.
8. What do you understand by torque developed by an engine? How its varies with speed?
9. What is tractive effort?

   A car engine is developing 15 kg-m torque at 3000 r.p.m. of the gear box ratio is 3.14 and final drive is 4.5 : 1. Find the value of tractive effort at driving wheels. The dia of driving wheels may be taken as 60 cms.
10. How compression ratio affects vehicle performance? How output of engine may be increased?
11. What is weight/power ratio? Describe the advantages of a greater W/P ratio.
12. Explain the following:
    (*i*) Pick-up (*ii*) Transmission resistance
    (*iii*) Rolling resistance (*iv*) Air-resistance.
13. Describe the procedure followed in Morse Test for determining I.P. of each cylinder of the engine.

14. A Morse test was performed on a 4-stroke, 4-cylinder petrol engine and following data were observed:

B.P. with 4-cylinder working = 21.7 kW

B.P. with first cylinder cut-off = 15.5 kW

B.P. with second cylinder cut-off = 15.6 kW

B.P. with third cylinder cut-off = 15.7 kW

B.P. with fourth cylinder cut-off = 15.5 kW

Fuel consumption = 0.07 kg per minute

calorific value of fuel = 40, 900 kJ/kg. Calculate the mechanical efficiency and indicated thermal efficiency. [**Ans.** 88.6%, 51.3%]

CHAPTER 8

Body and Chassis

A self-propelled vehicle is termed as an automobile vehicle. The wide application of automobile is seen in Land transport (roadways, passenger cars and trucks) and in Water transport (marines and ships) and in Air transport (aeroplanes, helicopters etc.).

CLASSIFICATION OF VEHICLES

I. On the basis of number of wheels

(*i*) Two-wheelers—Mopeds, scooters, motorcycles

(*ii*) Three-wheelers—Tempos, road rollers, transport vans

(*iii*) Four-wheelers—Cars, buses, trucks, tractors etc.

(*iv*) Six-wheelers—Truckes and buses

(*v*) Eights or more wheels vehicles—Special transport vehicles for heavy machinery and equipments.

II. On the basis of fuel used

(*i*) Petrol cars

(*ii*) Diesel vehicles

(*iii*) Gas (LPG) driven vehicles.

(*iv*) Hybrid or multi-fuel vehicles.

III. On the basis of engine/drive

(*i*) I.C. engine vehicles (petrol, diesel, gas)

(*ii*) Wankel engine (rotary piston engine)

(*iii*) Electric power driven vehicles (railway engines)

(*iv*) Battery operated vehicles (scooters and mopeds and same (*iii*))

(*v*) Solar energy operated vehicles.

IV. On the basis of weight of the vehicles

The Unladen weight or the weight of chassis without any load on it, is called KERB weight. The Gross Vehicle Weight (GVW) is the maximum load capacity of a vehicle. The difference between GVW and Kerb weight is called Payload.

PAYLOAD = GVW – KERB WEIGHT.

On the basis of Gross Vehicle Weight (maximum laden weight) the auto vehicles are classified as under:

(*i*) Light duty vehicles — GVW up to 1 tonne

(*ii*) Medium duty vehicles — GVW: 1 to 3.5 tonne

(*iii*) Heavy duty vehicles — GVW: 3.5 to 7.5 tonne

(*iv*) Extra heavy duty vehicles — GVW more than 7.5 tonne

GVW for family small cars varies from 1200 kg to 1350 kg

Kerb weight family small cars varies from 800 kg to 965 kg

GVW for trucks varies from 2600 kg to 3000 kg

whereas kerb weight fluctuates around 1800 kg for trucks

V. On the basis of type of wheel drive

(*i*) Front wheel drive — (2 wheel or 4-wheel driven by transfer case)

(*ii*) Rear wheel drive

(*iii*) Single wheel drive (2 wheelers)

(*iv*) All wheel drive [4 wheel drive or (4 × 2) or 6 wheel drive etc.].

VI. On the basis of steering wheel position

(*i*) Right hand drive

(*ii*) Left hand drive.

VII. On the basis of some utility

(*i*) Luxury cars (superior cars than ordinary ones), passenger cars and buses etc.

(*ii*) Sports utility vehicles (SUV)

(*iii*) Racing cars

(*iv*) Military vehicles, weapon carriers etc.

(*v*) Construction service vehicles Road rollers, Mixers, Cranes etc.

(*vi*) Goods transport vehicles (Trucks)

VIII. On the basis of the displacement volume of the engine:

A. Indian Vehicles

(*i*) Micro cars (500–700 c.c.)-Nano car, mini car, Stella etc.

(*ii*) Small Cars, Displacement up to 1000 c.c. Matiz, Spark, Maruti 800, Zen, Santro, Fiat Uno, Tata Indica etc.

(*iii*) Medium type cars (1000 to 2000 c.c.)

Contessa classic 2.0 DLX, Mitsubishi Lancer, Fiat UNO, Cielo, Tata safari, Sierra, Hondacity, Opel Astra, Ford Escort diesel, Fida etc.

(*iv*) Large cars (More than 2000 c.c.)

Mercedes Benz E 220 and E 250 diesel.

(*v*) Utility Vehicles or Diesel Jeeps (1500 to 3000 c.c.)

Mahindra MM 540 DP, Commander (NC 640 DP, Maruti Gypsy, Tata Sumo, C J 4-A Jeeps.

(*vi*) Transportation vehicles: (3000 Cc. onwards)

Ashok Leyland Comet and Panther, Allwyn Nissian, Eicher, Mitsubishi canter, Swaraj Mazda, Bajaj Tempo Excell-4, Tata LPT 1210 D Truck, Tata magic.

B. Foreign Cars

(*i*) Small cars (up to 1000 c.c): Renault type RZ 106.

(*ii*) Medium cars (1000–2000 c.c.): Volks wagon Golf GTI 16V, Mercedes Benz 200 and 200D, Spark 'USA 83'. Skoda Octavia.

(*iii*) Large cars (above 2000 c.c.): Rolls Royce Camargue, Audi 100 SE, Volvo 740, Hyundai X6 Saloon, Daimler Double-six Porsche 911 etc.

SPECIFICATIONS OF AN AUTOMOBILE

Whenever somebody intends to purchase a new vehicle or a second hand vehicle, he must see few specifications of that vehicle.

1. *Type:* It may be a luxury car or a small car or a mini car. If somebody is purchasing commercial vehicle such as a Bus, Truck, Mini bus, Mini/half truck, he should know about all types of vehicles of the required category.
2. *Make:* The reputation of a manufacturer also counts a lot. In most of the cases, name of maker and capacity of engine are indicated together such as Maruti 800, Contessa classic 1.8GL. The 1.8 indicates capacity in litres or it may be written as 1800 c.c. GL stands for Gasoline (petrol) engine.
3. *Seating capacity:* In case of cars, seating capacity (number of passengers that can sit in the car) is also indicated as 2 seater, 4 seater, 6 seater etc. For buses seating capacity-may be 30, 45, or 60, if doubles in case of double decker buses. In case of Trucks, loading capacity in seen *i.e.*, 3 ton, 5 ton, 10 tone etc.
4. *Steering drive*: In India, all the vehicles are fitted with right hand drive or steering wheel on right hand side. Many countries adopt left hand drive in which steering wheel is fitted on left side.
5. *Drive:* Most of the vehicles are either 2-wheel drive or 4-wheel drive (4×4 is the indication for 4-wheel drive). The number of wheels, which get power from engine

directly, show the number, *i.e.*, 2-wheel drive or 4 × 4 (4-wheel drive) or 4 WD. (4 × 2) shows that vehicles is 2-wheel driven but may be driven by 4-wheels at the time of need.

6. *Model:* It has importance when you are going to purchase a second hand vehicle. Manufactures also allot a code number such as Tata LPT 2213-6 × 2 Truck, Mahindra Jeep MM-540 DP, Tata Sumo etc., year of manufacture is also plays a keyrole in selection of used vehicles.
7. *Power steering*: Power braking, automatic door locking. Air-conditioning, Turbo charging, Music system etc. are also have their importance in the selection of a car.
8. *Fuel economy:* In this age of high prices of petrol and diesel oil, the mileage given by the vehicle also plays an important role in the selection of the vehicle. Every customer likes a fuel efficient vehicle with least pollution producing exhaust.

DIESEL CARS

Technological development and the increasing refinement of diesel engines have increased the demand for diesel cars, forcing manufacturers to provide a diesel variant of their models in order to increase their sales, such is Ford Fusion, Tata Indica, Ambassador, Zen and many others. The diesel cars suit to those who run longer distances frequently.

Requirements of a Good Car

The large wheels and good grand clearance make an ideal choice on city roads, especially during the rainy season when they are full of pot-holes. The tall design of a car gives a better road presence and also puts the driver at an advantage while driving in congested traffic where auto and taxi drivers try to 'bully' small cars.

The height from inside of a car gives more convenience to occupants. The driver's seat should also be well-positioned to slide easily. The seat should be quite comfortable with adequate support to the thighs. Once inside on the seats, the driver should have a commanding view on all sides with a minimum of blind spots. The steering wheel should also be adjustable according to the heights. The dash board should be well laid out and there should be enough pockets to take care of various titbits. The entry to the car seats of the rear side should also be very easy and comfortable and there must be quite sufficient leg room and shoulder room and head room to make a long journey more pleasurable.

The gear box should be short and well placed and the driver must feel very comfortable while using it for changing the gears to adjust speed requirements. The maintenance requirements should be minimum.

Brakes of a car must be very effective and should require least effort to operate. A good car also has generous luggage space. The instrument panel should have a digital clock and indicators to show distance-to-empty information that lets know the driver how far he can go on the fuel he has. There should be a powerful air-conditioning and heating system specifically designed for our country's hot and humid and cold climates.

There are some other additional features which make a car more friendly such as (*i*) the lights that come on, while remote unlocking, battery saver switches to switch them off, in case the driver forget to do so, (*ii*) a good music system and/or DVD player facility and USB port and (*iii*) mobile phone charging socket etc. Finally the cost of the car should manageable by the majority of customers and the car must be eco-friendly with attractive body outlook.

Specifications of Tata's 'NANO' Car

It is latest light car by Tata. See Fig. 8.1

1. Dealer price: Rs. 1 lac
2. Cost on road: Rs. 1.2250 lac
3. Top speed: 90 km per hour
4. Mileage/Fuel consumption: 23 km in a litre
5. Engine:
   (*i*) Fitted on rear side.
   (*ii*) Made of aluminium, twin cylinders
   (*iii*) Capacity 624 c.c.
   (*iv*) Max. power - 33 BHP
   (*v*) Multi-Point Fuel Injection (MPFI)
   (*vi*) Tubeless tyres
   (*vii*) 4-speed manual gears
   (*viii*) Seating for 4 persons
   (*ix*) Emission norms: As per Bharat-3 and Euro-4
   (*x*) Capacity of petrol tank-30 litre
   (*xi*) Models (*i*) Basic (*ii*) Deluxe (models).

Fig. 8.1 Tata's nano

MAIN COMPONENTS OF AN AUTO-VEHICLE

The auto vehicle may be divided in two parts:

1. **Basic Structure or Chassis:** It includes frame over which all parts are mounted such as suspension system, axles, wheels and tyres. The chassis may also be termed as the vehicle without body and fenders. The chassis consists of the following parts:

   (*i*) Engine

   (*ii*) Frame

   (*iii*) Transmission system: The system which helps in transferring power of engine to the driving wheels.

   (*iv*) Steering system

   (*v*) Braking system

   (*vi*) Suspension system

   (*vii*) Wheels and tyers axles

   (*viii*) Fuel Tank and supply system

   (*ix*) Electrical system, Air-conditioning system etc.

2. **Super Structure or Body:** It carries some parts of electrical and cooling systems but mainly contains seating arrangements. It gives an attractive look to the car, comfortable seats, safety belts, lock fittings, space for luggage, doors, windows, windshields etc. are integral parts of the body. A good design of body plays a key role in increasing the sales of the vehicle. Most of the cars are 5 seaters. Medium size cars and luxury cars are also provided with space for luggage on the rear side (called Dicky). Modern car bodies are generally part of the chassis at the floor in the frameless constructions.

Bodies of automobiles should have the following properties:

1. The body should provide comfortable sitting space to passengers with easy movements for entering and stepping out. The sufficient space for luggage is also a must.
2. The body has to bear so many loads and resistances. It should be strong enough to with stand all such forces and also the impact loads.
3. The car frame along with body, particularly in frameless construction, may be supposed like a simply supported beam at front and rear wheels. The body must have sufficient stiffness to counter sagging between the wheels or wheel base.
4. The body is subjected to twisting on rough and pitted roads. It should have adequate torsional stiffness to resist the twisting, vibrations and stresses such caused.
5. The design of the body should be aerodynamic so that the value of Air drag or air resistance should be minimum possible.
6. The body should be able to provide protection and safety to the passengers and luggage from bad weather/Sunshine/dirt and dust.

7. The material used in making body should be resistant to rust and corrosion and less dentable.
8. The body should be made strong enough to protects passengers to the maximum possible extent in cases of collision or serious accident.

Materials Used in Body Making

Zinc coated steel, aluminium alloys, plastics, thermoplastics for boot covers and grills, toughened glass for window panes and windscreens, laminated glass—it does not shatter into pieces an impacts. Paints of different colours to make the body attractive.

Boot-space: The space for keeping space wheel and some luggage in the rear portion of vehicle is known as boot-space.

TYPES OF VEHICLE BODIES

1. *Sedan:* Closed type body with 2 or 4 doors. Most of the cars are sedan type.
2. *Convertible body:* The top of the car or jeep is made of fabric sheets which could be folded down. Some cars have sliding hard tops.
3. *Limousine:* This is just like sedan but with a separate compartment for driver.
4. *Coupe:* It has one or two seats for driver and his companion. The rear body is used as luggage space with separate door. It is a closed body vehicle.
5. *Station wagon:* It may have folding seats to make space for luggage. Station wagon are built with 2 or 4 doors and are also provided with a large tail gate.
6. *Ambulance:* Body with separate driver cabin, a gate at rear side and side benches in the main compartment used for carrying patients.
7. *Van type:* Body with 2 or 4 doors, more seating capacity up to 8 passengers but without luggage space or it may have seating at front only, and luggage at rear side.
8. *Passenger bus body:* Enclosed body with 2 doors and multiple windows and a large seating capacity upto 60. It may also be double-decker type.
9. *Truck body:* Body with separate driver cabin, rest of the body for cargo. It may be full body or half body with provisions for supporting cover sheets to protect luggage from rain, fog, sunshine etc.
10. *Crew cab:* It is partly a saloon and partly a goods carrier. Front portion is a saloon with all luxuries and the rear portion is meant for carrying goods.

Automobile's Chassis

If we remove body or carriage portion from the vehicle, the remaining part is called CHASSIS. The basic or the main structure over which all the other parts are fitted or suspended is called FRAME.

Frame

Frame serves as the foundation of the chassis to which all the main parts are attached. The chassis frame must be stiff and strong enough to withstand shocks, twists, vibrations, transmission thrust, torque stresses and many other strains to which it is subjected on road while the vehicle is in movement.

Loads on Frame/Chassis

1. Self load or dead weight of chassis parts such as engine, transmission system, suspension system, steering system, fuel tank, body plus passengers and their luggage etc.
2. Torque produced by engine and drive axles.
3. Resistances due to Air, tyre rolling, friction gradient of road etc.
4. Impacts and vibrations when the vehicle runs over a broken path and rough roads.
5. Loads while the vehicle is negotiating a curve, and applying brakes, and sometimes sudden brakes.
6. Vibrations of engine and the transmission system.
7. Inertia loads due to brake application.
8. Overloading of the vehicle.
9. Cornering forces, effects of side wind and road camber at turns.
10. Torsional load when vehicle passes over a road bump. The twisting causes a shear stress in the frame.
11. Impact loads due to road obstacles. These result in distortion of the frame and disturbs the wheel alignment.
12. Collision impact when the vehicle collides with some other vehicle or object.

Frame Construction: The frames are usually made in channel section or box section or I-section of special pressed alloy steels containing 2 to 3% Nickel. Pressed steel sheets and sheets of Al-alloys are also used, as in case of integral frame body of several types of cars. Most of the modern cars use the body itself as a frame, that is, no separate frame is provided for mounting body. Such constructoin is called frameless construction or integral frame. Use of channel or box section frames is more popular.

Terms Related with Frame/Chassis

Refer to Fig. 8.2.

1. *Wheel Base:* It is the distance between centre lines of front wheels and rear wheels as seen in side view.
2. *Wheel Track:* It is the centre to centre distance between wheels seen from front side or rear side.
3. *Front Overhang:* The part of the chassis/frame projecting out of the axis of front axle is called front overhang.
4. *Rear Overhang*: It is the part of the chassis/frame projecting beyond the axis of the rear axle.

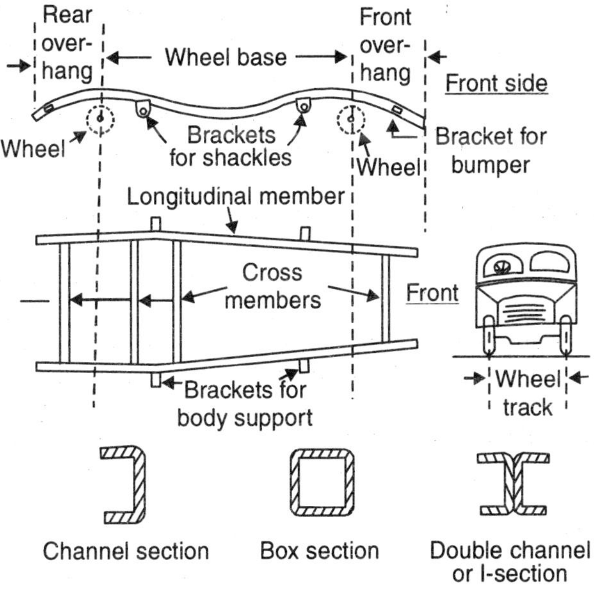

Fig. 8.2 Conventional frame and sections used in frames

TYPES OF CHASSIS FRAMES

There are three types of frames in use:

(*i*) Conventional Frame

(*ii*) Semi-Integral Frame

(*iii*) Integral or Unitised Frame.

(*i*) *Conventional frame:* This frame is simple type. It carries engine, transmission system and the body. It is supported on wheels and axles by means of suspension springs. A conventional frame is shown in Fig. 8.2.

It consists of two longitudinal members and few cross-members to provide strength and rigidity and base for mounting parts. The frame is bent upward at rear and front to accommodate suspension and movement of axles. The frame is narrowed at front side to have better steering lock, which gives a smaller turning radius. Brackets are provided for mounting of springs shackle bearings and bumpers. This type of frames are used an trucks and buses only.

The channel section or square box section steel members are used for making frame members. Double channel or I-section members are also used for greater strength and stiffness against torsion and bending. The frame members are made from mild steel sheet Nickel (3%) alloy steel and carbon (0.3%) sheet steel. In most of the designs of Truck chassis, the longitudinal members are kept straight and not bent on front and rear sides of the body on this type of frame is made of flexible material such as wood or some stiff material and may be isolated from the chassis with the help of rubber mountings.

(ii) *Semi-integral or half frame chassis:* These types of frame are made half of the full size of vehicle. The front half-frame supports the power unit, gear box and steering system. The remaining rear half is combined frame and body. The floor of the body acts as half frame partly mounted on the front half frame with stiff mounts such that a part of the chassis frame load is transferred to body structure also. Old imported Fiat cars were fitted with such structure. These days, this types of frames are not popular due to heavy weight and difficulties in fitting the two halves. After severe accident it becomes a very difficult task to correct the body portion for wheel alignment. Therefore such frame is not used in present day cars. A semi-integral frame is shown in Fig. 8.3.

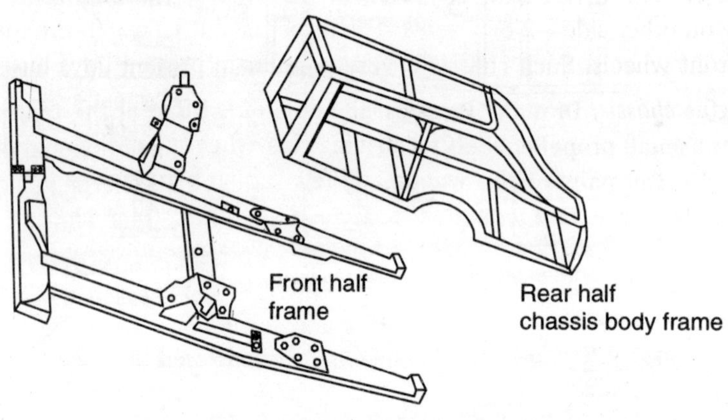

Fig. 8.3 Semi-integral frame and body

(iii) *Integral or Unitised frame or Frameless constructions*: In this type, there is no separate frame and all the parts are attached to the unitised body. The body and frame are a single unit firmly welded together. The body itself takes all the loads. The load is diffused through the whole structure and the body becomes lighter but stronger. This structure is widely being used on modern passenger cars.

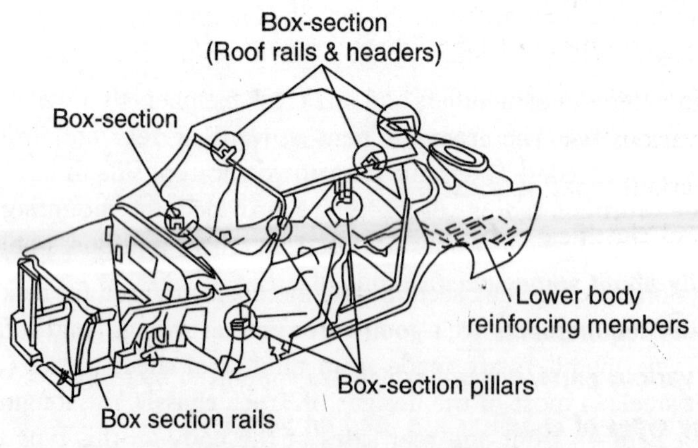

Fig. 8.4 Integral frame and body

Auto chassis may further be classified according to the mounting position of the engine at the frame.

(*i*) *Full forward chassis:* The engine is placed outside the driver seat as in case of cars and old trucks and buses. The driver is unable to see the road near the front wheels.

(*ii*) *Semi-forward chassis.* The driver seat is placed just above the engine. Bedford pickups and few Tata trucks are the examples. Motor cycles also have such chassis. It increases seating capacity without increasing space. The driver may see the road from very near.

(*iii*) *Bus chassis:* The driver seat is placed at the side of the engine. It increases seating capacity on other side—2 or 3 seats. The driver is able to watch the road from very near to the front wheels. Such chassis is very popular in present days buses and trucks.

(*iv*) *Rear engine chassis:* In many vehicles, the engine is fitted at the rear part of the frame. It utilises a small propeller shaft for driving rear wheels, gearbox and differential are put in one units. Germany's Volks wagon and few Indian Vikram type vehicles are examples. Scooters also have rear engine fitted chassis frames. Gear shifting and accelerating mechanism do not work smoothly in four wheelers otherwise rear engine chassis works quite satisfactorily.

QUESTIONNAIRE

1. What are various loads which a vehicle frame has to withstand?
2. Describe briefly the functions of a frame.
3. Why is the frame narrowed at front?
4. What is a body of an automobile? What purposes does it serve?
5. What are the components which are mounted on the frame?
6. At how many places the engine can be mounted on the frame?
7. What are the materials used in making frame and body?
8. Describe various types of frames.
9. Why unitised frame is so common in the present days Cars?
10. What are various requirements of a good body?
11. Describe various types of bodies.
12. Give a broad classification of auto vehicles.
13. Write briefly about some specifications of a car.
14. What are the requirements of a good car?
15. Name the various parts/systems which are supported by frame.
16. How many types of chassises are used on vehicles?

Chapter 9

Suspension Springs and Shock Absorbers

The chassis is mounted on wheel axles through suspension springs to isolate the body cabin from experiencing road shocks and route irregularities. The springs oscillate up and down due to road shocks and these oscillations are restricted within tolerable limits by a device called spring damper, the most commonly used damper is a shock absorber. When a vehicle moves on rough roads, side tilting and shocks are experienced. The purpose of springs is to absorb these shocks.

Springs may be either tension type or compression type. A tension spring is stretched under tensile load and returns back to its original position when the load is removed. A compression springs gets compressed under a compressive load and returns back to its normal position after the load is removed.

TYPES OF SUSPENSION SPRINGS

1. Leaf springs or Laminated springs: These springs are widely used in auto vehicles for suspension purposes. Four to eight flat alloy steel strips of varying lengths are placed on one another as shown in Fig. 9.1. They are then clamped with axle at centre by U-Bolt and Nuts. The strips are also called blades. These blades are bound together at 2 to 4 places by steel straps as shown in the figure. The upper longest leaf is called **Master-leaf** and has eyes (round bent ends) at both ends. The leaf spring is basically in a semi-elliptic form, but under load it is flattened and when it gets flattened under load, the leaves slide over each other. The sliding effect helps in bending and springing. Graphite is applied between blades to minimise friction during sliding. The one end of the leaf spring is attached with the frame's longitudinal member by using a simple cylindrical pin, while the other eye is used for mounting a small part called shackle which helps in increasing or adjusting the length of leaf spring while the vehicle is passing over road projections/pits. The blades of the leaf spring slide over each other for flattening, bending and springing action is under loads. The initial curvature provided in a leaf spring is called **Spring Chamber.**

When moving on rough road, leaf spring vibrates vertically up and down, the phenomenon is called **Yawning.**

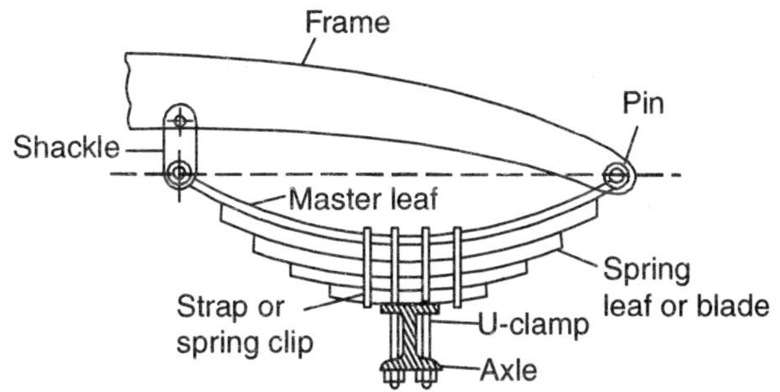

Fig. 9.1 Semi-elliptical type leaf spring

In vehicles having steering gear at rear of the front axle, the shackle is placed in the front eye, but if the steering gear is mounted forward of the front axle, the shackle is placed at rear end. The front suspension with leaf springs is limited to Trucks, Buses and other heavy duty vehicles. In cars, the leaf springs are commonly used only at rear axle only. The front suspension is independent type.

Heavy duty vehicles may have double leaf-springs, the upper one is called **helper or auxiliary leaf-spring** as shown in the Fig. 9.2. It is capable of taking violent shocks smoothly under heavily loaded vehicle.

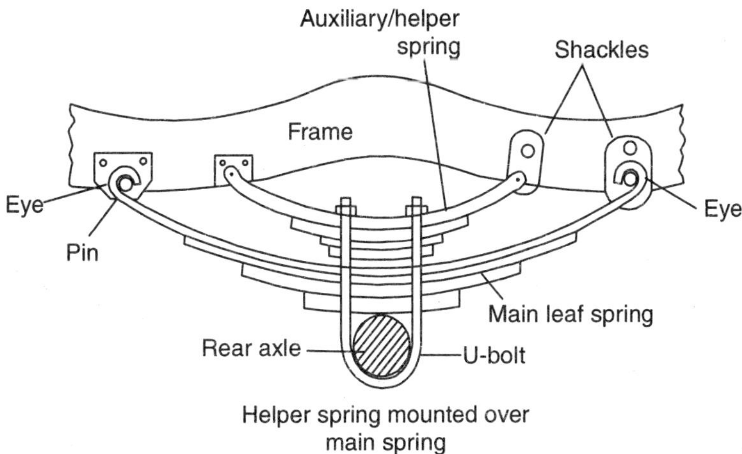

Fig. 9.2 Auxiliary or helper spring

The leaf spring blades are made from chrome-vanadium steel, Silico-Manganese steel and carbon steels.

2. Coil springs: Almost all the independent suspension systems use coil springs. In few models, coil springs were also used on rigid axle suspensions also. They acquire less space. The coil spring can store more energy per unit volume then leaf spring and they also have no friction and noise during springing. They can bear bending as well as shear stresses but cannot take side thrust and torque reaction. To overcome this weakness, some other devices are provided. A coil-spring is shown in Fig. 9.3.

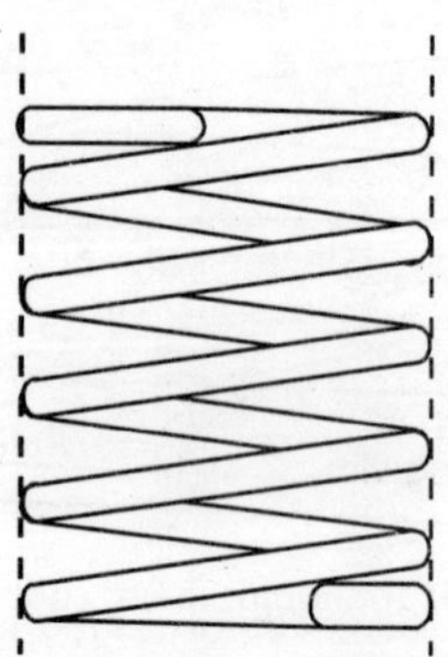

Fig. 9.3 Coil spring

3.(a) Rubber springs: Rubber is an excellent vibration damper. Rubber springs are shown in Fig. 9.4. It can store greater energy per unit weight than the steel. Due to this reason rubber springs are also used on some vehicles, supported with some mechanical device to guide the spring. Their life is quite longer. The compression-shear type spring takes the load partly by shear and partly by compression. Their vibration damping property is also very good.

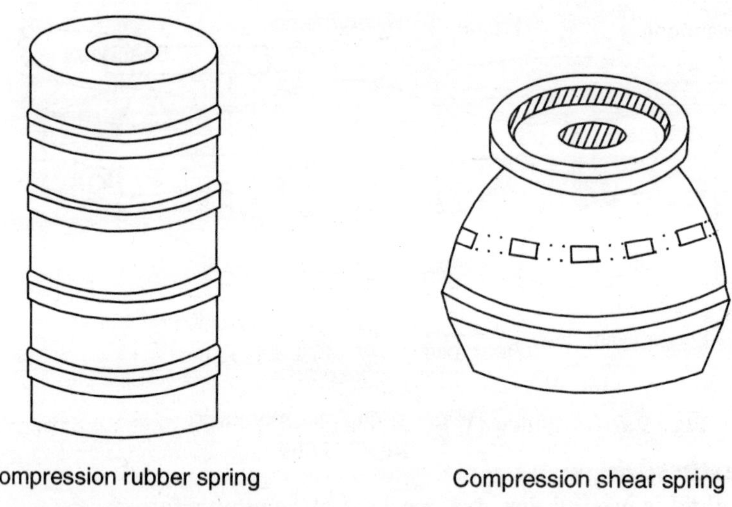

Fig. 9.4 Rubber springs

(b) Steel reinforced rubber springs: Some rubber springs are reinforced with steel helical springs bonded in their rubber body. It can take greater loads and vibrations. They also have very good vibration damping quality. Therefore, no shock absorber is required with them.

(c) Torsional shear spring: This spring consists of an inner metal rod bonded with rubber body and covered by an outer metallic shell. The spring action is produced when shaft is made to rotate slightly about its own axis relative to the outer shell. It is shown in Fig. 9.5.

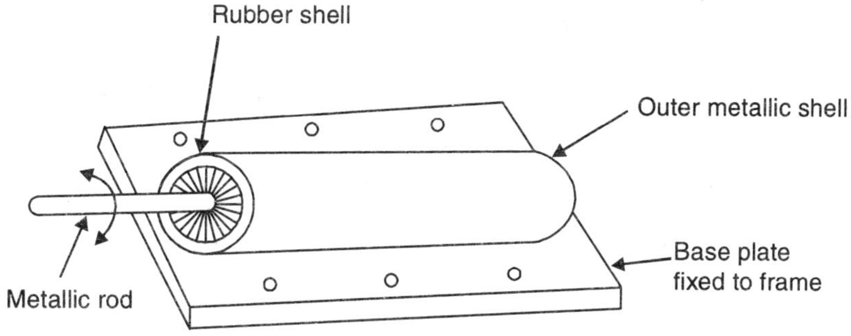

Fig. 9.5 Torsional shear rubber spring

4. Plastic springs: In some foreign vehicles, resilient plastic rings in compression are used. The plastic rings can slide over each other in a cylinder to give a spring action.

5. Air springs: Generally two types of air springs are used in independent suspension systems (Fig. 9.6).

(*i*) Bellows type, (*ii*) Piston type. They are used as road shock damping device. They are filled with compressed air. Both are used in wishbone type independent suspension system. The air springs are used to suspend axles and body. The springs absorb shocks and vibration caused by road irregularities.

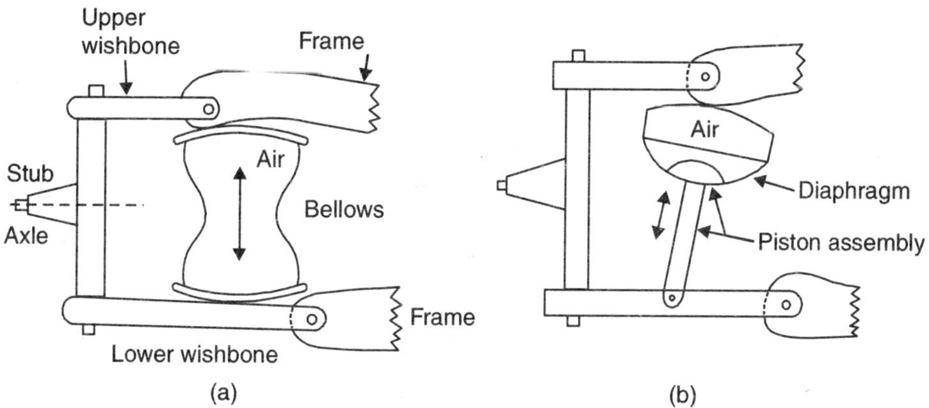

Fig. 9.6 (a) Bellows type air springs. (b) Piston type air springs

6. Hydrolastic springs: In this spring, rubber is used as a spring material under compression and shear and a fluid is used under pressure as a damping medium. It acts similar to a hydraulic shock absorber. Pure hydraulic suspensions are also used in some foreign models of luxury vehicles.

TORSION BAR

It is a steel bar used for absorbing torsional vibrations and shocks and it maintains stability of the vehicle. Its one end is fixed to the frame and the other end is connected to a swinging arm on whose outer end is mounted the wheel (Fig. 9.7). When wheel strikes a bump on the road, it starts vibrating up and down which exerts a torque on torsion bar. The bar acts like a spring and dampens the vibrations.

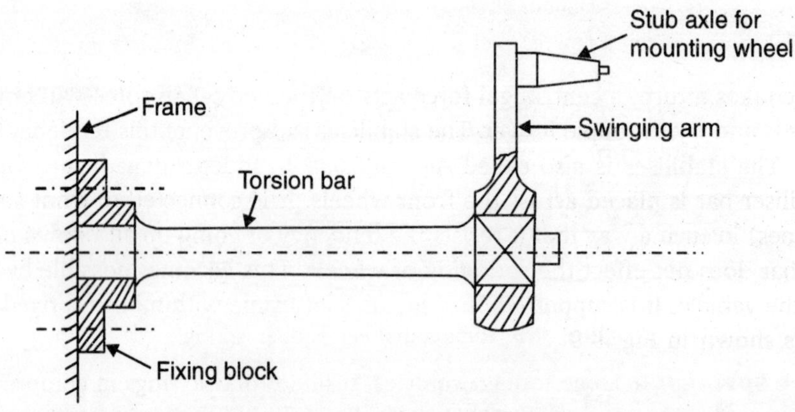

Fig. 9.7 Torsion bar as spring

The torsion bar does not take driving and braking thrusts for which additional devices are attached. Since it does not has any friction device, therefore, its damping process is not quick.

RADIUS RODS

The torsion radius rods are fitted to the rear axle casing (in rear axle driven vehicles) and cross member of the frame of the chassis. See Fig. 9.8.

The radius rods prevent the rear axle casing from rotating in clockwise direction as there is a tendency of the casing to rotate when vehicle is moving.

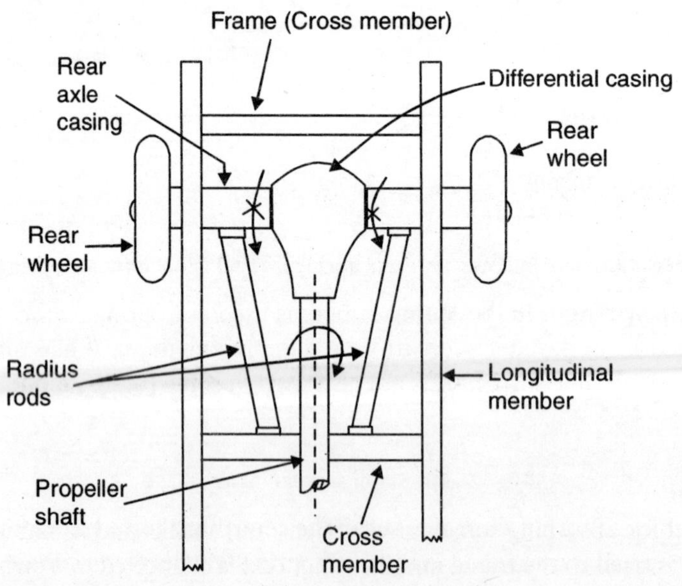

Fig. 9.8 Radius Rods (rear wheel suspension not shown)

STABILISERS

When a vehicle takes a turn, a centrifugal force acts on the body. This force will tend to move the vehicle outwards and try to topple it over. The stabiliser rod prevents this tendency of rolling over of the vehicle. The stabiliser is also called *Anti-roll Bar.* In independent front wheel suspension system, a stabiliser bar is placed across the front wheels. It is connected to front suspension links (lower wishbones) in such a way that it resists the tendency of going up and down of front wheels. The stabiliser bar does not effect the springing of wheels. This becomes possible by not fitting the bar rigidly to the vehicle. It is supported on X-member of frame within bushes fixed to frame. The arrangement is shown in Fig. 9.9.

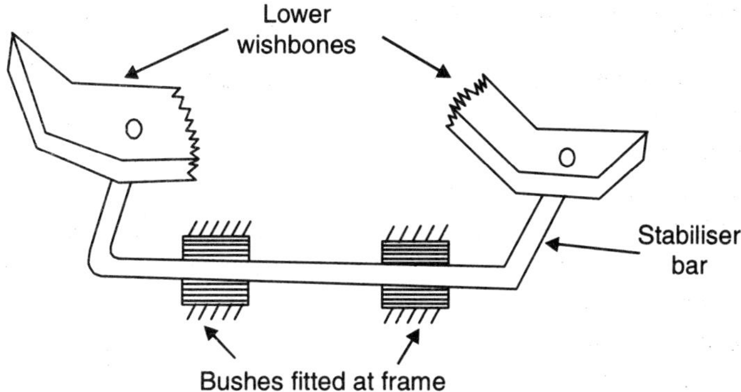

Fig. 9.9 Stabiliser bar at front side

The stabiliser bar is also fitted on rear suspension. The Fig. 9.10 shows a stabiliser bar fitted to a rigid axle rear suspension. The bar is also supported in two bearings fixed to the cross-member of the frame.

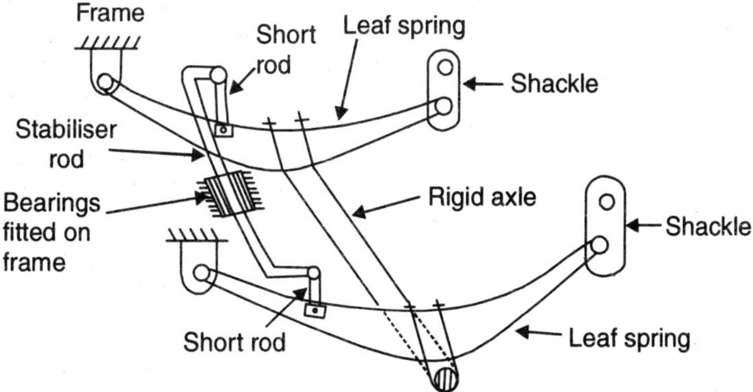

Fig. 9.10 Stabiliser bar at rear end

Suppose the vehicle is taking a left side turn. The centrifugal force will try to roll-out it, the load on the right leaf spring will increase and load on left spring will decrease, consequently the wishbone rod or short rod on left side moves down and the other moves up on the right side. Due to movement of two short rods in opposite directions, the stabiliser rod is twisted which counters the tendency of vehicle's rolling-out. The stability to the vehicle against lateral forces is thus achieved successfully.

SHOCK ABSORBER

To damp the vibrations of suspension springs, the shock absorbers are used. A shock absorber always used in conjunction with a spring. When springs start vibrating under rough road conditions, these vibrations are damped by shock absorber which enhances the riding comfort. So the shock absorber may also be called as *shock damping device.*

Many types of shock-absorbers are used on automobiles. But Telescopic Type, Direct Action, Double Acting shock absorber is widely used an almost all the vehicles. It is a Hydraulic type shock absorber. The fluid or oil filled in the shock absorber is prescribed by the manufacturer. Some of the types of the shock absorbers are:

Gas filled (N_2 at high pressure) type: Some modern cars are using *Electro-Rheological (E-R) fluid filled shock-absorbers.* These shockers are controlled by a microprocessor. The microprocessor senses the shock, sends electric pulse to ER fluid whose viscosity immediately increases and helps in better damping of spring vibrations. The E-R shock-absorbers are piston type.

Telescopic double acting shock absorber: This shock absorber is commonly used in automobiles. It is a hydraulic type shock absorber. If a vehicle is not fitted with shock absorber, the suspension spring will remain oscillating up and down for quite a longer time after a hit on the bump. The shock absorber enables the springs to settle down in a short time and thus the vehicle moves smoothly without vibrations. In hydraulic shock absorber, the damping is proportional to the square of speed of piston. The oil used in shock absorber is a mixture of transformer oil (60%) and turbine oil (40%).

In Figs. 9.11 (*a*) and (*b*) a telescopic shock absorber has been shown. The shock absorber stretches up increasing its length and returns back (downwards) when suspension springs oscillate up and down respectively. Since the length of shock absorber is elongated or reduced during spring oscillations, that is the reason that it is called a telescopic type shock absorber.

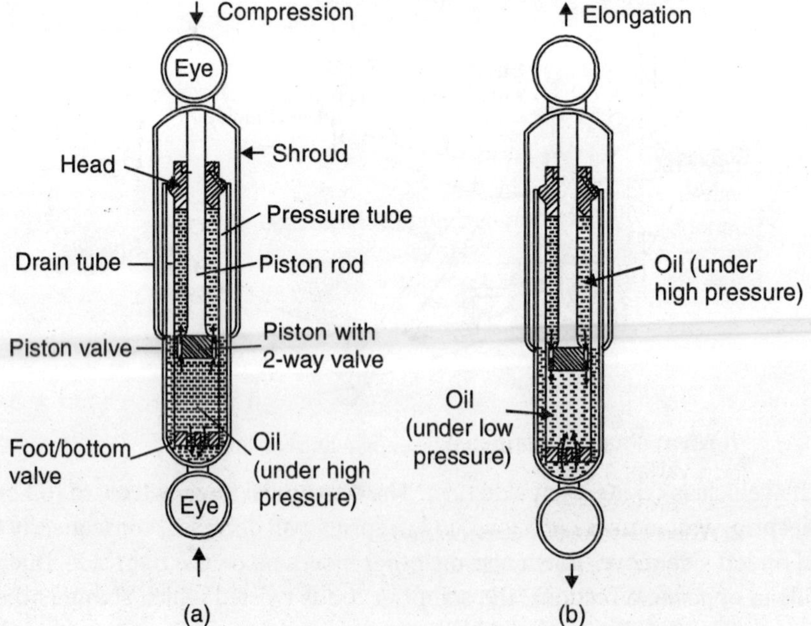

Fig. 9.11 (a) Shock absorber under compression (b) Shock absorber under elongation

The shock absorber is mounted between the axle and the frame in rigid axle suspension and between wishbones enclosed within coil springs in independent suspensions. A piston rod is attached at the top and carries a piston at its lower end. The piston reciprocates along with outer cover (shroud) inside a pressure tube concentrically contained in an outer tube called drain tube. Both the pressure tube and drain tube are oil-sealed. The pressure tube always remains full of oil but the drain tube, which surrounds the pressure tube, contains about two-third of oil and one-third air. The drain tube acts as a oil-reserve for the pressure tube. The quantity of oil in the drain tube must be sufficient to keep the foot valve constantly immersed in the oil.

When the shock absorber is compressed Fig. 9.11 (a), the fluid in the pressure tube passes upwards through restricted openings of the 2-way valve in the piston. At the same time, the oil, being under pressure by piston, passes down through openings in the 2-way foot valve. By this action the piston is able to move against the resistance of the oil at a very slow speed. This helps in absorbing the shocks on the vehicle.

Figure 9.11 (b) shows the action of shock absorber when it is lengthening, the oil filled in the top space of the piston, came under pressure and is forced downward through the opening in the piston valve. Simultaneously, the oil from the drain tube enters the pressure tube through openings in the foot 2-way valve; the oil pressure between piston valve and bottom valve falls down due to piston moving upwards during lengthening. Thus the process of lengthening becomes very slow which damps the spring vibration.

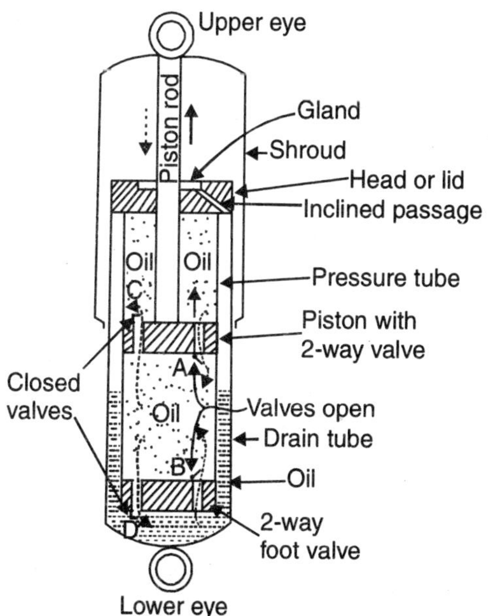

1. When shocker elongates:

Valves A and B ! open }
Valves C and D ! closed } Oil flows inside pressure tube

2. When shocker is compressed:

Valves A and B ! closed }
Valves C and D ! open } Oil flows out of pressure tube

Fig. 9.12 Principle of working of shock absorber

A gland is also formed at the top head of the pressure tube. If any oil is scrapped off by piston rod, it collects in the gland and then flows down into the drain tube through an inclined passage cut in the head. Since the oil acts on both sides of the piston valve, it is termed as **Direction Action** and **Double Acting** shock absorber. The working of the shock absorber may easily be understood by the line diagram shown in Fig. 9.12. The firm arrows show the flow of oil when piston moves up compressing the oil between head and piston. The dotted arrows show the flow of oil when piston moves downwards compressing the fluid between the two valves and the oil flows upward from piston valve and downward to drain tube from foot valve as shown in the figure.

RUBBER MOUNTINGS OF THE ENGINE

The engine is subjected to complex vibrations and stresses. As a result of vibrations, the following free movements are produced:

(*a*) Bounce and yawning about a vertical axis.

(*b*) Fore and aft movement and roll about a horizontal longitudinal axis.

(c) Sideways shake and pitch about a horizontal lateral axis.

The three inertia axes of the above six motions—(*i*) bounce, (*ii*) yawning, (*iii*) fore and aft movement, (*iv*) roll, (*v*) sideways shake (*vi*) pitch, intersect at the centre of gravity of the engine as shown in the diagram see Fig. 9.13.

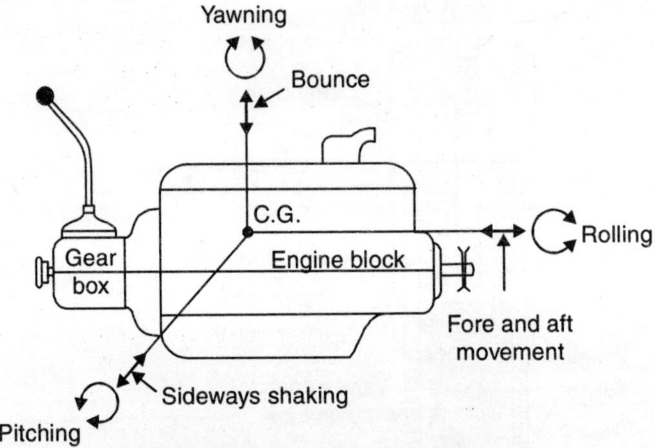

Fig. 9.13 Vibratory motions of engine

All this warrants a *resilient anti-vibration system* for mounting the engine so that the vibratory forces are reduced to the relatively small spring forces transmitted by the mountings themselves.

The mountings support the static load of the engine unit and isolate the chassis frame from engine vibrations. They also insulate the engine against vibrations of vehicle mechanism and structure while moving on road. They minimise undesirable movements of the engine and under all working conditions and reduce vehicle shake.

The engine mountings must be capable:

(*i*) to carry static or dead weight of the engine unit.

(*ii*) to withstand torque reactions.

(*iii*) to bear fore and aft thrust when the propeller shaft is pushed back and forth due to road irregularities.

(*iv*) to lower the natural frequency of the engine and the mountings to well below 0.7 times the engine frequency in whatever direction it occurs, and

(*v*) to provide sufficient flexibility to save the engine from any kind of stress when the chassis frame is distorted in an accident.

By virtue of high ratio of deflection to load, rubber is used as spring medium in modern engine mountings although coil springs were also used for sometime. Rubber is highly resilient and has a self-damping effect on the spring action of mountings. It is beneficial in preventing resonant vibrations. There are less chances of sound transmission through rubber mountings as there is no metal-to-metal path for the sound to travel along.

The engine mountings are usually made of rubber to metal bonding, which has greater strength than rubber alone. A widely used type of engine mounting is *sandwich unit* which generally

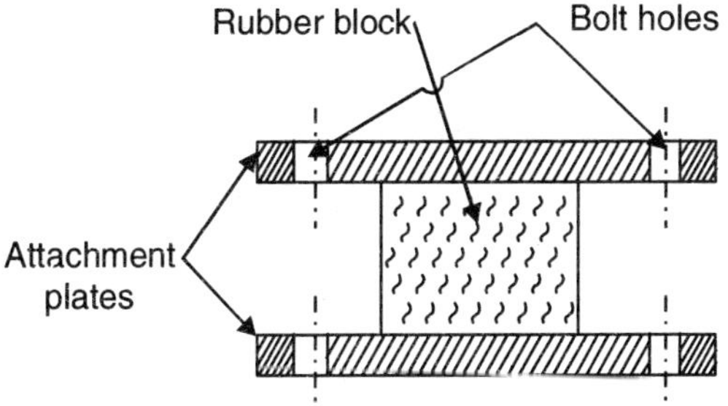

Fig. 9.14 Sandwich mounting of engine

consists of either a rectangular or a circular block of rubber bonded to metal attachment plates (See Fig. 9.14). In order to provide a combination of shear and compression loading, the limbs of U-shaped carriers are inclined in some designs (See Fig. 9.15).

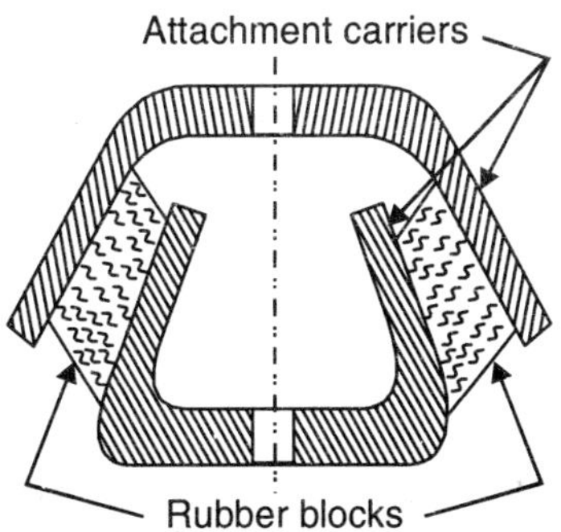

Fig. 9.15 Double sandwich mounting (inclined)

Another type of engine mounting is *cylindrical unit*. The earlier form consisted of upper and lower flanged rubber elements which were subject to only compression loading. The improved type consisted of a conical rubber bushing bonded both to a central sleeve and an outer attachment casing. This mounting may be loaded partly in shear and partly in compression. Refer to Fig. 9.16.

In early vehicles, the crankcase was usually cast with four integral supporting arms which were rigidly bolted to the chassis frame. The object being that the engine may act as a sturdy cross member to resist torsion at the front end of chassis frame. Later on flexible engine mountings were introduced to prevent the transmission of engine vibrations and noises to the body as well as to lessen the effect of road shocks upon engine. Since the engine is freed from the effects of chassis stress by the use of flexible mountings, the front portion of chassis frame is made of sufficient strength and rigidity to withstand any normal road shocks without distortion. All the externally connected parts such as exhaust pipe, cables, fuel pipe, etc., are attached flexibly to relieve them from any kind of stresses. *The various alternative methods of mounting the engines are as under:*

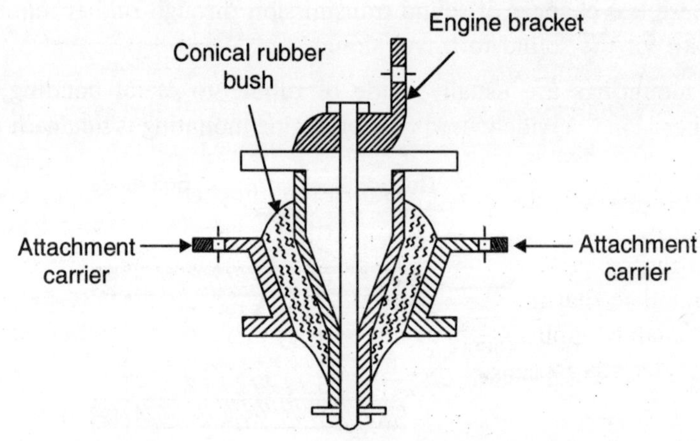

Fig. 9.16 Conical compression and shear mounting

(*i*) At two front and two rear places as in Tata bus or truck.

(*ii*) At one central position on the front cross-member and at two rear places as in Chevrolet and Oldsmobile 6 cylinder in-line engine cars.

(*iii*) At two forward places, one on either side of engine, and at one rear place corresponding to gearbox end as in Jeep.

It is the usual practice to employ *three-point mounting system* for the engine together with gearbox. Owing to greater mounting flexibility, this system is being preferred over *four-point mounting system.* The torque fluctuations induced by the power impulses from each cylinder are a major source of engine vibration. They tend to rock the engine unit in opposite direction of crankshaft rotation and thus establish the roll axis of the engine. The engine mountings must permit the greatest degree of freedom for engine movements about the roll axis. The roll axis slopes down towards the gearbox end of engine unit because the centre of gravity of engine is higher than that of gearbox. The engine mounting arrangement may be devised to produce *a centre of gravity* or *a centre of percussion* system of suspension.

In a typical *centre of gravity system* shown in Fig. 9.17, a pair of sandwich mountings support either the front or the rear of engine unit. They are installed in V-formation so that their projected normals meet on the roll axis under the static load of engine. The third mounting is of either the cylindrical or the sandwich type. It is placed at the other end of engine to intercept the roll axis.

In this arrangement, the engine is supported about its roll axis which passes through the centres of gravity of both engine and gearbox as in Jeep. As the engine unit rocks on its mountings, the sideways shake is reduced.

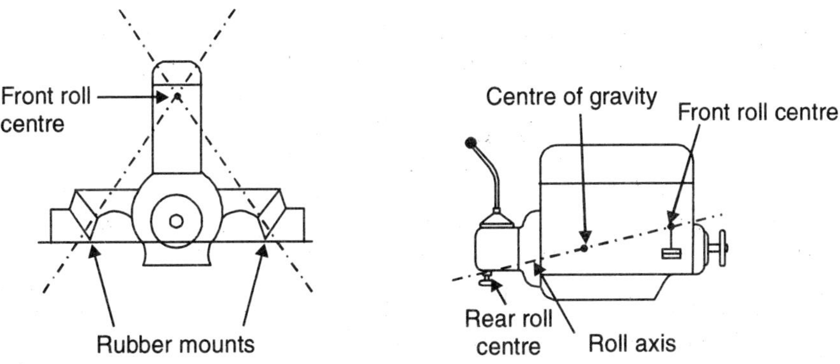

Fig. 9.17 Centre of gravity engine mounting system

In a centre of ***percussion mounting system*** too, the engine is supported about its roll axis but the transverse mounting planes are so disposed that no interaction from disturbing forces takes place between the front and rear mountings. In this arrangement, the rear mounting is located at the centre of percussion so that the disturbing forces may act upon the front mountings and *vice versa*. This system is mainly applicable to in line four cylinder engines where the vertically acting unbalanced secondary inertia forces are totally absorbed by the front mountings (See Fig. 9.18.)

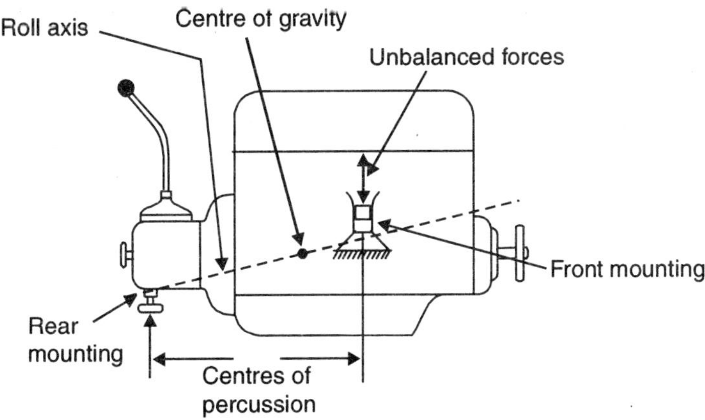

Fig. 9.18 Centre of percussion engine mounting system

In front wheel drive and rear engined cars, the torque reaction imposed upon the mountings is greater than the front engined rear wheel drive cars because the final drive assembly is integral with the engine unit. The engine mountings must be capable to accommodate the full torque applied to road wheels. In order to minimise the torque reaction loads upon the mountings, the distance between the front and rear mounting planes is kept as large as possible. In case of transversely mounted engines, it is difficult to increase the distance between the mountings due to space limitations in the engine unit compartment. For proper controlling of torque reaction movements, a fore and aft stay rod, in addition, is pivoted between the car body and engine cylinder head as shown in Fig. 9.19.

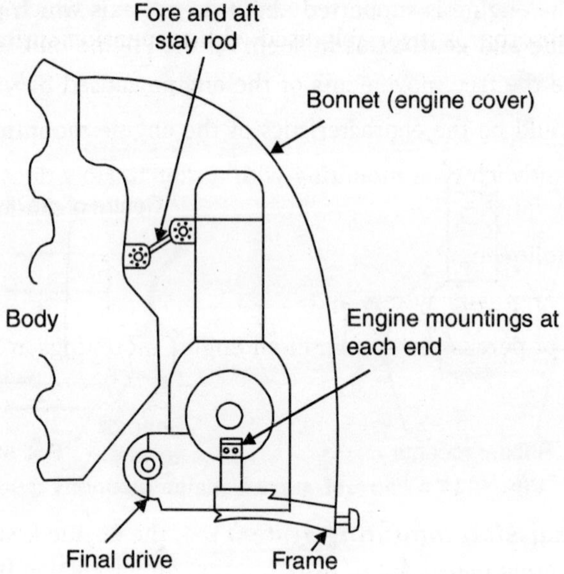

Fig. 9.19 Mounting in transverse engine and transmission unit (In front wheel driven vehicle)

QUESTIONNAIRE

1. What is the purpose of suspension springs? Classify various types of suspension springs.
2. Describe with a neat sketch a leaf spring suspension as used on auto vehicles. What is the function of shackle?
3. (*a*) What is function of helper leaf spring?

   (*b*) How spring leaves are tied together to form the leaf spring?
4. Describe briefly:

   (*a*) A coil spring

   (*b*) Rubber springs.
5. Describe air springs? What are their advantages?
6. Write short notes on:

   (*i*) Torsion bar

   (*ii*) Radius rods

   (*iii*) Stabiliser bar.
7. Describe with neat sketches, a telescopic type hydraulic shock absorber.

8. Why shock absorber is invariably used with suspension springs?
9. (*i*) What are the free movements of the engine caused by vibration?

   (*ii*) What should be the characteristics of the engine mountings?
10. Describe a sandwich type mounting with a sketch. How does it differ from a cylindrical type?
11. Explain the following:

    (*a*) Centre of gravity system and

    (*b*) Centre of percussion system in an engine mountings arrangement.

CHAPTER 10

Suspension Systems

INTRODUCTION

Wheels and axles are attached to the chassis frame through a system of springs and vibration damping devices. The suspension system helps in running the vehicle smoothly on pitted or bumpy road. Springs of different forms and types are fitted to the front and rear sides of the vehicle chassis. These springs absorb road shocks. The springing movement of the springs is dampened by a damping device called *shock absorber*. This system of shock and vibration absorbing devices is termed as suspension system. Some of the advantages of providing a suspension system are:

(*i*) It keeps vehicle movement on road stable by absorbing road shocks.

(*ii*) It dampers the vibrations and oscillations caused by rough roads.

(*iii*) It makes journey comfortable.

(*iv*) It increases useful life of vehicle parts by reducing stresses and strains developed by road shocks in various components.

Following are some important terms related with movement of vehicle on rough roads:

Rolling, Pitching and **Bouncing** (Fig. 10.1)

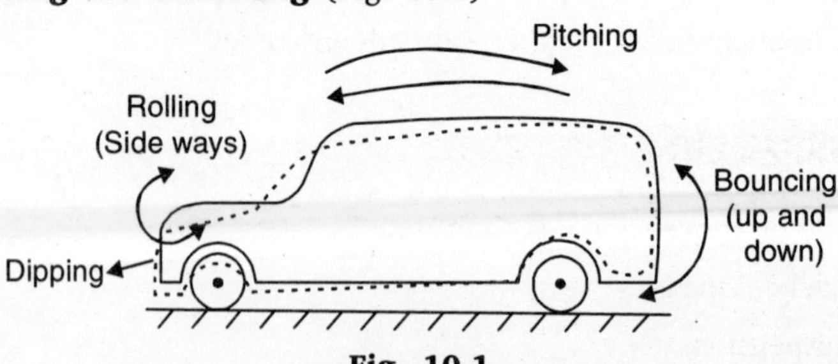

Fig. 10.1

Rolling is the movement of vehicle sideways or widthwise (left-right), pitching is the lengthwise movement between rear and front axles. Bouncing is the up and down or heightwise movement of vehicle body, Dipping or brake dip is the tendency to dive in front side during braking. Bottoming is the deflection of suspension springs on overloading the vehicle. Overloading reduces the ground clearance (shown by dotted lines profile of car).

LOADS ON SUSPENSION SYSTEM

Following loads are carried by a suspension system:

1. *Vertical load* of vehicle itself plus weight of passengers and luggage.
2. *Side thrust*: The centrifugal force developed during turning and the road camber, causes a side-thrust on the suspension, which is absorbed by springs and panhard rods. **Panhard** rods are employed to keep the rear axle casing in position against side thrust as shown in Fig. 10.2.

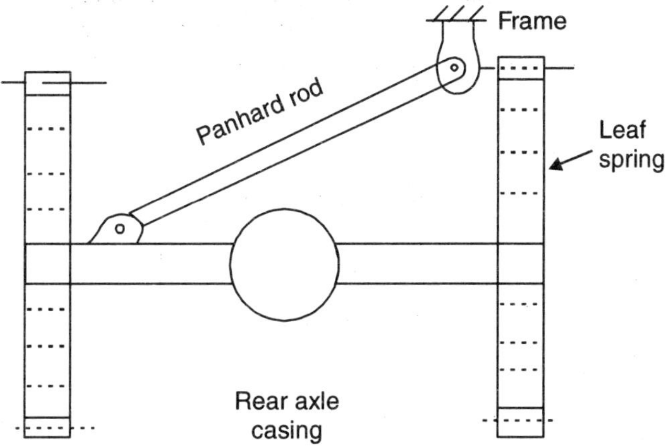

Fig. 10.2 Panhard rod

3. *Sprung load*: It is the load of frame, body, engine and transmission system.
4. *Unsprung load*: It is the weight of vehicle parts between suspension and road surface. It includes weight of wheels, tyres, brakes, front rigid axle, steering knuckle, rear axle mounting. It enhances the intensity of vibrations/shocks.
5. *Brake dip*: On braking the front nose of the vehicle dips down. This force is taken by springs, wishbones and/or by radius rods.
6. Rolling, bouncing and pitching—already discussed.

TYPES OF SUSPENSION SYSTEMS

1. Rigid axle or conventional suspension.
2. Independent suspension.
3. Mixed suspension.

1.(*a*) Rigid Axle Suspension: Generally semi-elliptic leaf springs are used in rigid axle suspensions. Buses and trucks employ this type of suspension on both sides-front and rear. But in light vehicles, the rigid axle suspension is ordinarily used on rear axle only. Front suspension is independent type on almost all the cars. The rigid axle suspension is shown in Fig. 10.3.

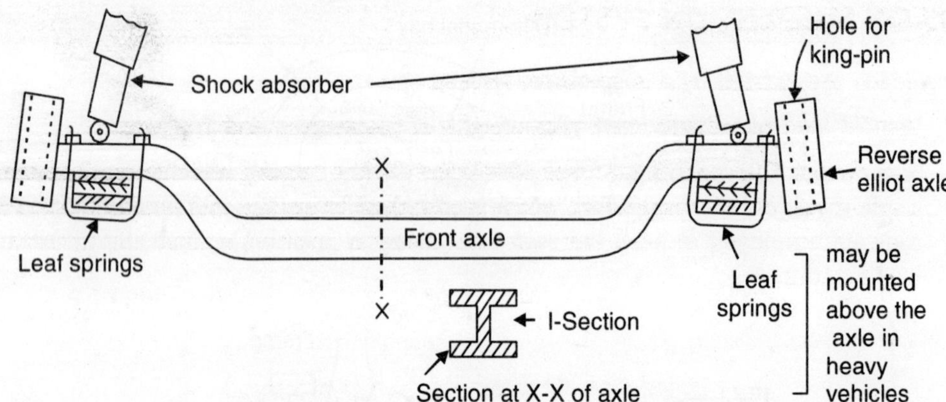

Fig. 10.3 Rigid axle front suspension

In rigid axle front suspension, the shackle on leaf springs which is made up of 6 to 20 leaves, is generally mounted on front end as shown in Fig. 10.4. In this arrangement the axle will be having its centre of rotation at the rear spring-eye, which will form the same arc as the radius action of the drag link of the steering mechanism. This arrangement reduces the tendency of wheels to wobble (side to side movement).

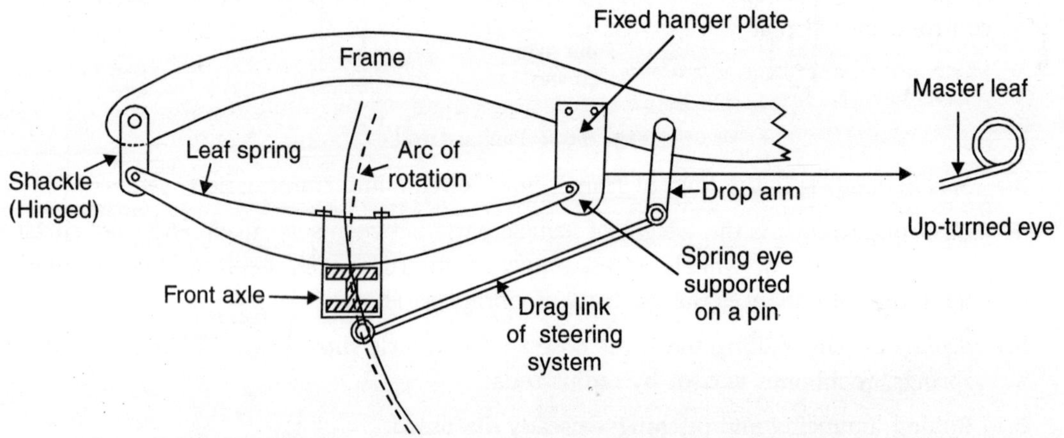

Fig. 10.4 Front rigid axle suspension showing position of shackle

In rear axle, the shackle is mounted on outer or rear side of the leaf spring. The spring is supported on axle, by means of a U-bolt. One end is mounted on the longitudinal member of the frame by a simple pin and the other end is attached to frame through a shackle. The shackle allows the change in length between the spring eyes during running of vehicle on roads.

The rigid axle is not suitable for front suspension because the engine is normally placed behind the front rigid axle. This reduces space for passengers. It also absorbs less energy in comparison to coil springs (about 1/3rd only), due to which the suspension is quite hard and uncomfortable for passengers, Steering control is also not satisfactory because of faulty steering geometry. On bumps, the changes in the camber angle of leaf spring may cause wheel shimmey or the wheel making rapid oscillations.

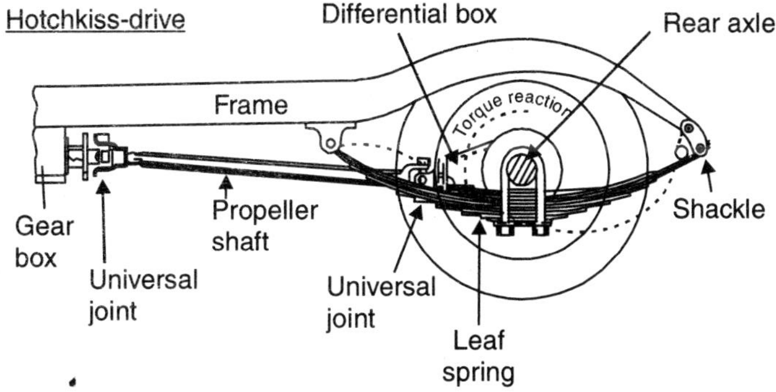

Fig. 10.5 Rear leaf spring suspension (Rigid axle)

(*b*) **Rigid Axle Rear Suspension:** Figure 10.5 shows a leaf spring type rigid axle suspension used on trucks and buses and on old cars. When rear wheels are driven (the rear axle is the driving axle), the rear axle casing has a tendency to rotate in a direction opposite to the direction of rotation of wheels. This twisting motion creates a rear-end torque which is absorbed by leaf spring. When vehicle is accelerated the rear end is tilted downward due to this torque. This torque is taken by panhard rod or control arm in case of coil springs. In hotchkiss drive the rear end torque is absorbed by panhard rod or control arm and rear spring together.

2. **Independent Suspension Systems:** In rigid axle suspension, the vehicle tilts on either side when wheel passes over any irregularity on the road and wheels also do not remain vertical. This causes lot of discomfort to the passengers. It also causes wheels to "wobble" or stagger, wheels-road adhesion also becomes weak. To overcome the above problems, independent suspensions are used on wheels, in which tilting of anyone wheel does not affect the other wheel or the level of car floor as shown in the Fig. 10.6.

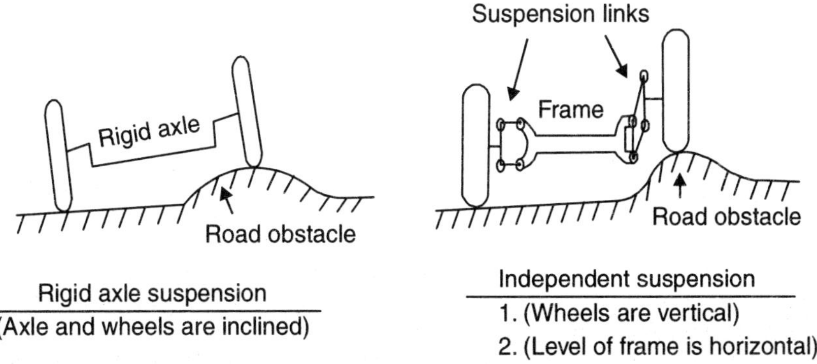

Fig. 10.6 Rigid and independents suspension

Advantages of Independent Suspension

1. Lighter springs (coil springs) are used which also increase comfort to passengers.
2. Unsprung weight is reduced, which increases tyre-life. Road adhesion of tyres is also increased.

3. Steering geometry is not altered with spring deflection as in case of rigid axle suspension.
4. Engine can be mounted at a lower level and may be moved sufficiently towards front side which results in more space for passengers.

On most of the passenger cars, the independent suspension is employed on front wheels only and rear wheels are mounted on rigid axle suspension.

2.(*i*) Independent Front Suspension: This system is universally employed on front wheels in almost all the passenger cars in combination with rigid axle suspension on rear wheels. Different types of front suspension system are used on different vehicles. Two types of front suspensions are commonly used:

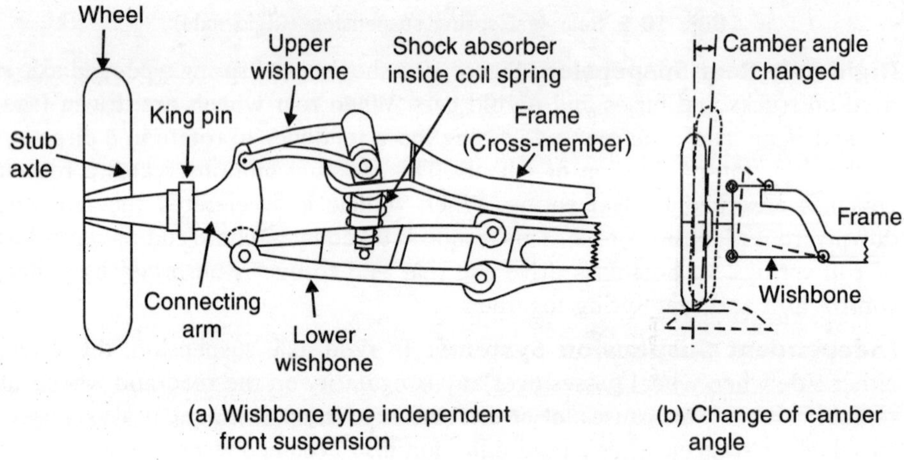

(a) Wishbone type independent front suspension

(b) Change of camber angle

Fig. 10.7 Wishbone type independent suspension

(*i*) *Wishbone type*: It is used on various models of Tata car company such as Sumo, Sierra and Estate and other vehicles of heavy and medium segment such as Scorpio.

This system consists of two wishbone arms (upper and lower) of V-shape. Upper wishbone is smaller in length. Both wishbone's wide open sides are connected with car frame's cross member and the closed sides are hinged to a connecting arm carrying king pin and a stub axle for wheel as shown in Fig. 10.7. The shock absorber enclosed in a coil spring is mounted between a seat on the lower wishbone and the underside of the cross-member of the frame as shown in Fig. 10.7.

The wishbone arms keep the wheels in position over the road and transmit the vehicle load to the coil springs. They also resist forces due to accelerator, braking and cornering or siding. The upper arms are made smaller in length than the lower arms.

This design helps to maintain the wheel track constant, thereby avoiding the scrubbing of tyres which also reduces tyre wear. In this system, car frame remains horizontal but wheels do tilt slightly from their vertical position causing a very small change in camber angle as shown in Fig. 10.7(b). Wishbone type suspension is also used on rear wheels on some cars such as Contessa Classic and Premier-Diesel cars.

(*ii*) **Macpherson Strut Type Front Suspension:** It is simple and lighter in weight. Most of the small and medium segment cars employ this suspension system such as Maruti 800, Matiz, Fiat Uno, Lancer etc.

This suspension uses lower wishbones only. The inner wishbone ends are hinged to the frame and keep the wheel in position. It also resists accelerating, braking and side or cornering forces. The other end is fastened to a channel shaped connector. The stub axle is attached at its centre and the upper end of the connector channel is attached to a strut containing shock absorber enclosed in the coil spring. The upper end of shock absorber is connected to the body frame. See Fig. 10.8.

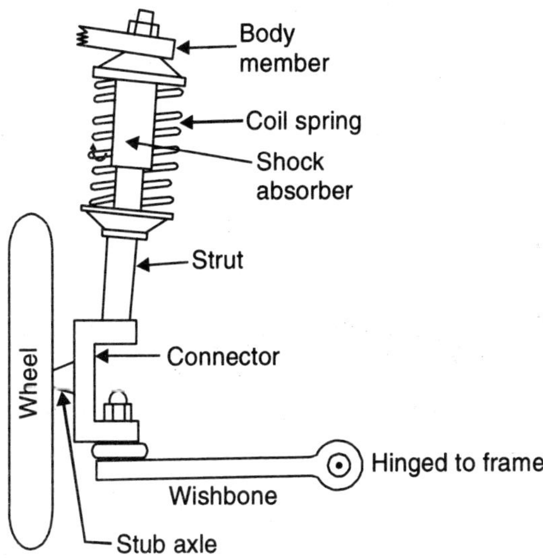

Fig. 10.8 Macpherson struct type front suspension

This system occupies less space, therefore, it is widely used on front wheel drive vehicles. The wheel alignment (camber angle) also does not change. The unsprung loads are reduced due to lighter construction.

On some cars, it is used along with an anti-roll bar, which gives self-stabilizing steering ability, *i.e.*, car remains running on its line of travel when the brakes are applied even through the road surface may be irregular Mac Pherson system is also used as independent rear suspension on many popular vehicles (Esteem, Corona).

(*iii*) **Mixed Suspension:** This system is commonly used on rear suspensions. The leaf spring and coil spring with shock absorber are employed. In some cars, quarter-elliptic type leaf springs are used. The coil spring takes the vehicle weight while leaf spring is subjected to drive and torque reaction. Almost all the front wheels drive vehicles are using combined or mixed suspension at the rear wheels. In small cars, the leaf spring is replaced by a lower oscillating arm attached to the car floor and rigid axle end as shown in the Fig. 10.9. A balance rod is also employed connecting both sides of the axle tube to keep the wheel track in position. This suspension is widely used on Maruti 800 and Matiz Cars.

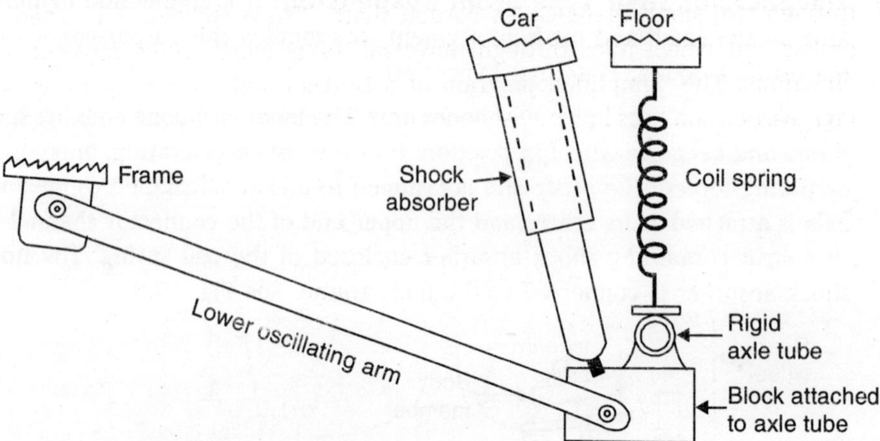

Fig. 10.9 Rear suspension of rigid axles in small cars

(iv) **Coil Spring Rear Suspension:** Coil springs with shock absorbers are also used on rear wheel drive vehicles in De-Dion rigid tube suspension. The De-Dion tube maintains the track at a constant width. It is not an independent suspension. Coil springs used in combination with radius arms make independent type suspension used on rear drive vehicles.

(a) ***Independent Rear Suspension Employing Radius Arms and Coil Springs:*** In this system, the two radius arms are employed each wheel. The frame of the car is fitted to the coil type compression spring enclosing a shock absorber. The two radius arms are attached to the wheel hub just like wishbones. The lower radius arm is larger than upper arm to the keep the wheel in vertical position at bumps or pits.

At bump or pit the wheel moves up or down without affecting the level of car frame. The coil spring helps to absorb the shock. The system is shown in Fig. 10.10.

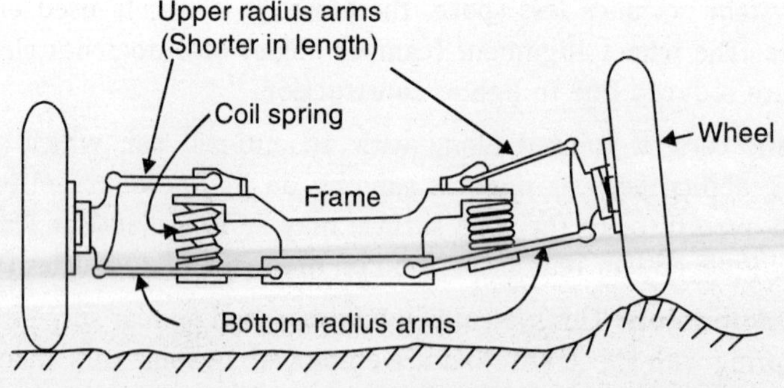

Fig. 10.10 Independent rear suspension with radius arms

(b) ***The De-Dion axle Method:*** It is generally not used on Indian cars. The De-Dion tube is rigidly attached to both wheels hubs which keeps the wheels parallel at constant track distance. The axle tube can move up and down but not to the sides. The side-movement is checked by the vertical slide attached to frame.

The motion and power from differential gear box is transmitted to wheels hubs through two small shafts called cordon shafts. These shafts are attached to inboard brakes and wheel hubs through universal joints which allow movements in many directions. The simplified diagram of a De-Dion axle rear suspension is shown in Fig. 10.11.

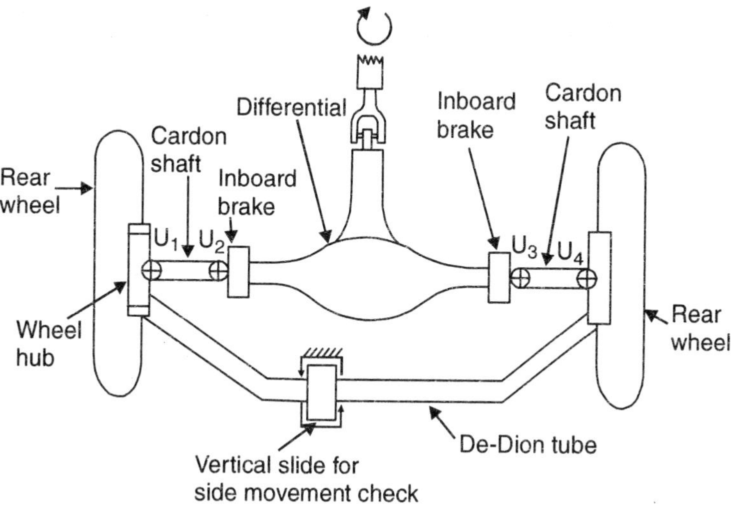

Fig. 10.11 The De-Dion rear axle suspension (top view)

It was used on old times racing cars. On Indian cars, generally rigid axle suspension systems or mixed suspension systems are used.

QUESTIONNAIRE

1. Why is it necessary to employ a suspension system on automobile vehicles?
2. What is function of shock absorber? How it differs from the function of springs?
3. Write short notes on:

   (*i*) Rolling, (*ii*) Pitching,

   (*iii*) Bouncing.
4. Describe the following:

   (*a*) Vertical load of vehicle (*b*) Sprung load of vehicle

   (*c*) Unsprung load.

5. What are side thrust and brake dip?
6. Describe a rigid axle suspension system with suitable and neat diagram.
7. What is the function of shackle? Describe its effects when it is mounted on front and rear sides of the leaf spring.
8. (*i*) What are advantages of independent suspension over rigid axle suspension?

   (*ii*) Describe a wishbone type independent suspension.
9. Describe with neat sketch, a Macpherson strut type suspension. What are its advantages?
10. Briefly describe the De-Dion axle type suspension on rear wheels.
11. Describe an independent rear suspension with radius arms.
12. Describe any type of a front independent suspension with a neat sketch.

CHAPTER 11

Braking Systems

INTRODUCTION

Braking system is a must in all the vehicles. A good braking system saves the vehicle, its passengers and the third party on the road. Brakes are provided to control the speed of the vehicle. The function of brakes is to stop or slow down the speed of the vehicle according to the needs of the driving on road. When vehicle is braked, each wheel develops a certain amount of a braking force. Greater than the number of wheels braked, higher will be the braking effort and sooner the vehicle will get slowed down or stopped. When brakes are applied, the clutch pedal is first pressed, so that the power unit is disconnected. When brakes are applied the kinetic energy of the vehicle is converted into heat. The removal of heat from the braking system is important. The proper air circulation is necessary for heat removal. A hot brake drum/disc results in brake fading or inefficient braking.

BRAKING REQUIREMENTS

1. The brakes must stop the vehicle within a very small distance *i.e.*, stopping distance should be minimum.
2. The braking system must come into action without delay as soon as the brake pedal is pressed by driver.
3. The system must be strong enough to sustain the sudden braking forces.
4. It must operate with least effort by the driver.
5. It must neither slip or should cause any skidding of the vehicle (in which wheel does not rotate due to firm braking but vehicle goes on slipping on the road.
6. The brake-lining should be durable with good braking property.
7. The braking system also controls the speed when the vehicle moves down a hill. It also holds the vehicle firmly when vehicle is braked on climbing a hill.

8. On braking, the vehicle should not drift to right or left.
9. The braking system should be light as possible. This minimises unsprung weight.
10. The braking system should have a longer life and minimum maintenance.
11. There should be a provision for a separate braking system which can be operated easily in case the main braking system fails.
12. Brakes should not make noise or vibrate when applied.
13. The system should be safe and remain unaffected by water, heat, road grit and dust etc.

Other Retarding Forces: The following forces/resistances also help to slow down the vehicle besides braking force:

1. Air resistance to the movement of vehicle.
2. Energy losses in friction in the transmission system.
3. Rolling resistance of tyres over the road. Greater the load on wheels, greater will be rolling resistance.
4. Gradient resistances on slopes (uphill or downhill).

Work done during braking, stopping distance and brake efficiency

Refer to Fig. 11.1.

If W = Weight of vehicle, N

V = Speed of vehicle, m/s,

F_r = Average friction at all the four wheels of car

$= (F_{r_1} + F_{r_2} + F_{r_3} + F_{r_4})$

d = Stopping distance or the distance moved by vehicle after applying brakes till the time of halting.

g = Acceleration due to gravity (9.81 m/s²).

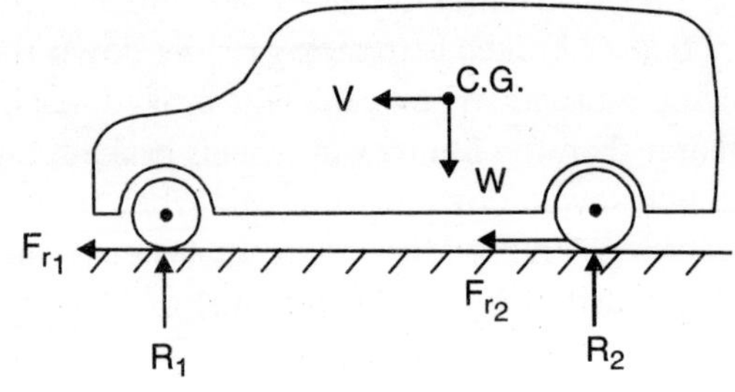

Fig. 11.1

The kinetic energy (K.E.) possessed by vehicle:

$$\text{K.E.} = \frac{1}{2}\frac{W}{g}V^2 \qquad ...(1)$$

$$\text{Braking work done} = F_r \times d \qquad ...(2)$$

Equating the two equations

$$F_r \times d = \frac{1}{2}\frac{W}{g}\cdot V^2 \qquad ...(3)$$

from which, stopping distance or braking distance may be determined.

If μ = coefficient of friction between road and wheel tyre,

then $F_r = \mu \cdot R = \mu \cdot W.$

Where R is the sum of values of resistances at all the wheels of the vehicle. It will be equal to the weight of the vehicle (W).

$\therefore$ Stopping distance
$$d = \frac{1}{2} \cdot \frac{WV^2}{g} \times \frac{1}{\mu \cdot W}$$
$$= \frac{1}{2} \cdot \frac{V^2}{\mu \cdot g}$$

If braking efficiency (η_{br}) is also taken into account,

then
$$d = \frac{1}{2} \cdot \frac{V^2}{\mu . g . \eta_{br}} \quad \text{...(4a)}$$

If $\mu = 0.6$, $\eta_{br} = 1$ and $g = 9.80$ then $d = \dfrac{V^2}{11.76}$ m ...(4b)

If the vehicle slows down from V_1 to V_2 speed over a distance of x meters

then,
$$F_r \cdot x = \frac{1}{2} \frac{W}{g} \left(V_1^2 - V_2^2\right) \quad \text{...(5)}$$

From this equation 'x' distance may be calculated.

If η_{br} = Braking efficiency. It is the rate of retardation which the braking system brings a vehicle to rest from a certain speed of the vehicle. It is expressed in m/sec^2 or as a percentage. It is also termed as braking ratio.

If f_{br} = Braking retardation in m/sec^2

= $\mu \cdot g$ for limiting retardation

= g if $\mu = 1$. This will be the maximum value of f_{br}.

Then,
$$\eta_{br} = \frac{f_{br}}{g} \times 100\% \quad \text{...(6)}$$

Its value ranges from 70 to 85%.

Highly efficient brakes with 100% efficiency, are ideal but practically not desirable, as they produce sudden greater deceleration which may cause:

(*a*) Rapid wear of tyres due to sudden braking and consequent skidding.

(*b*) Injury to passengers.

(*c*) Goods loaded collide with each other and also with the body of the vehicle and get damaged. Body may also get damaged by striking of goods due to sudden braking.

(*d*) There is a faster wear of brake shoe linings and the disc or brake drum.

(*e*) Driver may loose control over steering.

(*f*) Damage may also be caused due to braking system.

APPROXIMATE STOPPING DISTANCES

The brake performance test done on brake drum type of brakes (not the disc brakes) on the normal plane road at speed of 50 km per hour, the stopping distances were found as entered in the following table:

Test Speed of vehicle = 50 km per hour		
Stopping distance in meters	**Brake efficiency**	**Braking quality**
30.4	30%	Very poor
18.28	50%	Fair
12.80	70%	Good
10.67 and less	80% to 100%	Excellent

The stopping distance vary with the conditions of the road surface and the tyre tread. Some distance is also added to above values for (*i*) Driver's reaction time, (*ii*) Time taken in operating the brake system. This distance may be up to 6 meters or more depending upon the activeness of the driver and speed of vehicle. The braking distance is also affected by the brake effort applied by the driver.

Solved Example: A car is going at a speed of 60 kmph when brakes were applied, it took 25 meters to stop. Determine the braking efficiency and the value of retardation. Take coefficient of friction as 0.8 between road and the tyres.

Solution:

Given

$$V = 60 \text{ Kmph} = \frac{60 \times 1000}{60 \times 60} = 16.66 \text{ m/sec.}$$

$$d = 25 \text{ m}, \; g = 9.81 \text{ m/s}^2, \; \mu = 0.8$$

∴ Braking efficiency,

$$\eta_{br} = \frac{1}{2} \cdot \frac{V^2}{\mu \cdot g \cdot d} \times 100\% \qquad \text{from Eqn. } (iv\text{-a})$$

$$= \frac{1}{2} \times \frac{16.66 \times 16.66}{0.8 \times 9.81 \times 25} \times 100\%$$

$$= \frac{277.56}{392.4} \times 100\%$$

$$= 70.73\% \qquad \textbf{Ans.}$$

∴ Retardation,

$$f_{br} = \frac{\eta_{br} \times g}{100} \quad \text{(Refer to equation } (vi))$$

$$= \frac{70.73 \times 9.81}{100} = 6.94 \text{ m/sec}^2. \qquad \textbf{Ans.}$$

ROAD TYRE ADHESION

The adhesion between the wheel tyres and road depends upon the following factors:

1. Weight of vehicle acting on the wheels.
2. The tyre inflation pressure.
3. Type of tread pattern on the tyre surface.
4. The coefficient of friction between tyre and the road. Wider tyres (radial type) have good adhesion.

If the braking force on the wheel is less than the force of adhesion, the vehicle will stop gradually. When the braking force exceeds the adhesion force at any wheel, the wheel stops rotating but starts slipping or skidding on the road till the kinetic energy is dissipated in friction between tyre and road. Skidding causes rapid wear of tyre and also possess steering difficulties. During braking, the force of adhesion decreases at the rear wheels on account of WEIGHT TRANSFER during braking. This phenomenon may result in locking of the rear wheels and skidding. If only front wheels get locked due to braking, the vehicle will not skid but dipping (the tendency of vehicle to dive in its front side) will occur which may also be fatal. Some regulating devices or anti-skid devices in suspension system on rear wheels are provided to avoid skidding.

THERMAL ASPECTS

During normal application of brakes, the kinetic energy of the rotating wheel is converted into heat which is dissipated to the brake drum and the atmosphere. The rim of wheel should have sufficient open space for air circulation. In case of skidding (wheel is stopped when braked but goes on slipping on the road), the kinetic energy is dissipated in friction between the wheel tyre surface and the road. The tyre gets excessively heated up and the surface is badly scratched and worns out. The tyre, brake drum, brake lining and the rim must be strong enough to bear the heat such generated.

If braking or retarding force $= F_b = \dfrac{\text{Weight of vehicle} \times \text{Rate of decelerations}}{g}$

and Driver's effort = P Newton's

(*i*) Then for mechanical brakes, the mechanical advantage $= \dfrac{F_b}{P}$. The mechanical advantages in small passenger cars varies from 50 to 100, in LCV's it is 100 to 250 and in HCV's it may vary from 250 to 800.

(*ii*) **For Hydraulic Brakes**

$$F_b = p \times A \qquad ...(1)$$

where p = Pressure of brake fluid per unit area of the piston of the master cylinder.

and A = Area of the piston of master cylinder.

Heat generated = Work done in braking

$\therefore$ Work done in braking = Braking force × Stopping distance

$$= F_b \times d \text{ N.m}$$

$$\therefore \quad \text{Heat generated 'H'} = \frac{F_b \times d}{4200} \text{ kcal}$$

As 1 kcal = 4200 N.m or Joules

If n = Number of wheels braked

$$\text{Then total heat generated at one wheel (during braking)} = \frac{F_b \times d}{4200 \times n} \text{ kcal}$$

If V = Linear velocity of the wheel in m/min.

$$\text{then} \quad H = \frac{F_b \times V}{4200\, n} \text{ kcal/min per wheel.}$$

WEIGHT TRANSFER DURING BRAKING AND BRAKING RATIO

The braking effort of each wheel must be proportional to the weight of vehicle on the wheel to achieve maximum deceleration during braking. When the wheels are braked, there is a change in the distribution of weight between rear and front wheels. This change of weight, from rear wheels to front wheels, is proportional to the rate of deceleration. Referring to Fig. 11.2, the inertia force acting at centre of gravity (towards front side) and the retarding force (produced due to braking) acting at the road and wheel surface, form an *over turning couple* as shown in the Fig. 11.2. This over turning tendency increases some weight (w) on the front wheels and decreases the same amount on the rear wheels. The phenomenon is well-displayed when, on sudden braking, every passenger bends forward and the luggage is also tilted or slips towards front side. Thus the brake dip causes some weight transfer on front wheels. About 12% of vehicle weight is shifted to the front wheels. Therefore, it is necessary to apply more braking effort on front wheels (about 60%) for achieving efficient braking. For maximum retardation on braking, the ratio of braking force on the front the wheels to that on rear wheels should be equal to the ratio of the transferred weights. This ratio is called BRAKING RATIO.

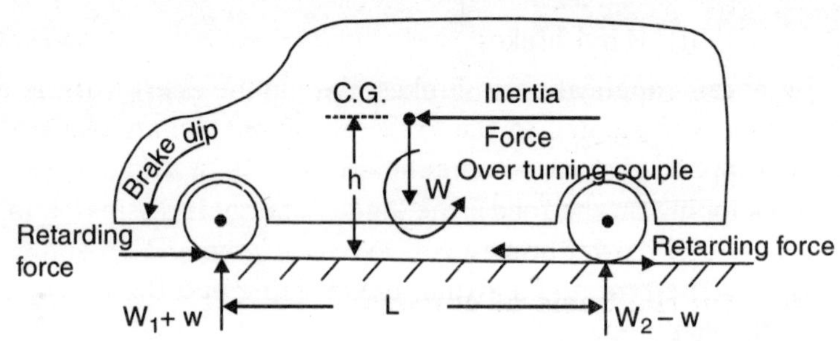

Fig. 11.2

CLASSIFICATION OF BRAKES

The brakes are friction creating devices which reduce the speed of the vehicle as and when desired by the driver. They may be classified on the basis of various factors:

1. **On the basis of method of operation:**
   (*i*) Mechanical brakes
   (*ii*) Hydraulic brakes
   (*iii*) Pneumatic or air brakes: (*a*) Compressed air brakes, (*b*) Vacuum brakes
   (*iv*) Electromagnetic brakes or Electrical brakes.
2. **On the basis of purpose served:**
   (*i*) Main operational brakes
   (*ii*) Parking brakes or hand brakes.
3. **On the basis of location:**
   (*i*) Wheel mounted brakes
   (*ii*) Transmission mounted brakes (Band brakes and Disc brakes)
4. **On the basis of driver's ergonomics:**
   (*i*) Hand brakes (*ii*) Foot brakes.
5. **On the basis of application of braking effort:**
   (*i*) Manual brakes.
   (*ii*) Servo or power assisted brakes (Hydraulic brake assisted by vacuum system)
   (*iii*) Power brakes.
6. **On the basis of construction:**
   (*i*) Drum brakes (*ii*) Disc brakes
   (*iii*) Band brakes.
7. **On the basis of movement of brake-shoes:**
   (*i*) Internal expanding brakes (*ii*) External contracting brakes.
8. **On the basis of combined brake systems:**
   (*i*) Drum type and disc type combination: Matiz, Ceilo, Esteem, Zen and Lancer cars have disc brakes on front wheels and drum type brakes on rear wheels.
   (*ii*) Hydraulic and Mechanical Combination: Some vehicles are provided above combination. Generally mechanical brakes are used as parking brakes.
9. **Special type of brakes:**
   (*i*) Engine exhaust gas operated brakes (*ii*) Pneumatic-hydro brakes
   (*iii*) Hill holding brakes.

MECHANICAL BRAKES

1. Parking Brakes or Hand Brakes

Mechanical brakes are widely used on 2-wheelers and tractors. They are also used on cars as parking brakes or hand brakes. Parking brakes are used for parking the vehicle and when starting and stopping a car on an elevation or hill to prevent backward movement.

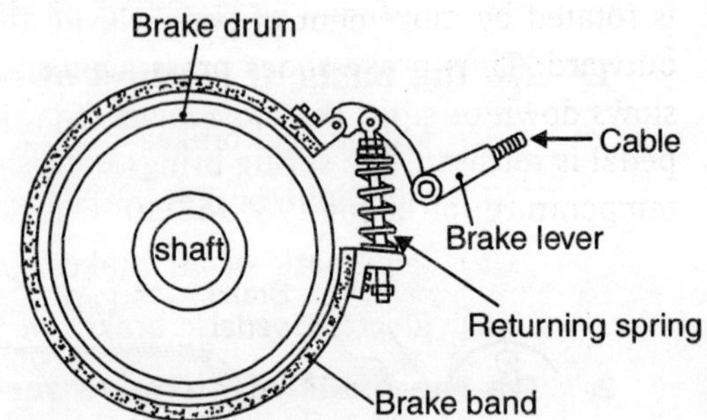

Fig. 11.3 (a) Parking/hand/band brake

Parking brakes utilise a band brake fitted on the counter shaft of the gearbox before the universal joint. The brake drum is mounted on the counter shaft of the gear-box, over which a brake band is wound round the periphery. This brake band is linked to the hand lever by means of a spring, a brake lever and a cable. See Fig. 11.3(a).

Location of hand brake on transmission line is shown in Fig. 11.3 (b).

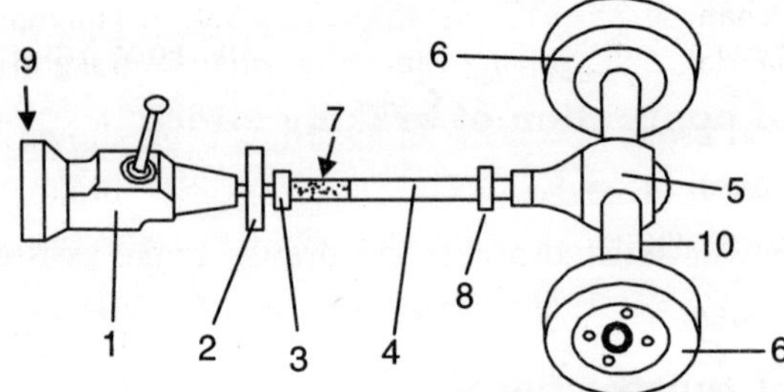

Fig. 11.3(b) Location of hand brake: Ref. 1. Gearbox; 2. Hand brake drum; 3. Universal joint; 4. Propeller shaft; 5. Differential; 6. Wheel; 7. Slip joint; 8. Rear U. joint, 9. Clutch, 10. Rear axle casing.

When the hand brake is applied, the brake band gets tightened on the drum and the motion is thus stopped. When hand lever is released, the brake band looses its grip on the drum. The vehicle now becomes ready for starting. In present day cars with drum brakes on rear wheels, the rear wheel drum brakes are used as parking brakes too.

2. Drum Brakes

Drum brakes need a round drum for creating friction between shoe-lining and the drum's internal surface. The brake drum is made on the axle hub and the wheel rim is attached to it by means of bolts and nuts to the hub. A brake plate of pressed steel sheet is attached to the axle casing. The two horse-shoe shaped brake shoes are attached to brake plate. The lower ends of the shoes are pin-jointed at the bottom over the plate and a cam is fitted between the two upper ends of the shoes. A retaining spring keeps the shoes pressed over the cam. The outer surface of shoes is lined with asbestos or special fibre material lining riveted to the shoes. It is made from asbestos, some filler materials and powdered resins. The mixture is moulded in dies to form the shape and then heated under pressure until a hard slate like board is formed. The cam is operated by foot or hand brake lever connected to the cam by means of cables and levers. See Fig. 11.4. When cam is rotated by movement of pedal, lever and cable linkage, it pushes the ends of the brake shoes outward. Thus brake-shoes press against the inner surface of rotating brake drum. This friction slows down or stops the movement of brake drum and the wheel gets slow down or stops. When pedal is released, the spring brings both shoes together relieving the pressure on brake drum. A temperature of about 175° to 250° C is developed between the lining and brake drum.

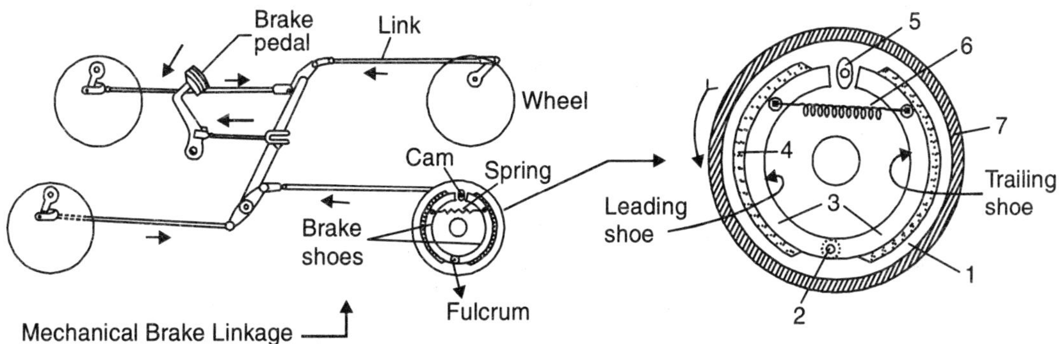

Fig. 11.4 Drum type mechanical brake: 1. Fixed brake plate; 2. Fulcrum; 3. Brake shoes; 4. Fibre material; 5. Revolving cam; 6. Spring; 7. Brake drum

A simplified mechanical brake operating linkage has been shown in the Fig. 11.5. The lever is turned by foot pedal or hand brake lever. The cables turn the brake cams from vertical position to horizontal position. This movement of cam expands the brake shoes.

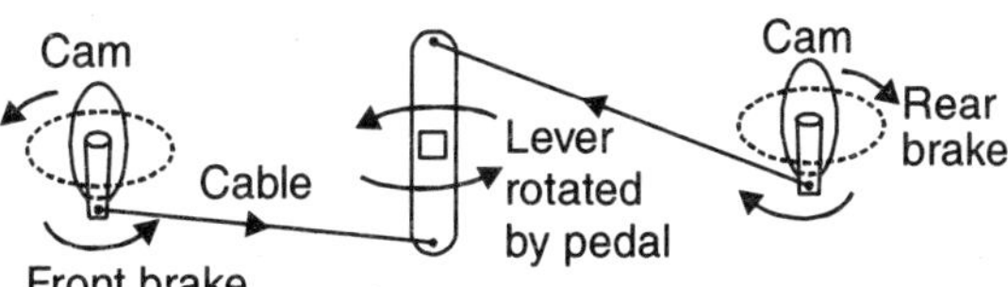

Fig. 11.5 Mechanical brake linkage simplified

(a) Leading Shoe and Trailing Shoe

The shoe which moves in the direction of movement of the brake drum is called a leading shoe. During braking it offers more braking torque and engages with the drum smoothly. The other shoe which moves against the direction of rotation of brake drum is called trailing shoe. It offers lesser braking torque and so it is less efficient. Such brakes are commonly used on two wheelers and on many vehicles such as Lancer Car and Tata LPT 1612 truck. See Fig. 11.4 for leading and trailing shoes.

(b) Two Leading Shoe Brakes

To make the trailing shoe more efficient, many provisions are made to ensure its leading movement; one of them is called Girling Mechanical Brakes.

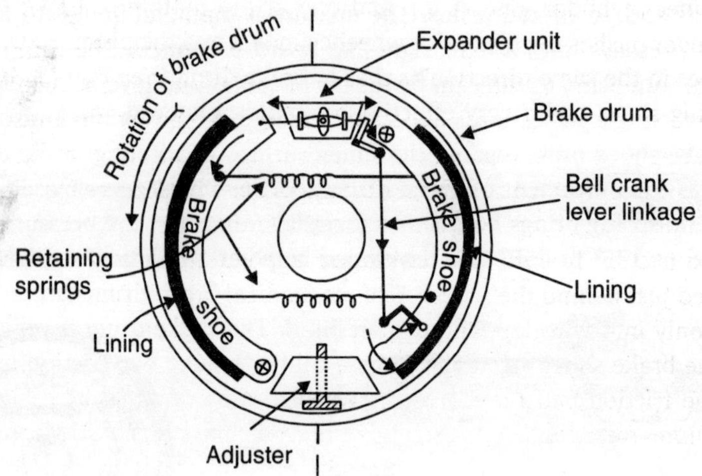

Fig. 11.6 Two leading shoes drum brakes or girling mechanical brakes

Girling Mechanical Brakes: These type of brakes use an expander unit which contains a cam with conical shape at both ends. When pulled by brake pull rod or pedal, the cam pushes two plungers in opposite directions as shown in the Fig. 11.6 in simplified form. One plunger rod pushes the leading shoe to press against the drum. The other plunger rod actuates a bell crank levers linkage which moves the bottom end of the other shoe in the direction of rotation of brake drum to act it as leading shoe and thus braking is enforced. The shoe connected with bell crank linkage is fulcrumed at the top. Thus both shoes move in the direction of rotation of the brake drum and become leading shoes which give very efficient braking effect in the forward direction of the vehicle. The two leading shoe brakes act as trailing shoe type when vehicle is moved backwards and become less effective. An adjuster unit is provided at the bottom, at whose sloped surface, the tip of the brake shoe slides. The wear-adjustment is also done by adjuster. Two leading shoes brakes are used in an Ambassadar diesel car and on front wheel in Bullet 500 C.C. bike.

(c) Two Leading Shoe Hydraulic Brake

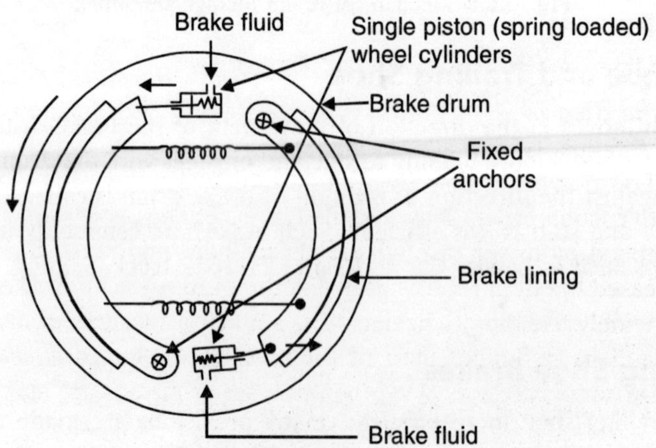

Fig. 11.7 Two leading shoe hydraulic brake

It utilizes to wheel cylinders-one at top and one at the bottom of brake plate. Brake fluid, through master cylinder pushes the pistons of wheel cylinders and the piston rods, attached to brake shoes move the shoes in the same direction as that of brake drum (See Fig. 11.7). Thus both brake shoes become leading shoes and a very effective braking is achieved.

DISC BRAKES

Disc brakes are also friction brakes. The brakes are applied at a metallic rotating disc attached to driving shaft fitted just behind the wheel. The mechanical brake drums are now being replaced by disc brakes not only in 4-wheelers but also on bikes. The brake drum is replaced by a circular metallic disc and the brake shoes are replaced by a caliper having two friction pads—one on each side of the disc. The friction pads are forced towards disc by braking force and grip the disc to retard it. The medium for actuating the caliper may be mechanical, hydraulic, pneumatic or electromagnetic.

Advantages of Disc Brakes

1. Lower inertia, less weight (about 20%).
2. More stability, do not wear easily.
3. Easy service and repair and replacement due to simple design.
4. More efficient cooling as only a small portion of disc remains in contact with friction pads. Pedal travel is almost constant for long durations.

Disadvantage

They do not have self-servo action, due to which greater operating force is needed. Generally, brakes are operated hydraulically. The principle of operation of disc brakes is similar to the brake system fitted an ordinary bicycles which are the common men's two wheelers running on roads since very olden times.

Types of Disc Brakes

1. Caliper type
2. Sliding caliper type
3. Swinging caliper type.

1. ***Caliper type disc brakes:*** In this type, a cast iron disc of suitable size is fitted to the wheel hub. A caliper type small housing is mounted over the axle casing in such a manner that a small portion of the disc always rotates between the caliper ends, each containing a piston with friction pad as shown in the Fig. 11.8. The pistons are actuated by hydraulic pressure applied by driver through master cylinder and brake fluid lines. When brake pedal is released the fluid returns back and the pistons are moved back by springs. These brakes are widely used in 4-wheelers. The standard thickness of disc on maruti 800 car is 11 mm and its minimum limit of thickness is 9.5 mm. Deflection is allowed upto 0.15 mm. On some vehicles, more than one set of pistons are employed.

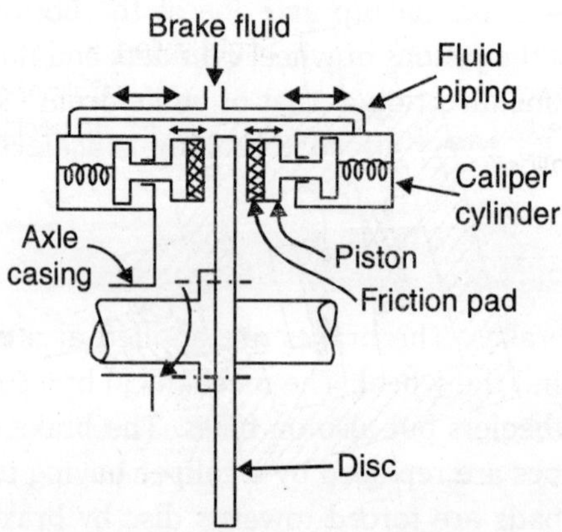

Fig. 11.8 Caliper type disc brake

2. ***Sliding caliper type disc brakes:*** In this type, the caliper body itself slides slightly to apply the brakes. This type also has two pistons, but in one cylinder. Both pistons are attached by springs. One piston pushes the friction pad towards the disc while the other piston pushes the caliper itself which slides towards the disc and the brakes are applied as shown in the Fig. 11.9. The pistons are actuated by fluid pressure. One of the friction pad is fixed at the caliper itself as shown on its left portion.

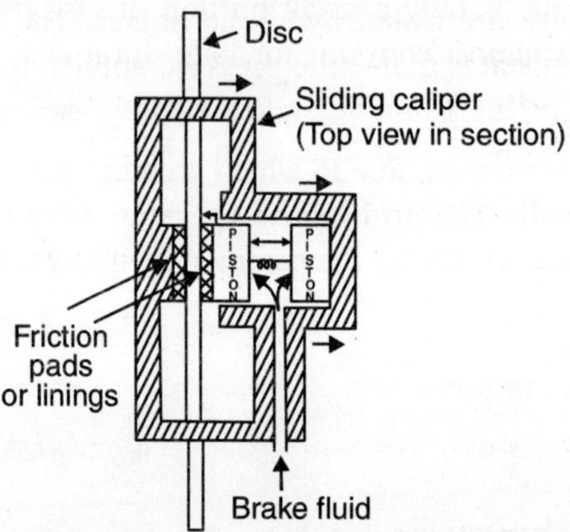

Fig.11.9 Sliding caliper disc brake

3. ***Swinging or floating caliper type disc brakes:*** In this type, the caliper is hinged at a fulcrum pin. The friction pads are made slightly slopy on braking side. One pad is attached to the piston and the other pad is attached to the caliper. The hydraulic pressure, applied by brake oil, presses the piston pad against the disc to apply the brake. The reactions on the caliper causes it to move the fixed pad slightly inwards, thereby applying equal pressure to the disc on its side. The caliper automatically adjusts its proper position by swinging about the pin. This type of brake is shown in Fig. 11.10.

   Some brakes employ ***ventilated disc*** consisting of two discs connected by vanes in such a manner that they form a single casting. These discs improve the cooling and keep the low temperature at brake surfaces.

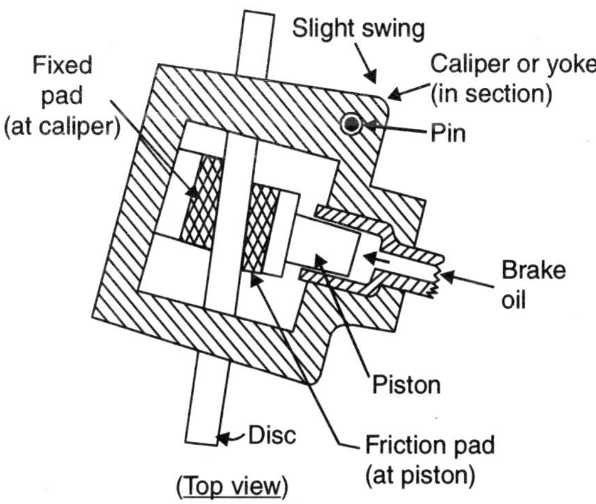

Fig. 11.10 Swinging type disc brake

HYDRAULIC BRAKES

Hydraulic brakes are used most commonly on almost all the passenger cars. The brakes which are operated by hydraulic pressure are called hydraulic brakes. The hydraulic brakes work on the principle of Pascal's law which states that—'the pressure applied to a liquid is transmitted equally in all directions'. In addition to the hydraulic brakes, hand brakes or parking brakes, mechanically operated by hand, are provided on the rear wheels. Hydraulic brakes have the following advantages:

1. Lesser braking effort is applied by driver.
2. There is equal braking force on all the wheels.
3. The operation is quiter as there are less moving linkages.
4. Brake pedal effort is multiplied many times due to hydraulic pressure.

The hydraulic brake system is used on both drum type and disc type brakes.

Main Components of Hydraulic Brake System

1. Brake pedal or foot pedal
2. Master cylinder
3. Oil pipe lines
4. Wheel cylinders in case of drum type brakes or caliper operating mechanism in case of disc brakes.

The hydraulic brake system utilises brake fluid or brake oil filled in the system for transmitting braking effort. A simple layout of hydraulic brake system is shown in Fig. 11.11. On some vehicles, a servo-mechanism is also provided to boost up the drivers effort for braking (vacuum or air assiated brakes).

The brake lines, master cylinder and wheel cylinders are always full of brake fluid containing polyglycols mixed with some mineral oil. Branded brake fluids are named by various manufactures as SAE-J-1703, DOT4, Shell brake fluid, Donax YB, Castrol Qstop, DOT4 etc.

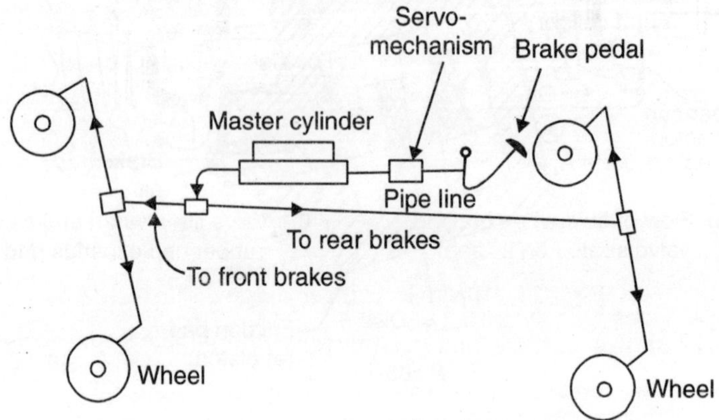

Fig. 11.11 Layout of a hydraulic brake system

Working

When the brake pedal is pressed by the driver, the rod attached to pedal, operates servo mechanism which multiplies the effort, the rod now moves a spring loaded piston inside the master cylinder filled with brake oil. The piston pushes the brake oil through a valve to the front and rear pipe lines. The pressurised brake fluid move the calipers (on disc brakes) or wheel cylinders in case of drum brakes. The piston rod of the wheel cylinder actuates the brake shoes to apply friction at the inner surface of the rotating brake drum within wheel assembly.

When pedal is released, the spring pushes the piston back to its position and brake fluid returns to the master cylinder and brakes are disengaged. The working of master cylinder and brake shoes with wheel cylinders is being described below:

1. **Master Cylinder:** It is the heart of the hydraulic brake system. It is connected to the brake pedal on one end (In some vehicles through a servo-mechanism) and to the brake pipe lines and wheel cylinders on the other end. It is a plunger type pump which pumps the brake oil at approximately 350 to 500 kPa pressure to the wheel cylinder during braking.

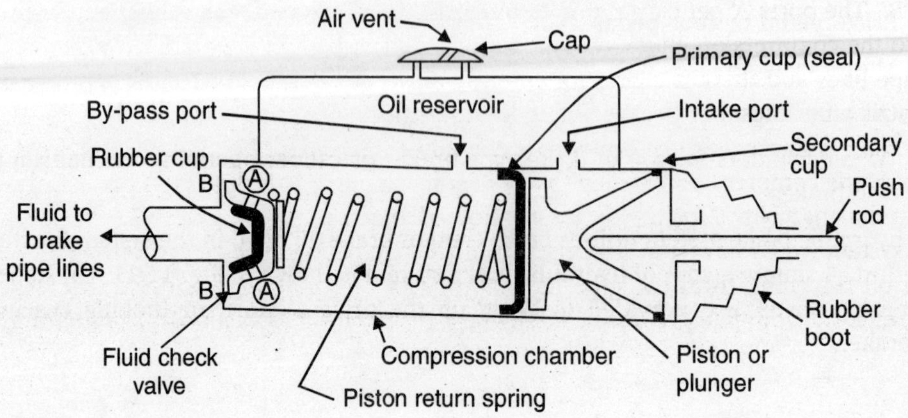

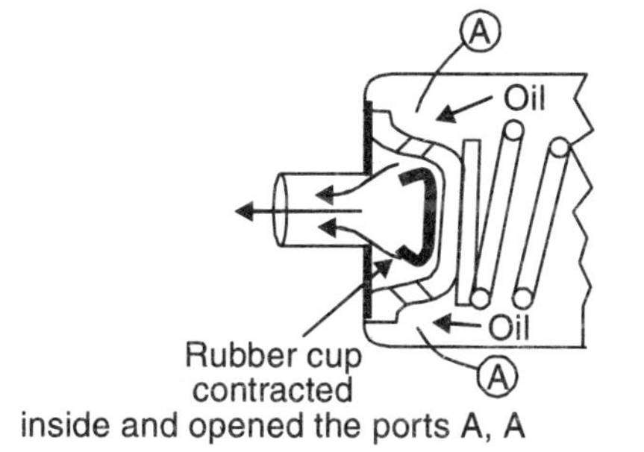

(a) Flow of fluid when braking, valve seated on its seat

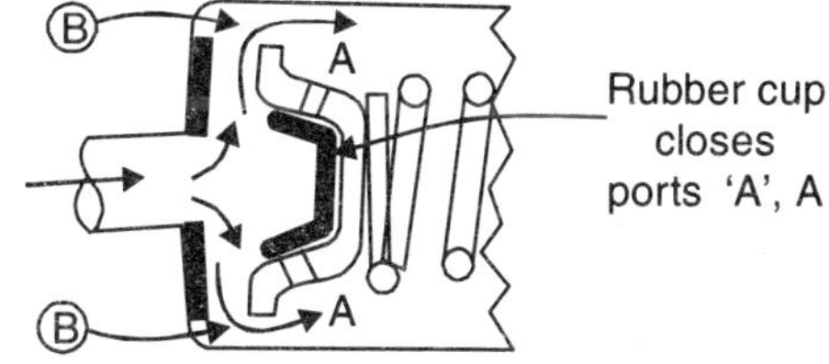

(b) Valve lifted when brakes are released, rubber cup expands and closes the ports

Fig. 11.12 Single plunger master cylinder

Figure 11.12 shows a single plunger master cylinder.

(a) Single plunger master cylinder: It contains a compression chamber in which a plunger is moved by a push rod operated by brake pedal. The brake fluid remains filled inside the compression chamber as well as space behind the plunger face. The compression chamber is also connected to an oil reservoir made at the top of the chamber. The brake oil is filled in the reservoir through a cap. 2/3 to 1/2 reservoir is always kept full of oil. The reservoir is connected to compression chamber through two ports. One by-pass port (about 0.7 mm. dia) opening in the front of the plunger face and another intake port behind the plunger face in the cavity. The cavity gets the oil from the intake port. There is a fluid check valve with rubber cup at the outlet of the compression chamber. The cavity behind plunger, compression chamber and brakes lines always remain full of brake oil under a small pressure of about 1.5 kgf/cm^2 in normal condition to avoid air leakage into the system.

During braking, the push rod pushes the plunger inside the compression chamber. The plunger moves against the spring and when it covers the by-pass port, the fluid is forced towards the check-valve. Ports 'A' open due to pressing down of rubber cup by fluid pressure. Thus oil from ports A enters the brake lines through the check valve. The hydraulic pressure, through brake lines, is transferred to the wheel cylinders which actuate the brake-shoes or caliper pistons (in disc brakes). When the driver lifts up his foot from the brake pedal, the push rod with plunger is pushed back by spring. The sudden return of spring with plunger causes a drop of pressure inside the compression chamber. The oil starts returning back to the compression chamber by lifting check valve through passages 'B'. The ports 'A' get closed due to drop of fluid pressure at the rubber cup's outer surface exposed to the compression chamber. The back flow of brake fluid is quite slow due to friction in narrow pipe lines and there is chance of entering air in the system due to low pressure or vacuum in the compression chamber because of sudden returning of spring with plunger. As the pressure in the chamber is lower than the pressure in the oil cavity behind the plunger, the oil from plunger cavity enters the compression chamber by deflecting the primary cup. Some oil from by-pass port also enters the chamber. Thus the vacuum is immediately destroyed, otherwise, air leakage inside the brake system would have made the system ineffective because the air is more compressible than oil. When oil returning from the pipe lines approaches its normal quantity in the chamber, the excess oil, entered from plunger cavity and by-pass port to destroy vacuum, passes back to the fluid reservoir through by-pass port. The returning oil from wheel cylinders also gets up heated during braking which increases its volume. The increased volume is also accommodated inside the fluid reservoir by forcing out some air to atmosphere from reservoir through the air vent made in the cap. The push rod is protected from dust and water by covering it with rubber boot. In single plunger master cylinder, if brake pipe lines get damaged, the whole brake system becomes ineffective.

(b) *Tandem master cylinder:* This is used in medium duty vehicles (1 to 3 tonnes rating) and heavy duty vehicles. In this arrangement, two separate pipe lines are used for front brakes and rear brakes. If front brake lines are damaged the rear brake will remain working, which makes it easy to have control on the vehicle during brakes failure of either side.

A simplied diagram of the tandem master cylinder is shown in Fig. 11.13.

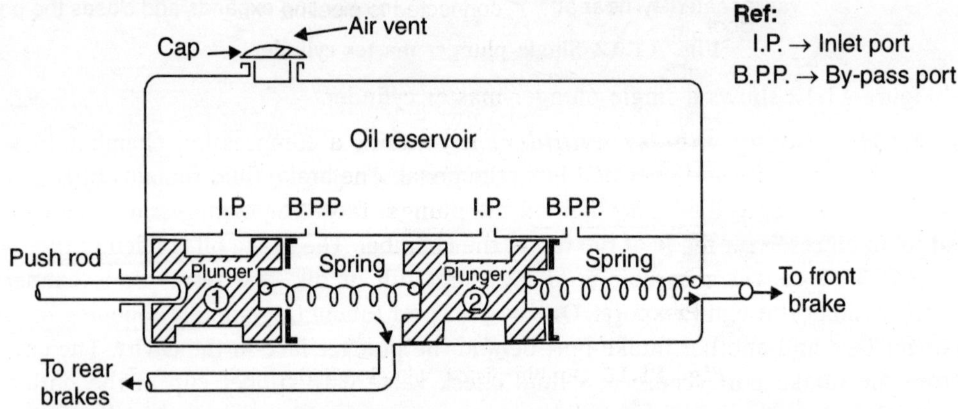

Fig. 11.13 Tandem master cylinder

The compression chamber has two plungers inter connected by springs. The push rod forces the first plunger to move inwards. The first plunger, through spring, pushes the second plunger to move brake oil towards delivery pipes lines. Both pipes lines are separate — one for front brakes and the other for rear brakes. Rest of the working is similar to the single plunger master cylinder. A pressure of 1.5 kgf/cm^2 is always maintained in the brake lines to avoid air leakage into the system.

2. **Wheel Cylinders:** A drum type hydraulic brake shown in Fig. 11.14 is operated by the movement of two pistons fitted inside the wheel cylinder. The pistons move outwards due

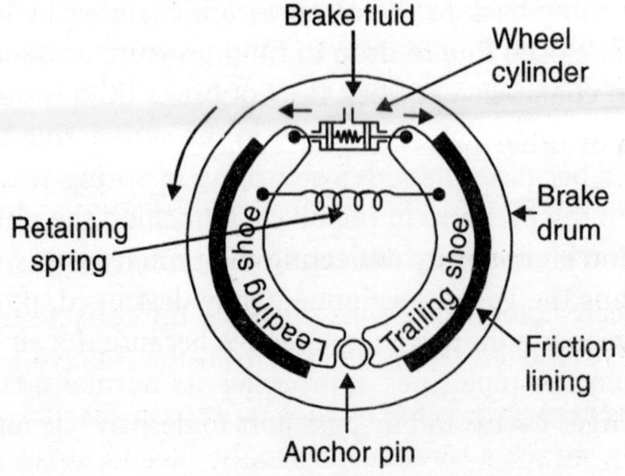

Fig. 11.14 Hydraulic brake (Drum type)

to pressure exerted by brake fluid filling in the wheel cylinder. The piston rods which are attached to brake shoes, move the brake shoes towards the revolving brake drum and the braking is affected. When brakes are released the piston slide back to their original position under spring force and push out the excess brake oil to the brake pipe lines. A detailed view of the double piston wheel cylinder is shown in Fig. 11.15. It is also called a *Slave cylinder.* One wheel cylinder is fitted on each of the four wheels of the vehicle. It is mounted on the back plate or brake plate inside the wheel drum. The wheel cylinder is fitted with two pistons connected by a light coil spring. There are two rubber cups which are located in front of each piston and act as oil seals.

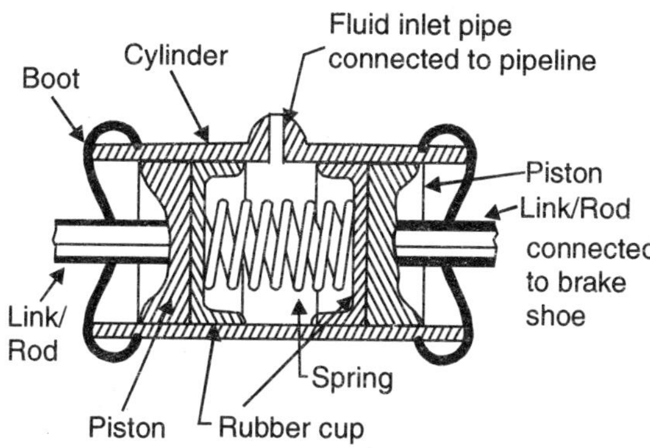

Fig. 11.15 Double piston wheel cylinder

Two rubber boots are also fitted outside on btoh sides of the wheel cylinder to protects them from water and dust. The brake pipe line is connected to the cylinder in the upper middle portion.

Brake pipe lines and hoses are made of special material to with stand the high pressure of brake fluid. Damaged or leaking pipe lines should immediately be replaced. A bleeder valve is also provided in each wheel cylinder to facilitate the bleading of brakes. The wheel cylinders of the front brakes are usually of larger diameter to provide more braking effort to face the weight transfer during braking.

Bleeding of Hydraulic Brakes

The process of draining out or removing air from the braking system of an automobile is termed as **bleeding of brakes.** Some amount of air gets entry into the system whenever;

(*i*) Repair of pipe line or other parts (master cylinder, wheel cylinder etc.), is carried out.

(*ii*) When there is leak in the pipe line or wheel cylinders.

(*iii*) When fluid level in the reservoir falls too low or the reservoir becomes empty.

Since the air is more easily compressible, its presence in the system makes the braking very poor or ineffective. The air is removed from the system in the following manner:

The hydraulic brake system may be bled by using a pressure bleeding equipment or manually. To bleed the system manually, attach a bleeder tube to the bleeder valve of the wheel which is to be bled. The bleeder tube's free end should be submerged in a clean jar which should remain partially filled with approved brake oil throughout the operation, otherwise air may be sucked in from the free end if left open in air. Two persons are required to bleed the system. One person will press and release the brake pedal while other will work at bleeding wheel.

Pump the brake pedal and force the fluid to flow through the bleeder tube/hose into the jar. For this, open the bleeder valve, fully press the pedal and allow the brake fluid to flow out into the jar and then close the valve. Slowly release the pedal, allowing it to return to its normal position. Continue operating the brake pedal and opening and closing the bleeder valve until air bubbles stop coming out with brake oil from the bleeder tube end which is dipped in the oil of jar. When air bubbles get completely vanished, close the bleeder valve and remove the bleeder tube and jar. The brake fluid collected in the jar should be poured back in the reservoir to correct its level up to half filled. Repeat the process on all the four wheels for complete elimination of air from the system. Keep it in mind that level of fluid in the reservoir must be kept half full during the bleeding process and should not fall below half-level.

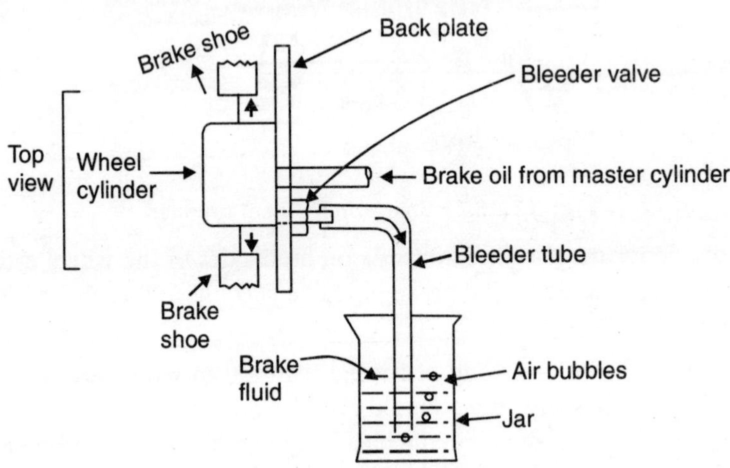

Fig. 11.16 Bleeding of brakes

The process of bleeding of brakes has been shown in the Fig. 11.16.

Trouble Shooting for Hydraulic Drum Brakes

1. Braking inefficient — Lining defective or greasy—Replace brake shoes/Clean the grease.

2. Brakes remain applied even after releasing the pedal —
   - (*i*) Springs of brakes shoes weak or broken—Replace.
   - (*ii*) Spring of pedal defective—Replace.
   - (*iii*) Wheel cylinder pistons siezed—Repair and clean.
   - (*iv*) By-pass port of master cylinder clogged — Clean the port.

3.	Pedal is felt spongy when pressed	— (*i*) Air in the system—Bleed the brakes. (*ii*) Secondary cup of master cylinder damaged — Replace the cup. (*iii*) Leakage in the system—Rectify it.
4.	Noisy brakes	— (*i*) Lining rivets touching the drum — Change lining and rivets. (*ii*) Thick lining—Rub-off excess thickness. (*iii*) Drum out-of-round—Rebore/replace.
5.	Pedal travel more than normal	— (*i*) Leakage from secondary cup—Clean or replace it. (*ii*) Low oil level in the reservoir—Level it up. (*iii*) Brake shoes out of proper position — Adjust the clearance between shoes and drum.
6.	Brake grab	— (*i*) Broken lining—Replace it. (*ii*) Drum badly scored—Rebore it.
7.	Wheel locked on braking	— (*i*) Broken shoes—Replace (*ii*) Broken lining/loose lining—Replace.
8.	Brake drum worn out	— (*i*) Brake pedal touches the foot board continuously — (*i*) Correct the pedal lift. (*ii*) Rebore the drum. (*iii*) Some drivers also keep foot on brake pedal, they should change their habit.

QUALITIES OF A GOOD BRAKE FLUID

1. Viscosity should not be affected by temperature.
2. The boiling point should be in the range of 250°–300°C to avoid vapour formation.
3. It should have lubricating property so as to lubricate master cylinder and wheel cylinder.
4. Should not spoil rubber seals of the system.
5. Should not corrode the metals.
6. Should not expire early and should have sufficient storage period, at least 3 years.

Always use brake fluid according to the vehicle manufacturer's recommendations for the vehicle.

TYPES OF SPECIAL HIGH POWER BRAKES

1. Vacuum Servo Brakes

A mechanism which adds to the driver's effort in applying brakes is known as servo-mechanism. In large and heavy vehicles, driver's foot effort is not sufficient to apply brakes, and therefore a device to increase the braking effort becomes a must.

Out of many servo mechanisms, one mechanism is called vacuum brakes which use power of vacuum, created by engine suction, for braking purposes. The Fig. 11.17 shows a vacuum servo brake system in a simple line diagram. The system consists of a small vacuum reservoir which is connected, through a non-return check valve, to the inlet manifold between the carburettor or fuel injectors and the air cleaner. The vacuum reservoir is also connected to servo cylinder to provide sufficient vacuum for repeated brake applications even after the engine has stopped.

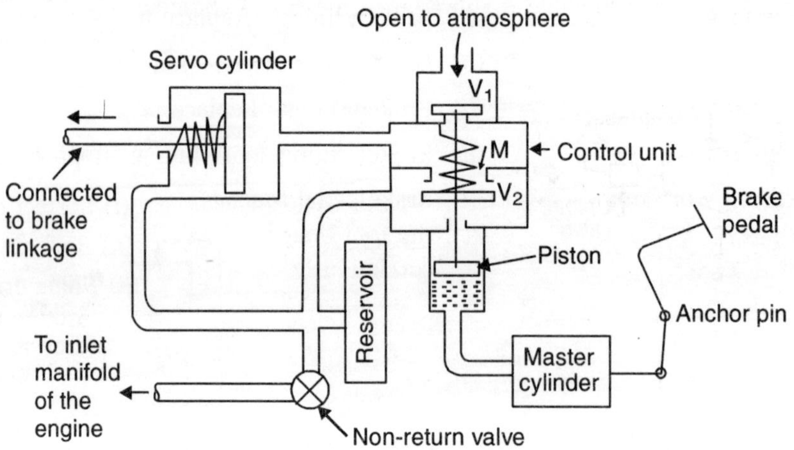

Fig. 11.17 Vacuum servo brakes

In a servo cylinder containing a spring loaded piston, the engine vacuum acts on both sides of the piston in the brake released position. The push rod of the piston is connected to the brake linkages piping of the mechanical or hydraulic type brakes. The control unit has a piston on whose rod two valves V_1 and V_2 are attached. The valve V_1 controls the connection between the atmosphere and the right side of the piston in the servo cylinder. The valve V_2 controls the connection between the vacuum reservoir and the left side of the servo piston. The piston of the control unit is actuated by the brake fluid forced through master cylinder by pressing brake pedal.

Since servo cylinder and control unit are connected to suction line or the inlet manifold of the engine, most of the air is sucked out from reservoir, Servo cylinder and the control unit. The valve V_1 remains closed under atmospheric pressure in normal condition and valve V_2 remains open. When brake pedal is pressed, the master cylinder forces the brake fluid to lift the piston in control unit. The valve V_1 lift up to open the passage for atmospheric air to enter in the upper part of control unit and act on the right side of the piston in the servo cylinder. The valve V_2 closes the middle passage M. Thus the left side of the servo piston remains connected with vacuum whereas the right side is subjected to atmospheric pressure. The servo piston quickly moves towards left side and the piston rod pulls brake linkage to apply brakes in the wheels. When the pedal is released, the fluid returns to master cylinder, the piston in the control unit moves downwards closing valve V_1 and opening valve V_2. When valve V_1 closes, the right side of the piston again gets connected to vacuum and the servo piston moves back to its position by spring and thus the brakes are released.

2. Electric Brakes

Electric brakes powered by battery current are also being used.

3. Air Brakes or Pneumatic Brakes

Air brakes are used on heavy vehicles such as buses and trucks. Air brakes provide heavier braking effort than the effort exerted by driver's foot pressure. These brakes are actuated by compressed air acting over a flexible diaphragm in the brake chambers.

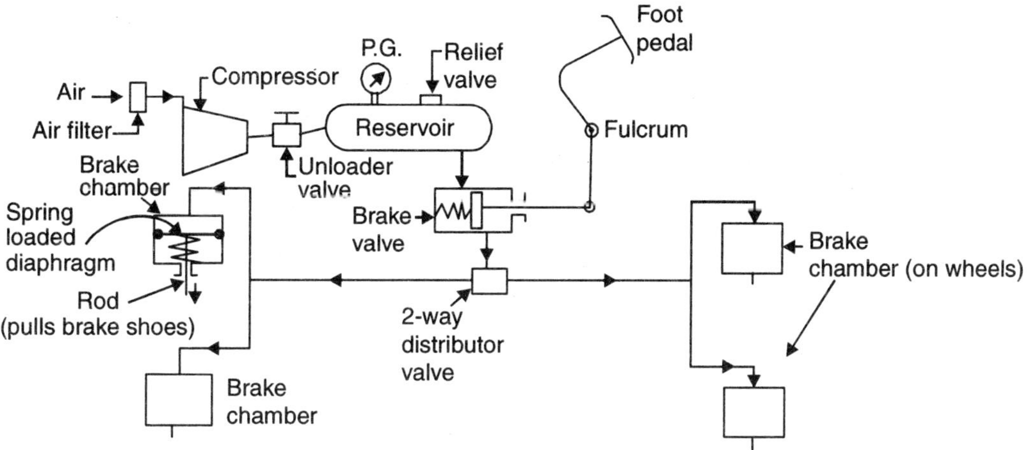

Fig. 11.18 Air brakes

The air brake system has a compressor which is driven by engine. It supplies compressed air. The compressed air is filled in a reseroir. Compressor outlet is fitted with an unloader valve which regulates the air pressure in the reservoir. When air pressure in the reservoir increases the prescribed maximum limit (about 900 kPa), the unloader valve relieves the compressor by discharging air to atmosphere. The reservoir is connected to a brake valve which is actuated by brake pedal. The compressed air (in the range of 700–900 kPa) from brake valve reaches at brake chamber fitted with spring loaded diaphragm. The air exerts pressure over the diaphragm, cupping it downwards. The diaphragm rod, attached to a cam, pushes the brake shoes apart to touch the brake drum and such the brakes are applied.

A relief valve for safety purpose is also fitted on the reservoir. The pressure gauge (P.G.) indicates the air pressure in the reservoir. Engine exhaust gas brakes are also used as an auxiliary brake on vehicles running in hill areas. They are used while the vehicle is comming down hill on slopy roads for slow downing the speed continuously when passing through heavy traffic.

Trouble-shooting chart for air brakes

Complaint	Cause	Remedy
1. Compressor does not unload	Unloader valve not working properly	Dismantle the unloader valve. Clean it and check for fault.
2. Pressure builds up very slowly	(*a*) Choked air filter	Check the air filter, clean it, and refit. If it is spoilt, replace it with a new one.
	(*b*) Leakage in the pipe line between the compressor and the reservoir or between the reservoir and the trailer brake (less effective brake)	Find out and rectify the points of leakage
	(*c*) Defective unloader valve	Check for escape of air; if it is present, the unloader should be dismantled, cleaned and rectified properly.
	(*d*) Faulty compressor	Dismantle the compressor, overhaul it and check the valves and springs.
	(*e*) Leaking past brake valve	Overhaul the brake valve.
3. Brake application is very slow	(*a*) Stuck brake control valve	Dismantle the brake control valve and overhaul it; see that the piston is moving freely.
	(*b*) Leaking brake cylinder	Overhaul the brake cylinder.
	(*c*) Worn out brake linings	Replace or adjust the brake linings.
4. Brakes work inefficiently even though the reservoir pressure is sufficient	(*a*) Wrongly adjusted brake linings	Replace or adjust the brake linings.
	(*b*) Worn out drums	Check the worn out parts of the drum; if necessary, true it
	(*c*) Incorrectly adjusted brakes	Adjust the brakes correctly
	(*d*) Leaking brake lines	Test for leakage; rectify the leakage if it is found out.
5. Brakes bind, *i.e.*, they grip, but do not release the brake drum even after braking process	(*a*) Brakes not adjusted properly	Adjust the brakes properly. Also check the brake diaphragm.
	(*b*) Defective brake valve	Overhaul the brake valve and then fit it.

4. Power Brakes

In the modern age of fast advancement, the speed of vehicles is being increased alarmingly. The power brakes give additional safety factor. Servo assisted power brakes provide instant stopping with least pressure on the brake pedal. Application of low effort on pedal also reduces driver's fatigue.

The system which assists the driver for swift braking with least effort, may be either compressed air or vacuum. When driver applies brakes, the compressed air or vacuum supplies greater part of the braking effort for powerful braking with negligible stopping distance. Vehicle with power brakes always has a caution at rear side—'Power brakes, keep distance' for the safety of vehicles following it. Air brakes, air hydraulic brakes, vacuum brakes and electric brakes come under the category of power brakes.

ADJUSTMENT OF BRAKES

The adjustment of brakes becomes necessary whenever the driver feels any type of defficiency. The repair may be divided in two categories:

(*i*) *Minor repairs or adjustments*: It is done for compensating the brake lining wear and is done without removing the wheels. The wear adjustment is done by adjusting a screw or cams. To check the proper running clearance between brake shoes and the drum, the wheel is to be lifted by jack. All the wheels must have same clearance between brake shoes and the wheel drums or the clearances of the pistons of disc brakes, otherwise, the vehicle will pull to one side on braking. The mechanical linkage used for parking and emergency brakes, should also be adjusted whenever the service brakes are adjusted for brake shoe lining wear. Hydraulic brakes may need bleeding if air is trapped in and proper fluid level is also to be maintained in the reservoir.

(*ii*) *Major repairs*: In this process, all the wheels are removed and the condition of brakes is checked throughly.

- (*a*) Worn-out lining needs replacement. Lining may be glazed or oil soaked or rivets may be protruding. Should be cleaned and repaired.
- (*b*) Brake drum may be showing signs of scoring of the surface, wear, distortion and ovality. The worn out drum has to be slightly skimmed on lathe machine, but maximum permissible limit should not exceed.
- (*c*) Inspect hydraulic system—Wheel cylinders, pipe lines for leakages and the working of master cylinder etc.
- (*d*) Brake shoe springs, operating cam, adjuster units etc. should also checked and repaired or replaced.
- (*e*) *Pedal free play* must be adjusted. Pedal free play ensures complete return of piston of the master cylinder and wheel cylinders. If piston of master cylinder does not return to its position, the by-pass port may be blocked and create hurdles to the returning fluid from the brake lines.

QUESTIONNAIRE

1. (*i*) Why a good braking system is necessary on automobiles?
   (*ii*) Describe various braking requirements.
2. How many types of brakes are in use on auto-vehicles?
3. Describe any mechanical brake system.
4. Differentiate between drum type and disc type brakes.
5. Describe various types of disc brakes.
6. Describe the main components of a hydraulic brake system with the help of a neat sketch.
7. 'Hydraulic brakes are better than mechanical brakes'. Justify the statement.
8. Describe the working of a master cylinder with a neat diagram.
9. (*i*) What is the advantage of using Tandem master cylinder?
   (*ii*) How does the Tandem master cylinder work?
10. What are parking brakes?
11. Describe an Air-brake system with suitable diagram.
12. Explain the following:
    (*i*) Braking force (*ii*) Weight transfer (*iii*) Stopping distance.
13. What is brake efficiency? On what basis does it determined?
14. Explain the following:
    (*i*) Bleeding of brakes (*ii*) Brake dip (*iii*) Two leading shoe brake.
15. Give a broad classification of brakes.
16. Describe a vacuum servo brake system.
17. What are the functions of the following in a vacuum brake system:
    (*i*) Unloader valve (*ii*) Control units (*iii*) Vacuum reservoir.
18. Write short notes on:
    (*i*) Tyre adhesion (*ii*) Thermal aspects in brakes.
19. What are the qualities of a good brake system? Why 100% brake efficiency is not desirable?
20. Describe various retarding forces acting on a vehicle besides the braking force.
21. Describe various types of adjustments done on used and less effective brakes.
22. (*i*) What is free play of the brake pedal? Why is it necessary?
    (*ii*) Describe with sketch, the working of a 2–piston wheel cylinder of hydraulic brakes.
23. Give a brief description of brake troubles and their remedies.
24. What do you understand by power brakes? What are their advantages?

CHAPTER 12

Front Axles and Steering System

FRONT AXLE

In rear wheel drive vehicles, the front axle does not rotate or transfer power of engine to the wheels. It is fitted with wheels through stub axles at its ends. About 40% of vehicle weight is transmitted to the wheels through the front axle. The steering and braking linkages are also supported on this axlc. The rigid front axle is shown in Fig. 12.1 (a). It is also called **Dead Axle** as it does not rotate.

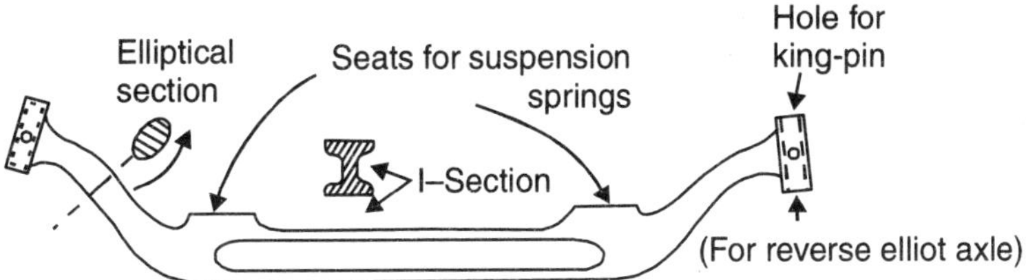

(a) Front rigid axle in rear wheel drive

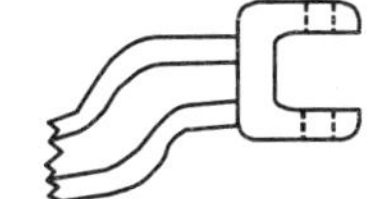

(b) Front rigid axle end for elliot stub axle

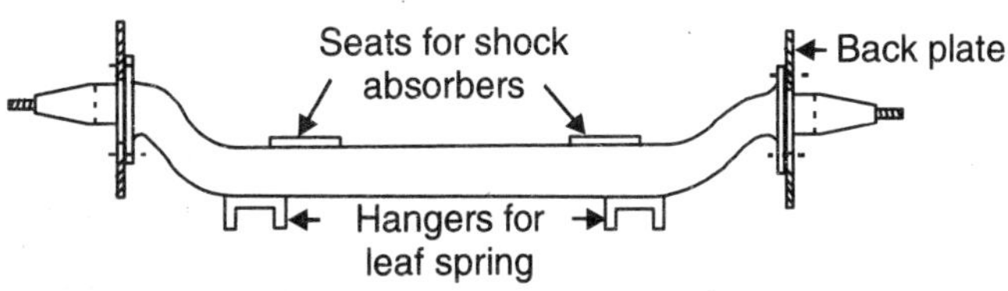

(c) Rear rigid axle in front wheel drive

Fig. 12.1 Rigid axles

It is used with leaf spring suspension on buses and trucks and in old models of cars. It is made in I-section to successfully resist bending due to vehicle load. The ends are made hollow to carry king-pins for stub-axle mountings. Stub-axle assembly facilitates steering and the wheel is mounted on it. Figure 12.1 (a) shows axle end for reverse Elliot stub axle and Fig. 12.1 (b) shows axle end for Elliot stub axle. The middle portion is made with down sweep to keep a lower height of the chassis. A lower height of chassis provides better directional stability and good braking. One spring seat is made near each end to support leaf springs. Front axles are made of carbon steel by drop forging process.

In front wheels drive vehicles, a TRANSAXLE is attached to the engine. Transaxle is a combined gear box and differential with a clutch on manually shifted transmission. Refer to Fig. 5.9 in chapter 5. In front wheels driven vehicle a rigid axle is used on rear side whose diagram is shown in Fig. 12.1 (c). It has fixed stub axles as there is no movement of wheels for steering purposes. Provision is also made for seating leaf spring suspension and shock-absorber. At both ends, back plates for brake-shoes and stub-axles for mounting wheels are fitted.

Front rigid axle, stub-axle and wheel assembly is shown in Fig. 12.2.

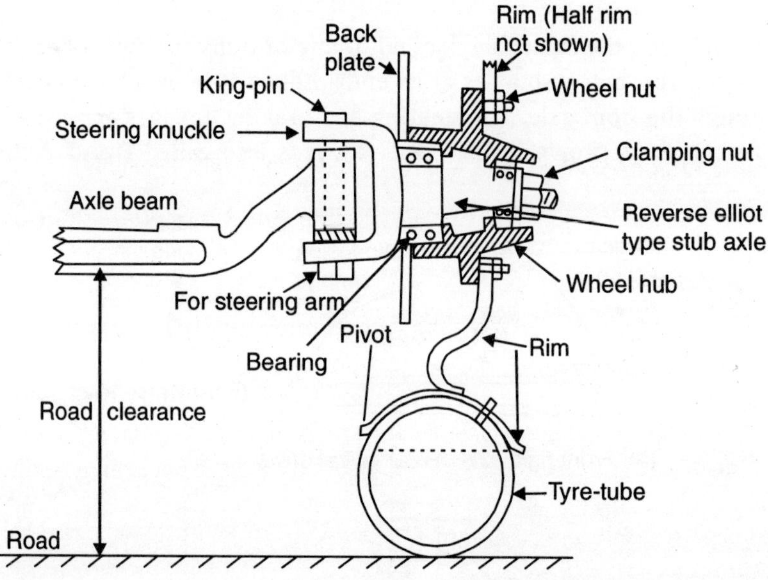

Fig. 12.2 Front rigid axle, reverse elliot stub axle and wheel assembly

Functions of Front Rigid Axle

1. Takes on load of vehicle (about 40%).
2. Bears braking torque.
3. Supports suspension system.
4. Carries steering linkage.
5. Stub axles and front wheels are mounted on it.

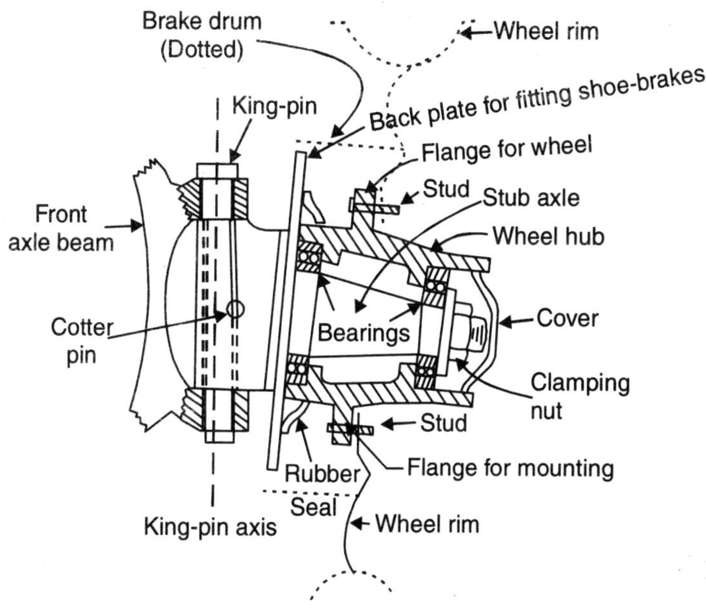

Fig. 12.3 Elliot type stub axle and front wheel assembly

A detailed diagram of assembly of wheel, elliot type stub axle and front solid axle beam is shown in Fig. 12.3. The wheel hub for mounting wheel rim is fitted firmly on stub axle and is supported on two bearings.

The wheel hub over stub axle is firmly held by clamping nut. The wheel rim with tyre-tube assembly is fitted on the four studs projecting out of the flange disc. The wheel hub freely rotates on the stub-axle. The back plate is fitted with brake-shoes assembly and is covered by wheel drum. Just behind back plate, an attachment is given for steering arm. The back plate does not rotate, it is fixed at the root of the stub-axle. The above assembly is used on rear wheel driven vehicles.

STUB AXLES

In rear wheel drive vehicles, the front rigid axle's both ends are fitted with stub axles by means of a ball joint or a king-pin. In Fig. 12.2, a front rigid axle using king-pin has been shown.

The wheels are mounted on stub-axles with ball or roller bearings interposed between the two. Various types of stub axles are used on automobile vehicles. Four types of stub axles used on Rigid axle (Front) are given in Fig. 12.3. The stub axle can swing around king-pin to facilitate steering. The king-pin holes in axle are fitted with bronze bushes. The king-pin's longitudinal movement is prevented by inserting a cotter pin in a slot cut on king-pin. A thrust washer is also placed at the load-carrying side between axle end and the stub-axle.

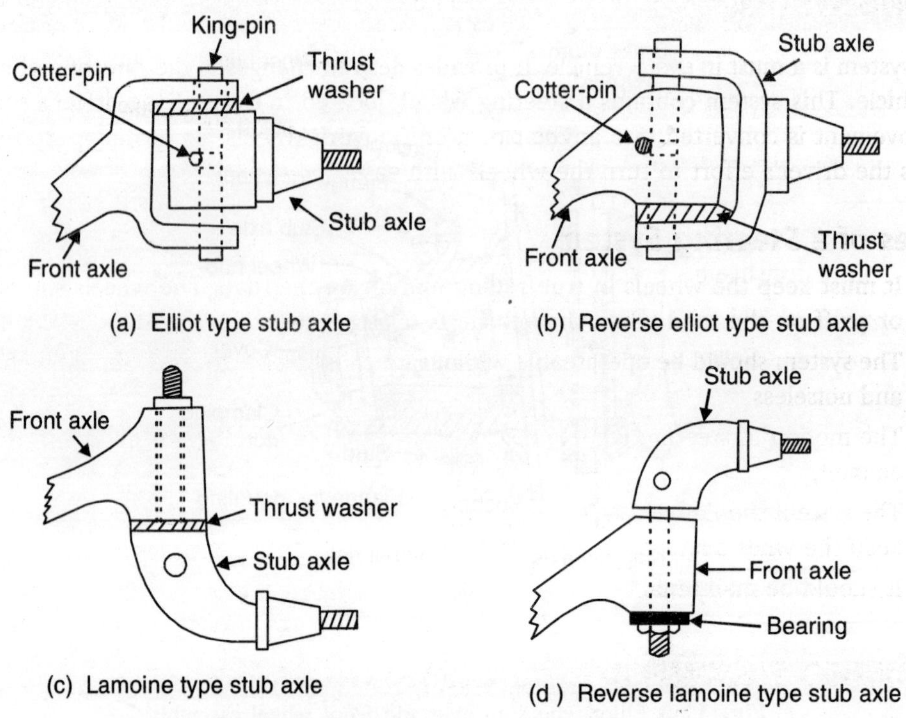

Fig. 12.4 Different types of stub axles

The reverse elliot type stub axles are commonly used because of its simple design and strength. Lamoine type stub axles are employed on front axles of tractors. Modern vehicles employ a ball joint type stub axle, particularly on independent suspension. Ball joint eliminates king-pin and provides swift and easy swinging movements for steering purposes. Ball joints supported stub axle is shown in Fig. 12.5.

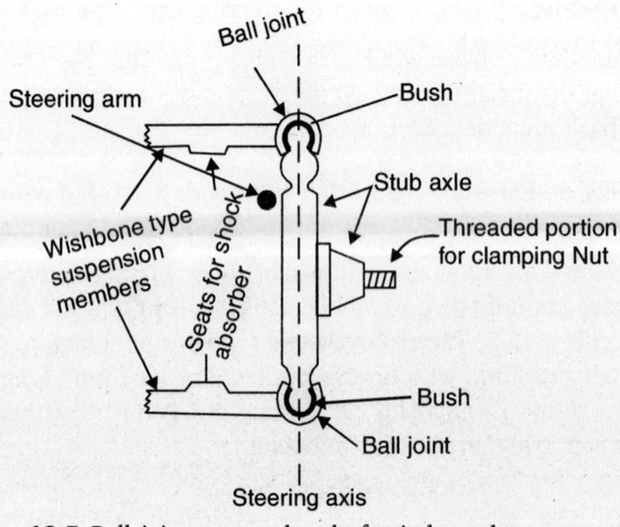

Fig. 12.5 Ball joint type stub axle for independent suspension

STEERING SYSTEM

Steering system is a must in every vehicle. It provides desired changes in the direction of movement of the vehicle. This system contains a steering wheel, located in front of the driver's seat, whose rotary movement is converted into angular turn of the front wheels for turning purposes. It also multiplies the driver's effort to turn the wheels with ease.

Qualities of a Steering System

(*i*) It must keep the wheels in true rolling motion on the road. The wheels should not rub or scuff on the road when the vehicle is taking a turn.

(*ii*) The system should be operateable without much effort by driver. It should be frictionless and noiseless.

(*iii*) The motion of steering wheel must be transferred to wheels very quickly in the desired amount.

(*iv*) The system should have a certain degree of self-centring action or self-righting torque to keep the wheels straight after turning is over.

(*v*) It should be maintenance free.

FRONT WHEELS GEOMETRY OR STEERING GEOMETRY

Wheel Alignment: This term is applied to such a proper setting of front wheels and steering system that provides an easy directional stability to the vehicle and a good steering control. If proper wheel alignment is there, there will lesser tyre wear, smooth turning when negotiating a curve and during straight running, the front wheels must run rolling exactly in parallel lines.

Front wheel alignment depends on the following factors—which are the part of the steering geometry. It is also termed as **Front End Geometry.**

1. Toe-in and Toe-out

***Toe-in*:** In the plan view of front wheels, the wheels are kept usually drawn inward slightly towards front side so that the distance between the front ends of wheels is slightly smaller than the distance between the back ends of wheels as shown in Fig. 12.6 (a).

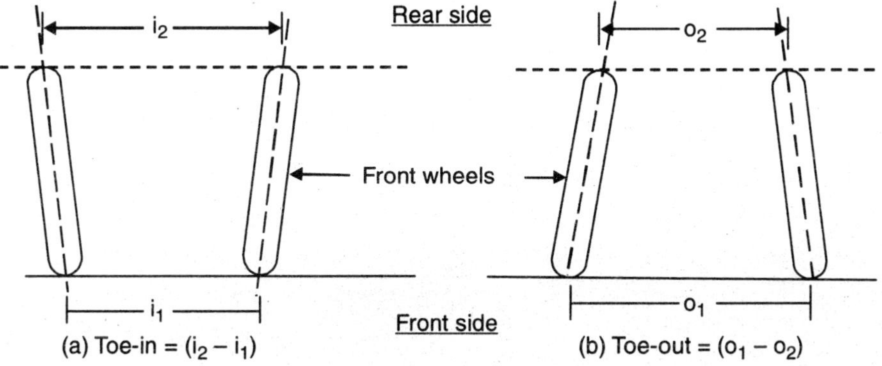

Fig. 12.6 Setting of front wheels at no load and at rest

The difference between these two distances is called Toe-in. The actual amount of toe-in is a fraction of a centimeter (1.5 to 6 mm). When vehicle with load is moving forward, there is a tendency of front wheels to move-outwards at their stub-axles under the effect of loading. The toe-in on the front wheels is provided to check this tendency and it ensures parallel rolling of the front wheels under loaded condition. It provides stability to steering, prevents side slipping and lowers the tyre-wear. It is provided on all vehicles except tractor which has toe-out for properly working in fields.

***Toe-out (on turns)*:** To prevent tendency of wheels to drag during turns, the toe-out is necessary. On turns, inner wheel travels a lesser distance than outer wheel, therefore, inner wheel assumes a larger angle than outer wheel. These different turning angles result in a toe-out of front wheels on turns. Generally, for 90° turn, the inner wheel turns at 23° (max.) to the frame centre line while the outer wheel takes an angle of 20° (max.) and such their axes are toe-out by (23–20) = 3° apart. Some car manufactures may provide different specifications such as 20° and 18° respectively. *For perfect steering, the centres of Arc A and B must coincide with each other and the common centre 'O' must lie on the line passing through the centre line of the rear wheels.*

The toe-out also counters the tendency of inward rolling of front wheels due to (a) the condition of soil on agricultural land (Tractors are provided Toe-out in place of Toe-in due to this reason), (b) side thrusts and cross-wind effects on vehicle body.

If the specified angles of front wheels are true, the car will run smoothly and properly both in straight driving and on turns. When a vehicle is taking a turn; suppose towards left side, as shown in Fig. 12.7.

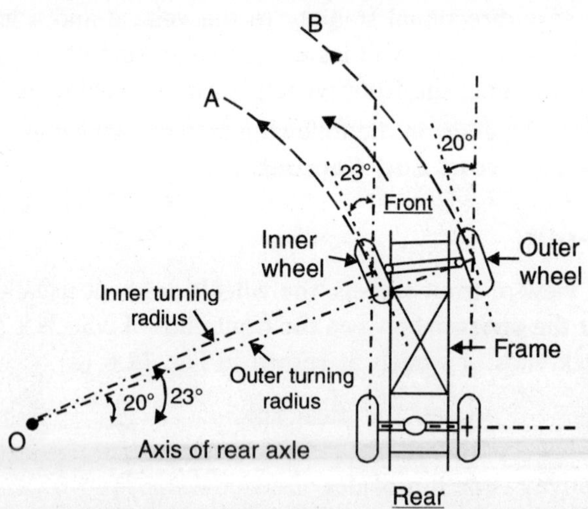

Fig. 12.7 Toe-out on turn

The inner and outer both wheels will move along arcs A and B. The centre of arc A of inner wheel path is at point O which lies on the line passing through axis of the rear axle/rear wheels. Point 'O' is called *instantaneous centre of rotation*. The angle of inclination of the inner wheel from vertical is 23°. At the same position, the outer wheel is following a larger arc 'B'. For proper steering and rolling of wheels, the centre of Arc A and B must coincide at common point O, lying at centre line of rear wheels extended towards left (in the direction of turn). The distance, seen from front side, between inner and outer wheels is greater than the distance at rear side of front wheels. This is the actual position of Toe-out, correctly known as **Toe-out on turns.**

This toe-out on turns helps smooth turning and the king-pins do not experience any unwanted stresses or friction.

2. Camber or Wheel Rake

Camber is the inclination of the front wheels from the vertical. The inclination angle should not exceed 2°. The camber is positive if the tilt is outwards at the top of wheel. It is termed as negative, if the tilt is inward. The camber, as shown in Fig. 12.8 is provided to prevent tilting of wheel top inwards when the vehicle is loaded excessively or there is play in wheel bearings or in king-pins. The camber in both front wheels is kept equal. The larger amount of camber will cause excessive tyre wear on sides—on outer side in positive camber and on inner side in negative camber. Commonly positive camber is given on most of the vehicles. Negative camber has been provided on Honda 4-wheel drive car and some models of Fiat 1100 car.

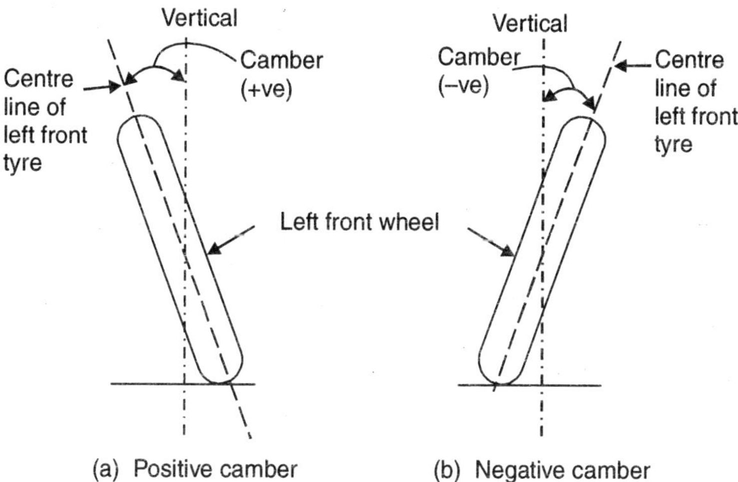

Fig. 12.8 Camber angle

3. Castor Angle or Castor

It is the angle (which ranges from 2° to 7°) between the king-pin centre line (or steering axis) and the vertical, in the plane of wheel. The castor is shown in Fig. 12.9. If axis of king-pin meets the ground ahead of the wheel centre line, the castor is positive. If it meets behind the vertical wheel centre line, it will be negative castor. It provides directional stability, prevents wheel wandering and shimmy. It helps to pull wheels forward and balances the tendency of front wheels to toe-in.

Uses of Castor

(*i*) To maintain directional stability and control.

(*ii*) To reduce steering effort.

(*iii*) To facilitate the steering returnability (easy return of front wheels in forward direction after the turning is over).

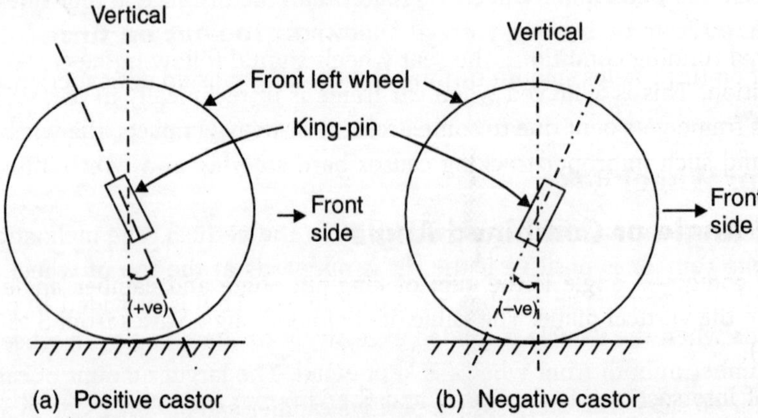

Fig. 12.9 Castor angle (seen from beneath the frame)

4. King-pin Inclination and Steering Axis Inclination

The inclination of king-pin from vertical is called the king-pin inclination. It is also termed as **King-pin rake.** If the king-pin is allowed to remain vertical, the cambered wheels would exert hard steering.

The modern vehicles employ ball joints in place of king-pin. Ball joint axis also called steering axis, is an imaginary centre line drawn through the centres of the upper and lower ball joints. The steering axis inclination is the angle between the ball joint axis and vertical.

The two inclinations are shown in Fig. 12.10. The king-pin axis or ball joint axis is slightly inclined inwards to keep the front wheels always pointing forward. It also helps in bringing back the wheels in straight forward position after negotiating a turn.

The value of king-pin angle varies from 3° to 9° and that of steering axis angle from 5 to 13°.

Vauxhall Carlton car employs a negative king-pin inclination where as, Maruti 800 car has 12°50′ king-pin angle.

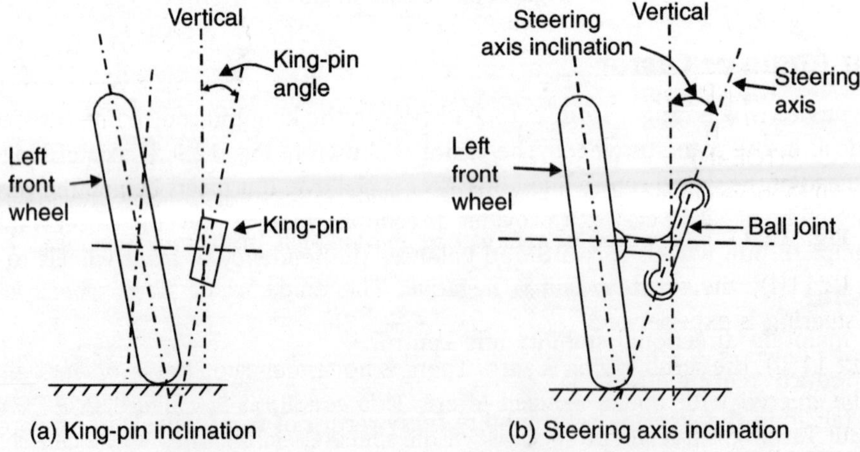

Fig. 12.10 King-pin and steering axis angles

5. Tracking

Under normal running conditions, the rear wheels should follow in the tracks of front wheels in a parallel position. This is achieved when car frame is in technically in perfect shape and size. When the chassis frame gets bent due to some accident or heavy impacts, the wheels do not follow the same track and such improper tracking causes hard steering and tyre scuffing on road.

6. Included Angle or Combined Angle

Included or combined angle is the sum of king-pin angle and camber angle of wheel. This angle is formed in the vertical plane. The value of combined angle varies from 5 to 12°. It is shown in Fig. 12.11 (ii).

The point of intersection of wheel axis and the king-pin axis may occur at three positions:

(*i*) Above the ground, (*ii*) At the ground, (*iii*) Below the ground.

All the above three situations are shown in Fig. 12.11 (i), (ii) and (iii).

The tractive force, through king-pin axis, acts at point 'K' and road resistance at tyre acts at point T. C. is the point of intersection of the two axes. The distance between these two points K and T is known as SCRUB RADIUS (in m.m.). Scrub radius = KT (–ve) or TK (+ve) as shown in Figs. 12.11 (i) and (iii) respectively.

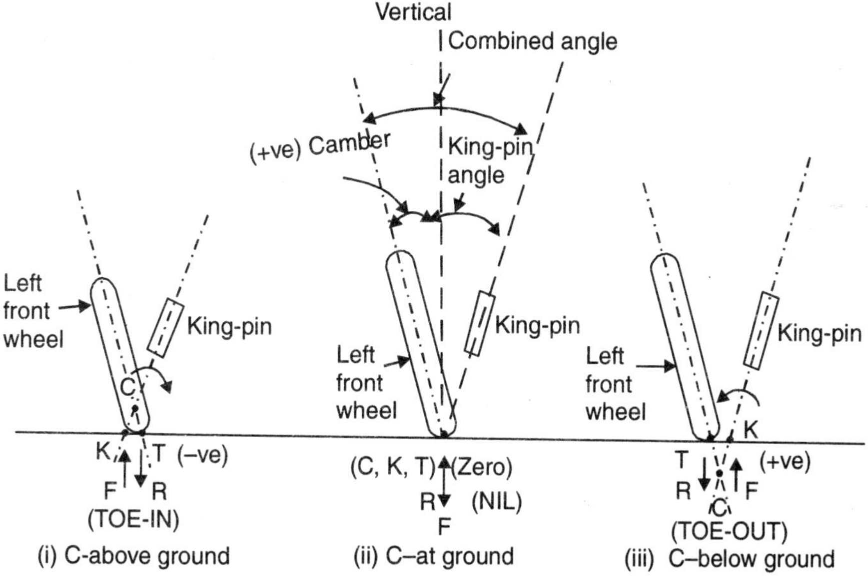

Fig. 12.11 Intersection of axes of tyre and king-pin (seen from front side)

In Fig. 12.11(i), the scrub radius is negative. The tendency of front wheels is to toe-in. Difficulty in steering is experienced.

In Fig. 12.11(ii), the scrub radius is zero. There is no tendency of Toe-in or Toe-out. The front wheels can be steered with much reduced effort. This condition is called TRUE CENTRE LINE STEERING. But Turning is not smooth because of the speed effect on tyres which causes the wheels to scrub and such the hard steering is experienced. Tyre wear also increases.

In Fig. 12.11(iii), the scrub radius is positive and a tendency of toe-out is produced in the front wheels. It is considered desirable.

At point 'K', tractive force (F) acts in the upward direction. At point 'T', Road resistance (R) acts in the downward direction. Both F and R form torque of opposite nature—clockwise in negative scrub and anti-clockwise in positive scrub radius. Clockwise torque produces tendency of Toe-in where as anti-clockwise torque produces Toe-out.

Scrub radius of 8 to 12 mm is desirable. Positive scrub radius is preferred as it require lesser torque to turn the wheel. Larger (+ve) scrub radius is also not desirable as it increases load on steering linkage and suspension system.

How to Check Wheel Alignment?

The sequence of checking wheel alignment is as follows:

1. First, check king-pin inclination or steering axis inclination.
2. Secondly, check camber angle on front wheels.
3. Now check the caster angle of front wheels.
4. In the last, check toe-in or toe-out.

King-pin angle, camber and caster are checked by an equipment called turn-table, level and gauge. The toe-in is checked by optical toe-in gauge.

ACKERMANN MECHANISM

As we have already seen in case of toe-out at turns that for perfect steering, all the wheels of the vehicle must rotate about a point, called instantaneous centre, located at the common intersection of tilted front wheels arcs and the line passing through the centre line of rear wheels axle. To achieve this set-up, two types of steering mechanisms are employed—

(*a*) Ackermann Mechanism—Universally employed.

(*b*) Davis Mechanism—Not in use now.

The Ackermann mechanism is universally employed on almost all types of automobiles because of its simplicity in construction and easy and low maintenance.

An Ackermann mechanism is a four-bar mechanism and is shown in Fig. 12.12. The two steering links are firmly attached to stub-axles over which front wheels are mounted. These two links are connected with each other through the track rod, CD.

When vehicle is moving straight, these links make equal angles 'α' with the centre line of vehicle frame in the vicinity of rear universal joint for achieving perfect rolling without slipping of any of the 4 wheels. In this straight running condition, the ball joints at C and D, are equi-distant, a, from verticals drawn at A and B respectively (shown by dotted lines).

When the vehicle negotiates a turn, say towards left side, the new configuration of links AC, BD and CD has been shown by dotted lines. The radii of both arcs OA and OB, intersect each other at point 'O' lying at the line passing through centre line of rear wheels axle.

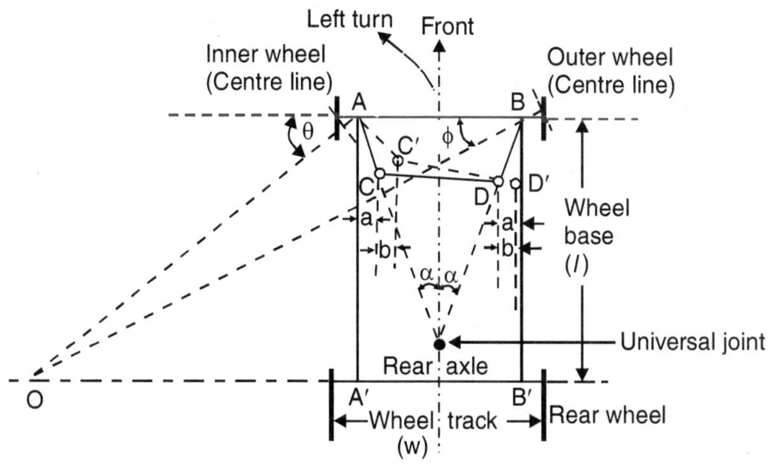

Fig. 12.12 Ackermann mechanism

If we neglect obliquily of the track rod, CD, in the turned position, the movements of the joints C and D in horizontal direction may be taken as equal, say 'b'. The radii of arcs OA and OB make turning angles θ and ϕ respectively at O with a horizontal line passing through front wheels axis.

(*a*) In ΔAOA′, $\cot\theta = \dfrac{OA'}{AA'} = \dfrac{OA'}{l}$ where l = AA′ = wheel base length

(*b*) In ΔBOB′, $\cot\phi = \dfrac{OB'}{BB'} = \dfrac{OB'}{l}$

$$\therefore \quad \cot\phi - \cot\theta = \frac{OB' - OA'}{l} = \frac{A'B'}{l} = \frac{w}{l}$$

or
$$\cot\phi - \cot\theta = \frac{\text{Wheel Track}}{\text{Wheel Base}} \qquad ...(1)$$

The Ackermann mechanism provides three positions for correct steering:

(*i*) When θ and ϕ are zero. The vehicle will run straight.

(*ii*) Turning to left at some particular values of θ and ϕ, satisfying the equation (1).

(*iii*) Turning to right at some particular values of angles θ and ϕ satisfying the equation (1).

Generally, values of turning angles are taken as:

$$\theta = 23°, \ \phi = 20° \text{ and } \alpha \simeq 15°.$$

and value of link AB = 0.85 to 0.90 of wheel track (w)

and length of steering arm AC or BD $\simeq \dfrac{1}{4}$ of AB.

On account of improvements in suspension system and tyres, small deviations of turning angles and links of Ackermann mechanism are now possible. Some vehicle take θ as 20° and value of ϕ varies in between 18–19°.

It is easier to steer a vehicle in reverse gear than in forward gear because the rear wheels turn on smaller radii than the turning radii of front wheels. Any small deviation from the true turning angles for perfect rolling of wheels, can readily be adjusted by tyres side wall flexibility and tread distortion.

Turning Radius: When a vehicle negotiates a turn, the radius of the circle on which the outer front wheel moves is termed as Turning Radius. It is usually proportional to wheel-base of the vehicle. It's extreme value ranges from 4.5 m to 8 m for passenger cars and about 14 m for heavy duty vehicles like trucks and buses; on motor cycles it lies between 2 to 4 m.

STEERING LOCK AND WHEEL LOCK ANGLES

When a front wheel is fully turned, it is turned by different angles in left or right direction with respect to its straight position. Figure 12.13 (a) shows a left front wheel's maximum movement in left side (θ°) and in right side (ϕ°).

Thus, a front wheel can be turned by a maximum angle of (θ + ϕ) before its motion is constrained. This angle is known as **Wheel Lock Angle.** In Ackermann linkage, its value is 20° + 23° = 43° (Fig. 12.13(b)), the maximum turning of left front wheel towards left (23°) during left side turning and maximum turning of front right wheel (23°) during right side turning. Thus, it is clear that the two front wheels will be locked when each of them is turned by 23° outside from their straight position. The combined angle of both front wheels (23° + 23° = 46°) is called **Steering Lock Angle** [Refer to Fig. 12.13(b)].

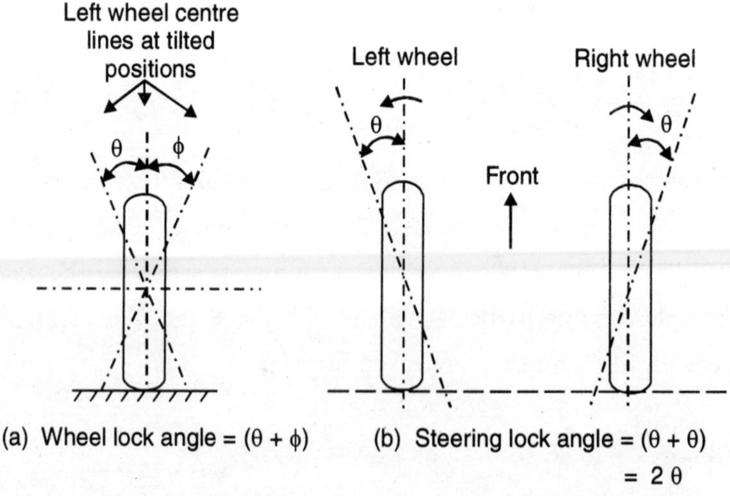

Fig. 12.13 Wheel lock and steering lock angles

Slip Angle: When a vehicle takes a turn at a curvature of radius '*r*', a centrifugal force acts on the wheels which produces a side thrust acting at the centre of wheel. The magnitude of the side thrust is same as that of centrifugal force.

$$\text{The centrifugal force} = \frac{W}{g} \times \frac{v^2}{r}$$

where W is the weight of vehicle and v is its velocity.

The side thrust causes the flexible tyre to distort slightly on turning side. The wheel then follows a path which is at an angle to its vertical plane of rotation. This angle is known as Slip Angle. It causes a force, called cornering force, to resist the side thrust. The cornering force acts perpendicular to the plane of wheel as shown in Fig. 12.14. The side thrust and cornering force form a couple of forces which help in bringing back the wheel in the direction of motion. This couple produces a **self-righting torque 'T'** at wheel.

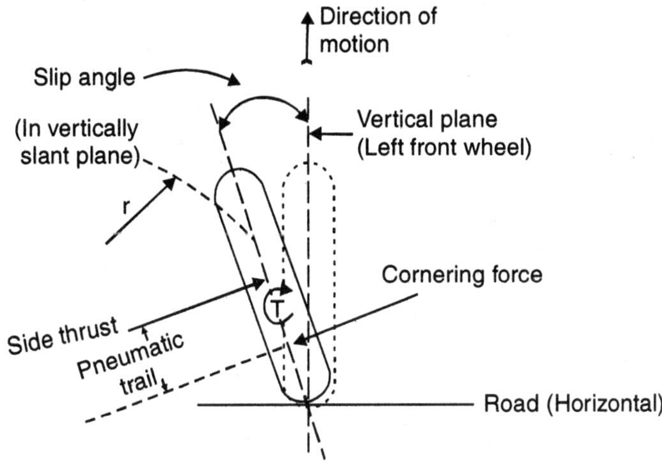

Fig. 12.14 Slip angle and cornering force

CAUSES OF POOR STEERING

1. Backlash in steering gear—should not be too much.
2. Excessive or lesser steering gear ratio (Normal ratio is 12 : 1).
3. Condition of king-pins and their bushings.
4. Play in wheel bearings or worn-out bearings.
5. Inflation pressures in the tyres are an equal or higher/lower than the prescribed limit.
6. Wheel alignment is not correct.
7. Ball joints worn-out.

OTHER TYPES OF STEERING SYSTEMS

1. Scooters and motor cycles are steered by single front wheel directly through handle and fork.
2. In 3–wheelers, the front third wheel is used for turning purposes.
3. On some vehicles a **fifth wheel steering system** is employed. In this system, front axle and wheel assembly is pivoted at centre. A fifth wheel, which is fitted between chassis frame and axle, serves as a pivot or turn-table around which axle assembly may be tilted. The 5th wheel has a ring-gear at its rim and is moved by steering wheel by driver, which in turn moves the front axle and wheel assembly in the desired direction.

Over steer and under steer: Generally both these problems occur on turns.

Over steer: If the vehicle gets steered 'more' than the normal desired steering, it is called 'over steer'.

Under steer: When the vehicle gets steered 'less' than the normal desired steering, it is called 'under steer'. These situations are created by—

(*i*) Slip angle, (*ii*) Cornering force, (*iii*) Inflation pressure, (*iv*) Load on vehicle, (*v*) Suspension springs particularly leaf springs which get flattened under load (*vi*) Camber of road surface this is provided on turns of the roads to help vehicles to move smoothly in their desired direction without side-skiding.

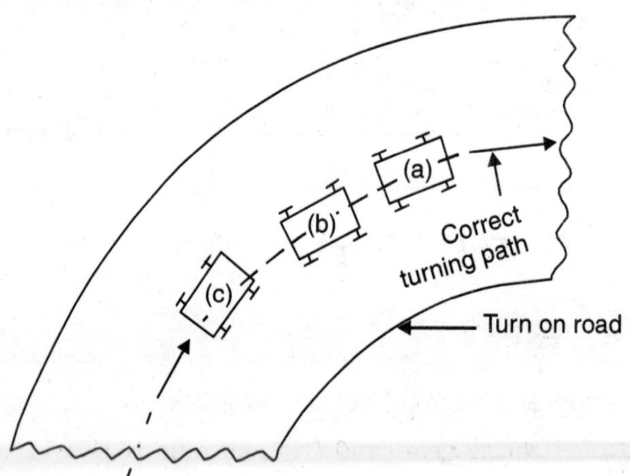

Fig. 12.15 (a) Normal steer, (b) Over steer, (c) Under steer

Vehicle heavily loaded at rear side cause over steer due to greater slip angle and lesser ability of rear wheels to turn than front wheels. Over steer is also produced due to change in effective length of the leaf springs due to load over rigid axle. Over steer causes greater chances of accidents at high speeds on turns especially.

When slip angles of front wheels are more than that of rear wheels (may be due to lower inflation pressure in front wheels than in rear wheels) then radius of turn is slightly increased which cause under steer. The vehicle, under steered, moves away from the normal path of turning. To keep the vehicle on normal path, the steering wheel has to be turned more than theoretically needed. Under steer is lesser undesirable than over steer.

The over steer and under steer may also occur during straight running of the vehicle when it is subjected to side forces due to cross-winds and road camber on straight road.

PARTS OF A STEERING SYSTEM

The steering system of a vehicle is composed of the following parts:

(*i*) Steering wheel

(*ii*) Steering column

(*iii*) Steering gear

(*iv*) Steering linkages

(*v*) Steering knuckle.

(*i*) **Steering Wheel:** It is a round ribbed wheel mounted on steering column in front of the driver's seat. It is attached to steering column or steering shaft. In its centre a push ring is fitted for blowing horn. It may also contain selector lever for controlling automatic transmission on cars having automatic transmission system.

(*ii*) **Steering Column:** It is normally a hollow shaft enclosed in a casing. It carries steering wheel at the upper end and is attached with steering gear at the lower end. In some cars, a gear shifting lever is also provided at its casing. Many designs of steering columns are in use, such as energy absorbing steering column or collapsible steering column etc.

(*iii*) **Steering Gear:** It serves the following functions:

(*a*) It provides mechanical advantage for easy steering of the wheels.

(*b*) It changes the rotation of steering column or shaft into to and fro motion of the pitman arm or drop arm or rack or drag link.

Several types of steering gear systems are employed on various automobiles. Some of the steering gears will be discussed in the coming pages.

(*iv*) **Steering Linkages:** These linkages connect the steering knuckle arms of the front wheels and drop arm of the steering gear. They swing backward and forward by drop arm, which is firmly connected to cross-shaft to actuate the steering knuckles for tilting front wheels. Many types of linkages are in use; a Pitman Arm type steering linkage is shown in Fig. 12.16 for rigid axle and in Fig. 12.17 for independent suspension using wishbone type suspension.

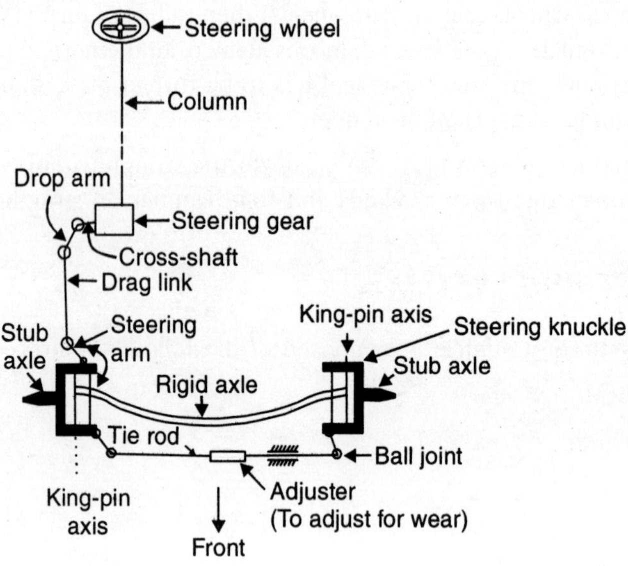

Fig. 12.16 One of the steering linkage for rigid axle

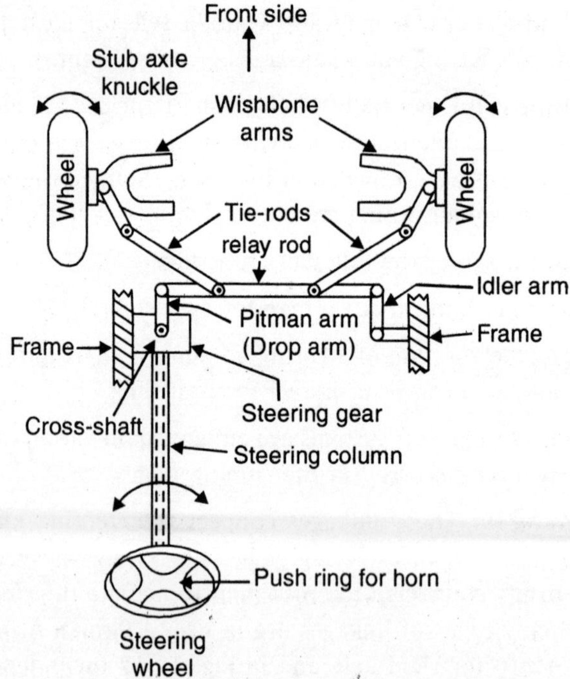

Fig. 12.17 Simple diagram of a steering linkage (Pitman arm type) (or parallelogram-type steering linkage) for independent suspension (front)

(*v*) **Steering Knuckle:** It is also called steering spindle. It is an integral part of the stub axle and moves around the king-pin. The motion to steering knuckle is conveyed by steering arm of the linkage. It turns the front wheels to right or left as desired by driver. It should be very tough and strong enough to bear various stresses developed during steering at high speeds. It is made of 3.5% Nickel steel or low Ni-Cr steel or molybdenum steel.

Reversible Steering Gear, if we turn stub axle, or front wheel, and the steering wheel is also turned—the steering gear used is called Reversible steering gear. If it does not turn, the gear will be called **irreversible**.

Irreversible Steering Gear does not transfer road shocks to steering wheel, making it comfortable to the driver. The irreversible gear employs very less pitch angle of the screw or worm used for driving rack or sector gear. But it makes steering quite heavy.

Present designs of steering gears are almost semi-reversible on all vehicles. Semi-reversible steering is the compromise between the above two types.

Comparison between Different Steerings

Sl. No.	Various Activities	Reversible Steering	Irreversible Steering	Semi-Reversible Steering
1.	Control of steering	Not easy	Difficult	Better
2.	Transfer of road shocks to steering wheel	Fully	Zero	Very less
3.	Load on joints of steering linkage	Less	More	Fair
4.	Fatigue to driver	More	Less	In between the two

Steering Gear Ratio (SGR): It is the ratio of the angle turned by steering wheel to the corresponding turning angle turned by the crosss shaft of the steering gear unit. Pitman arm is attached to the cross shaft. Its value varies from 12:1 to 20:1 for cars and about 35:1 for heavy commercial vehicles.

On an average, about one and a half turns of the steering wheel tilts front wheels by 41° to 45° on either side from centre line. The wheels get locked after about 45° turn usually.

Overall Steering Ratio (OSR): OSR may be defined as the angle turned by steering wheel to the corresponding angle turned by the front wheels. It is higher by about 20% than the value of SGR because, in determining the OSR, all the steering linkages are considered.

(*i*) In small cars 1.5 turns are required to turn the front wheels by 43°.

(*ii*) In diesel Ambassador cars 2.3 turns are required for complete steering lock (43°) of front wheels.

(*iii*) In sports cars 2.7 turns are needed for complete steering lock on either side—left or right.

BALL JOINT

The drop arm of steering linkage is connected to cross-shaft of the sector gear at its upper end and the lower end of the drop arm is connected, through a ball joint, to the drag link or link rod. (see Fig. 12.16). The other end of the drag link is also connected through a ball joint with the steering arm or link rod arm. The steering arm deflects the stub-axle for steering. The ball joints are made in such a way that the expanding upper spring compensates for wear or mis-adjustment in the ball joint. The ball joints permit swivel movement of two joining links without offering any significant friction.

A simple construction of the ball joint is shown in Fig. 12.18. A spherical ball is kept in position by 4-part split socket or four ball cups which act as bush for the ball. Any wear in the ball and socket is continuously adjusted by expansion of upper spring held by circlip. The lower spring takes on varying load induced in the joint during steering process. A dust cover is provided to protect the ball joint from dust, dirt and moisture.

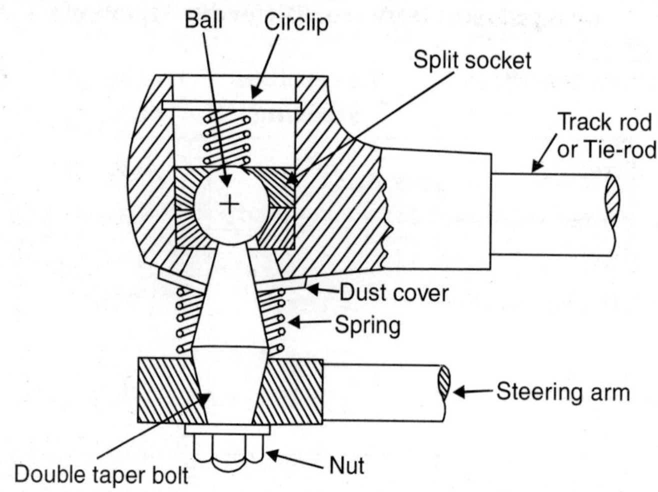

Fig. 12.18 Details of a ball and nut joint

TYPES OF STEERING GEARS

The steering gear is a very important part of the steering system. It performs two main functions:

1. It provides mechanical advantage to the driver and enables him to steer the vehicle smoothly.
2. It changes the rotation of steering column at right angles which in turn is again converted into linear motion of drag link by pitman arm or drop arm.

The steering gear assembly is enclosed in a casing, called steering gear box.

Usually the following type of steering gears are used on various vehicles:

(i) Worm and Sector type

(ii) Rack and pinion type

(iii) Recirculating balls type

Besides above three types, the following gears were also previously employed by some vehicles:

(a) **Worm and Worm-wheel type Steering Gear:** It is not being used now. It is replaced by worm and sector type.

(b) **Cam and Lever type:** It is also not very much popular.

(c) **Screw and Nut type:** It was improved by providing recirculating balls to minimise the friction.

(d) **Worm and Roller type:** It is also not much popular. It was used on Premier Padmini Cars and some Jeeps.

(i) **Worm and Worm Sector Type Steering Gear:** It is similar to worm and worm-wheel type. The worm wheel was replaced by a toothed sector. It is cheaper and smaller in size and easy to install. The construction is shown in Fig. 12.19. The steering column or steering shaft is attached to a worm and a sector gear is meshed with it. The sector gear is mounted firmly on cross-shaft. It is used on some small cars. Worm is a large screw with several threads of various types, used for transmitting motion and power.

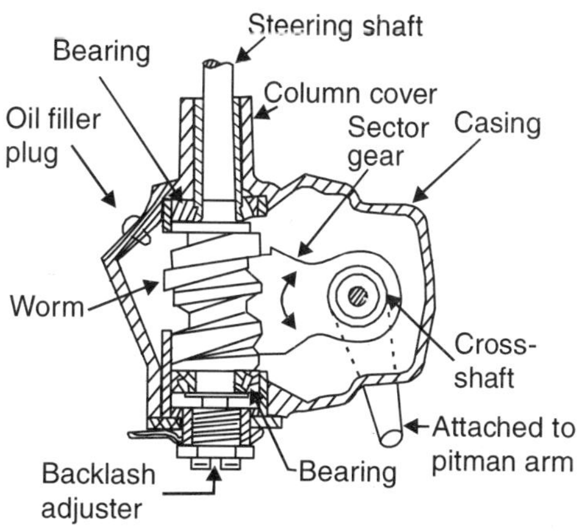

Fig. 12.19 Worm and sector steering gear

(ii) **Rack and Pinion Type Steering Gear:** It is used on quite a good number of vehicles such as Maruti 800, Cielo, Fiat UNO, Zen, Ambassador (Diesel version), Standard Herald etc. A pinion is mounted at the end of the steering shaft and meshed with a straight bar rack with Acme teeth on some portion in it's middle. The rack at both ends is connected through ball joints with tie rods as shown in Fig. 12.20. The tie-rods move steering arms which tilt the stub-axles for steering.

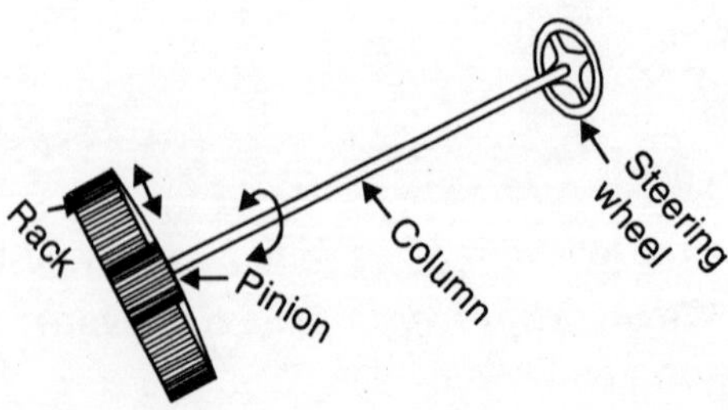

(a) Steering wheel, column and gear assembly (rack and pinion type)

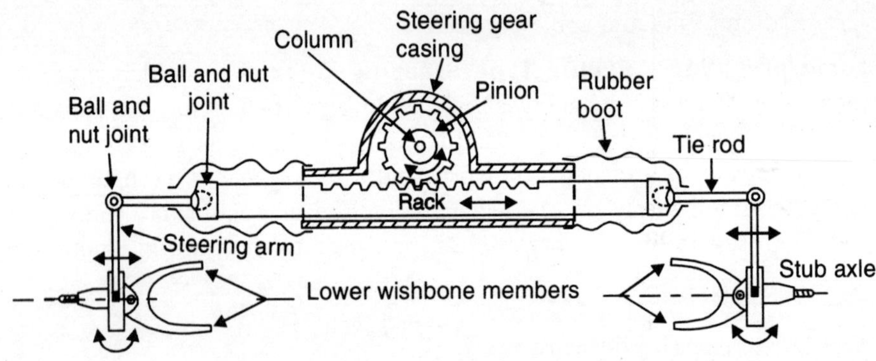

(b) Steering gear and linkage assembly

Fig. 12.20 Rack and pinion type steering gear

(iii) **Recirculating Balls Type Screw and Nut Steering Gear:** It is an improved design of screw and nut type steering gear. The screw is attached to steering shaft. A nut with splines cut on its outer surface moves over the screw. The splines of the nut are meshed with sector gear mounted on cross-shaft. To reduce friction between the nut and screw, a recirculating ball race is provided between them. The balls roll between the screw and nut and as they slide upto the upper end of the nut, the balls enter the return guide tube and then roll back to a lower outlet where they re-enter the groove between the screw and nut. See Fig. 12.21. This gear is employed on Omni car, Tata 407, Tata 1210, Eicher 10.70, Dodge/Fargo 89, M4. The screw and nut gear without balls is used in Swaraj Mazda. It is also used with power steering system.

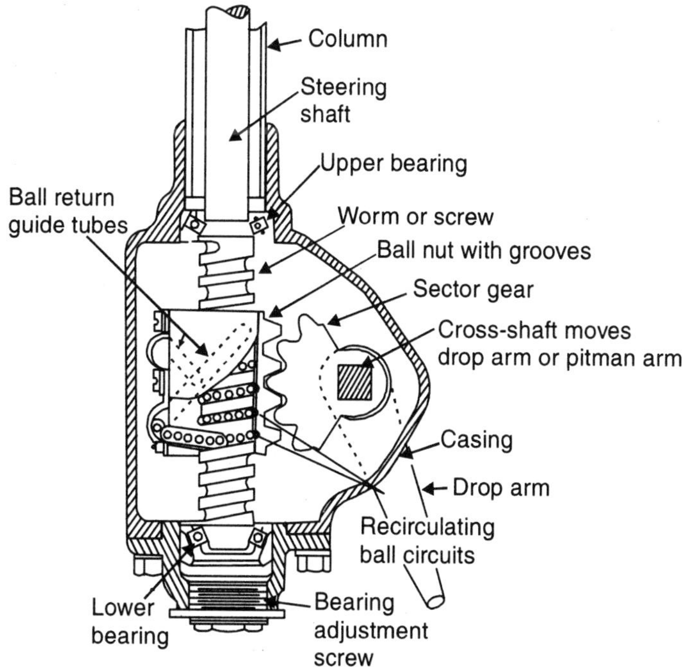

Fig. 12.21 Recirculating ball type screw and nut steering gear

FOUR WHEEL STEERING (4 WS)

Usually two front wheels are involved in steering process. Daihatsu's Terios and Honda 4 WS system steers all the four wheels. When steering wheel is turned, the output pinion shaft for rear wheels is rotated by rack and pinion system, incorporated in the front steering gear box. This rotation is transmitted, through a centre shaft, to an offset shaft fitted in the rear steering gear box which steers the rear wheels as shown below in Fig. 12.22. Thus all the four wheels are steered simultaneously. In this system, power steering is essential.

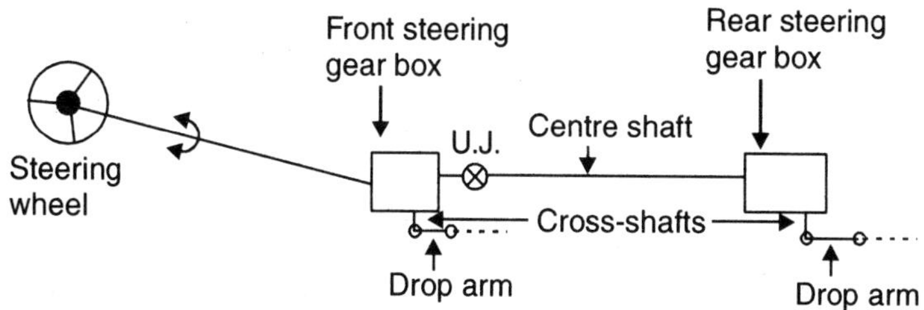

Fig. 12.22 4-Wheel steering

4-wheel or all wheel steering is used on Honda Siel Car along with power steering system.

POWER STEERING

Power steering is a power assisted steering in which a booster arrangement is set in operation when the steering wheel is turned by driver. The external power, utilised in the operation of power steering system, may be compressed air, hydraulic pressure or some electrical mechanism. Most of the power steering systems use hydraulic pressure for steering of vehicles. The power steering becomes operative when the driver's effort exceeds 10 newton force. The power steering system is designed in such a way that if power system fails, it can be operated manually.

Advantages of power steering

1. It requires lesser effort for steering.
2. Driver's fatigue on long drives is reduced.
3. There is efficient absorption of road shocks.
4. It gives better directional stability.
5. It is more safe and easy to operate.
6. The performance of steering gear is improved and vehicle gives a joyful driving.
7. It provides higher maneuverability or driving the vehicle on not so easy paths.

Power steering employed on vehicles is usually of two types:

(*i*) Integral type

(*ii*) Linkage type

(*i*) In **integral type,** the power unit or booster assembly is built into the steering gear box itself. It is generally integrated with recirculating ball steering gear. It consists of two parts:

(*a*) Power cylinder and piston unit.

(*b*) Oil controlling rotary valve unit.

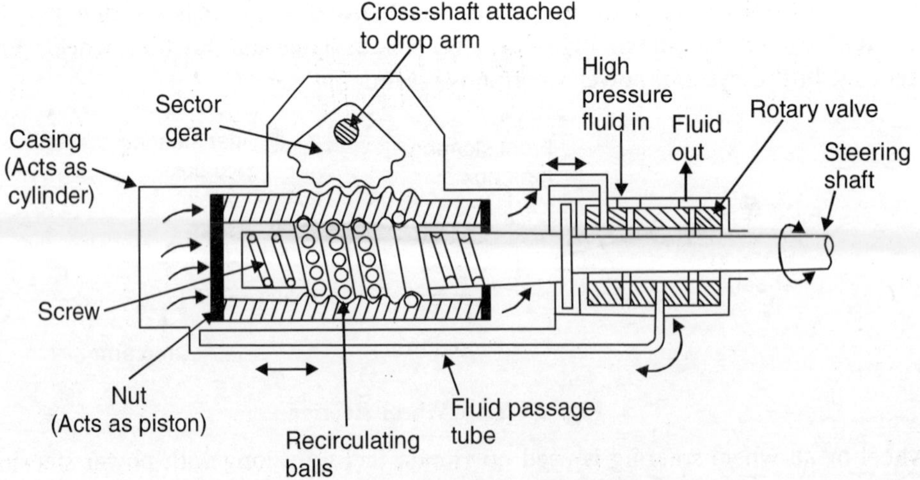

Fig. 12.23 (a) Integral type power steering

Integral type power steering is used on chrysler cars.

Referring to the Fig. 12.23 (a), the high pressure fluid from a hydraulic pump, is always remain filled up in the steering gear casing. When the steering shaft is rotated for steering purpose, the spool below the rotary valve turns and allows the flow of fluid to either back side of nut or to its front side. The pressure of fluid is applied at the end (right or left) of the nut, which acts as power piston. The outer teeth at the nut surface are engaged with the teeth of sector gear. The fluid pressure at nut helps in turning the sector gear and in the movement of the screw which is being turned by driver. The effort applied by driver is thus lessened. The detailed diagram of a rotary valve is shown in Fig. 12.23 (b). It is a cylindrical shape, two piece valve with passages for flow of fluid. The inner spool is mounted on steering shaft end. Flow of oil is controlled by inner spool valve.

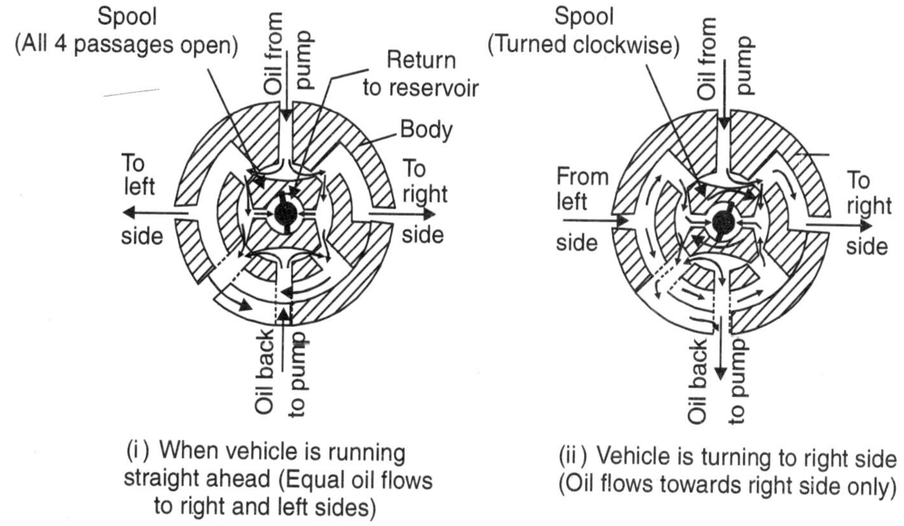

Fig. 12.23 (*b*) Working of rotary valve

In **linkage type**, the power or booster assembly is a part of linkage. The car engine drives a pump to develop fluid pressure. The fluid pressure acts on a double acting power piston inside a power cylinder. The right or left movement of the piston is transmitted to toothed rack operated by pinion manually. Then power piston rods boost the movement of rack and minimise the driver's effort to rotate pinion for steering. The variation in fluid pressure is achieved by a control valve linked with the steering shaft or column.

When the valve is in horizontal position (see Fig. 12.24), it closes both passages to cylinder and the fluid circulates freely in the pipeline leading to and from pump. It is the neutral position of valve and is achieved when vehicle is running straight.

When the driver turns the wheel towards right, the valve is also turned (see Fig. 12.24 (a)). The fluid now flows to the cylinder and pushes the piston to right side. The piston rods from both ends push the rack or the drag link for wheel turning.

When the valve is turned to left side, the direction of flow of fluid changes and piston moves to left side, turning the wheels to left through the linkage. See dotted valve and arrows in Fig. 12.24 (b).

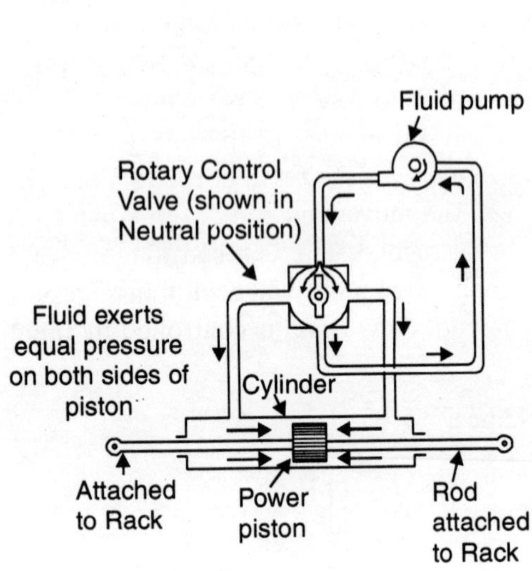

Vehicle running straight path

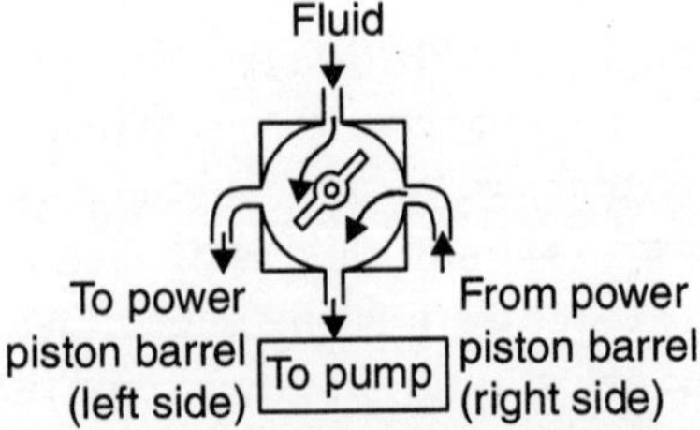

(a) **Control valve turned to right side.** Fluid flows to left side of power piston and pushes it to right side.

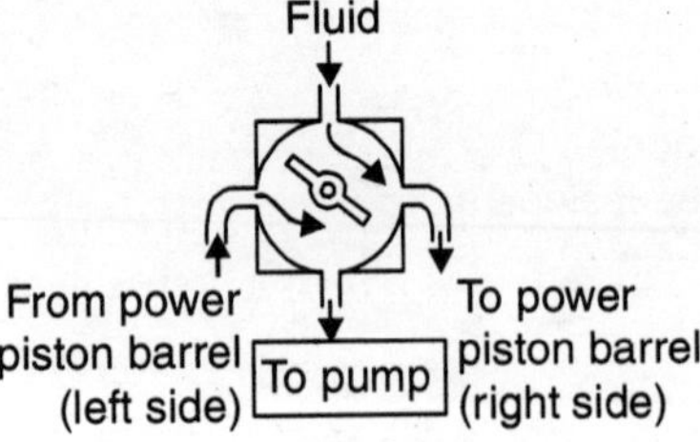

(b) **Control of valve turned to left side.** Fluid flows to right side of power piston and pushes it to left side.

Fig. 12.24 Linkage type power steering

SOME TROUBLES RELATED WITH STEERING

A. Poor Returnability of Steering Wheel

After steering is over, the steering wheel should return to its straight driving position. If the returnability is not proper, the reasons may be:

(*i*) Low inflation pressure—inflate the tyres to the specified limit.

(*ii*) Steering gear and steering column may be not in proper alignment—Align the steering column.

(*iii*) Due to hard steering:

(*a*) Lubricate the steering gear and linkages at joints. Defective joints should be replaced.

(*b*) Shaft bearings may be tight—replace.

(*c*) Steering gear too tight—adjust properly.

(*iv*) Front wheel alignment might has been disturbed, check and correct it.

B. Increased Backlash in Steering

Due to wearing out of ball joints and teeth of steering gear, there appears a slackness in steering. Check for backlash and the causes of slackness. Some of the causes may be:

(*i*) Drop arm may be loose on splines—correct it/replace.

(*ii*) Steering gear box may be loose fitted in the chassis frame—tight it and adjust the alignment of gear box.

(*iii*) Adjustment of linkage may not be proper—correct it.

C. Steering Becomes Hard to Operate

Due to hard steering, driver has to apply more efforts for steering, which causes early fatigue to the driver. Following items require checking and repair:

(*i*) Wheel alignment may be faulty—correct it.

(*ii*) Castor angle increased—Adjust it.

(*iii*) Tyre pressures may be low—check and correct.

(*iv*) Adjust steering gear and align the steering column if the same has disturbed. Steering gear may be too tight, correct it.

(*v*) Lubricate the steering gear and linkage joints after thorough cleaning, to reduce friction.

QUESTIONNAIRE

1. Draw a neat sketch of a rigid front axle. How wheels turn on it for steering?
2. What is a steering system? Which qualities it should possess?
3. Explain with sketches the following terms:

   (*i*) Castor (*ii*) Camber

   (*iii*) King-pin inclination.
4. What are the main components of a steering system? Describe them briefly.
5. How many types of steering gears do you know? Explain working of a rack and pinion type steering gear.
6. Which factors influence the steering?
7. Describe wheel alignment. What are the factors related to it?
8. Explain the following:

   (*i*) Irreversible steering (*ii*) Reversible steering

   (*iii*) Turning radius (*iv*) Self centring steering action.
9. Explain the following:

   (*i*) Toe-in (*ii*) Toe-out

   (*iii*) Steering axis inclination.
10. Write short notes on:

    (*i*) Over steer and under steer (*ii*) Slip angle

    (*iii*) Cornering force.

11. Explain the Ackermann Steering principle.
12. (*a*) What is self-righting torque?
    (*b*) What are the effects of low inflation pressure and high inflation pressure?
13. What is power steering? Describe any power steering system with neat sketch.
14. Describe:
    (*i*) A ball joint (*ii*) Drop arm
    (*iii*) Tie-rod (*iv*) Drag link.
15. Describe the working of a worm and sector type steering gear.
16. With the help of neat sketch, show how balls recirculate in a recirculating ball type steering gear.
17. (*i*) How 4-wheel steering is achieved?
    (*ii*) What is combined angle? What are its three positions?
18. Describe a steering linkage with sketch for independent suspension.
19. Write notes on:
    (*i*) Hard steering (*ii*) Difficult or poor returnability
    (*iii*) Backlash in steering system.
20. Describe with the help of neat sketch the working of worm and sector type of steering gears.
21. (*i*) Differentiate between Ackermann steering mechanism and a Davis steering mechanism.
    (*ii*) What are various requisites of the steering system?

CHAPTER 13

Final Drive and Differential Gears

The motion and power from gear box is transmitted through final drive to the driving wheels. It consists of the following parts:

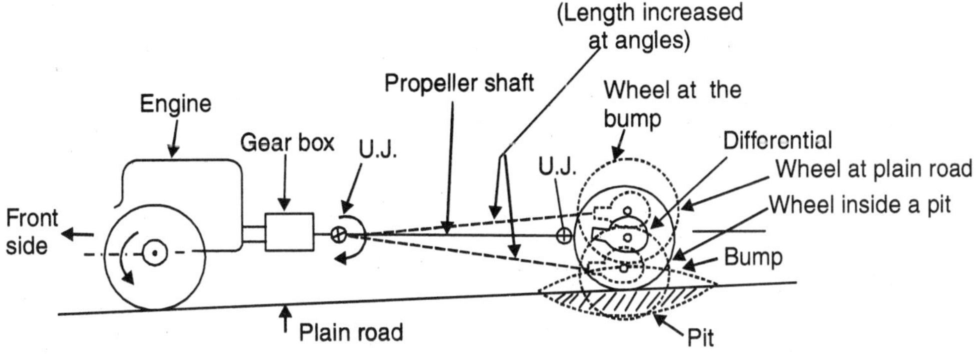

Fig. 13.1 Changes in length of propeller shaft while passing over pit or bump

Propeller Shaft: In case of front wheel driven vehicles, the propeller shaft or drive shaft is not required. In rear wheel driven vehicles, a long hollow propeller shaft is commonly used. In some cases solid shafts are also used. The propeller shaft connects the output shaft or main shaft of the gear box to the differential gear box fitted at Rear axle. The two universal joints are also fitted at the two ends of the propeller shaft. The propeller shaft transmits the motion and power from gear box to rear axle.

NEED OF PROPELLER SHAFT

The vehicle sometimes has to climb a hill or moves downward a slope. Even on plain roads, the vehicle has to pass through many pits or projected portions (BUMPS) on the road. One or two of the wheels may be inside a pit or at bump and other remaining wheels on the plain road. The changes in the angle of drive between gear box shaft and propeller shaft, and between propeller shaft and differential, need a variable length of propeller shaft for smooth driving, see Fig. 13.1. The movement of suspension springs due to jerks and bumps also makes it necessary that the length of propeller shaft must be self-adjustable.

To adjust for changes in driving angles, the universal joints are used. To adjust for changes (up to 3 cms) in the length of propeller shaft, a slip joint or a telescopic joint is provided within the shaft itself. The splined end of the shaft slides into the other internally splined end to increase or decrease the length. The splines require lubrication for frictionless sliding movement.

TELESCOPIC PROPELLER SHAFT

The propeller shaft is made of high strength alloy steel. It should be well balanced to avoid whirling at high speeds. In Hotchiss drive, there is fitted one universal at each end of the propeller or drive shaft. But in torque-tube drive, there is only one universal joint at the gear box side. In Hotchiss drive, the propeller shaft is open (no casing) whereas in torque-tube drive, the drive shaft is enclosed inside the torque tube. The torque reaction, braking torque and driving thrust are born by torque-tube and side thrust is taken by leaf springs. A smaller solid shaft may be employed in torque-tube drive. A telescopic type propeller shaft has one slip joint situated near to the gear box side. It serves to adjust the length of the shaft whenever required by rear axle movements. The sliding of shaft inside slip joint is known as telescopic action. The 4 × 4 wheel drive vehicles use divided propeller shaft or two shafts. In between the two shafts, the transfer case is fitted for (4 × 2) drive. The splines of the slip joint are properly lubricated. Some slip joints are provided with a rubber element incorporated in between the two sliding tubes surfaces for noiseless movement.

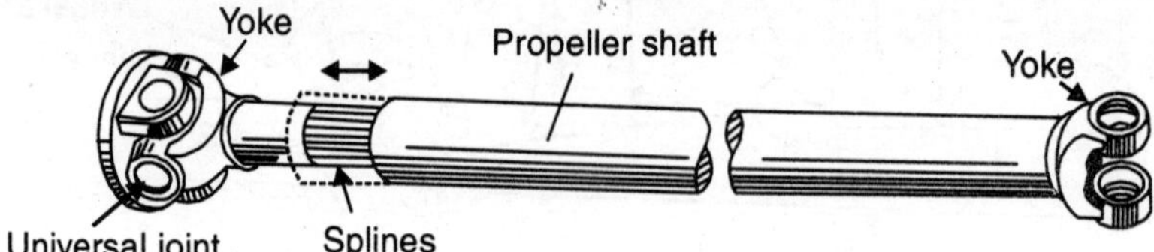

Telescopic propeller shaft with yokes on ends

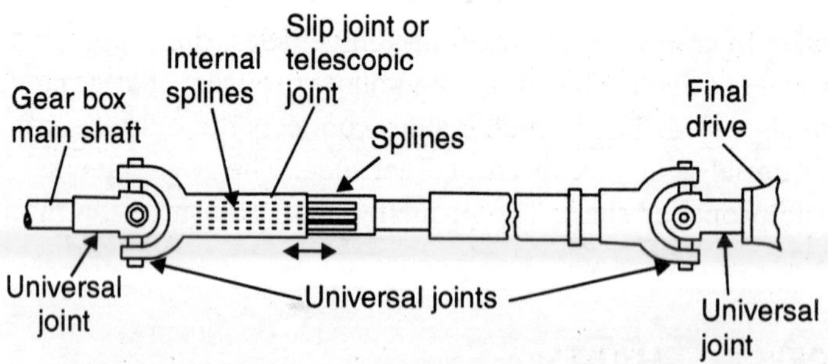

Fig. 13.2 Telescopic propeller shaft drive line

The slip joint causes the two half shafts to rotate together at the same speed like a solid shaft. The splines (External on left part and Internal on right part as shown in the diagram) facilitate the sliding movement between the two shafts whenever the need for adjustment of length arises Fig. 13.2.

UNIVERSAL JOINTS

The universal joint allows the transmission of power at an angle which varies according to the position of rear axle and wheels. The rotation and transmission of power is quite efficient at small angles of propeller shaft movements of up and down, normally up to 18° on either side.

Hooke's Universal Joint: It has been shown in Fig. 13.3. It is the most simple type of universal joints and is widely used on vehicles. It is also known as *cordon or cross-joint.* Its exploded view is shown in Fig. 13.4.

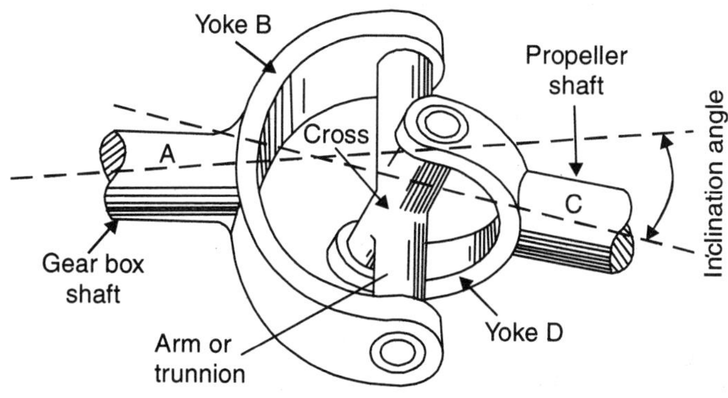

Fig. 13.3. Hooke's universal joint

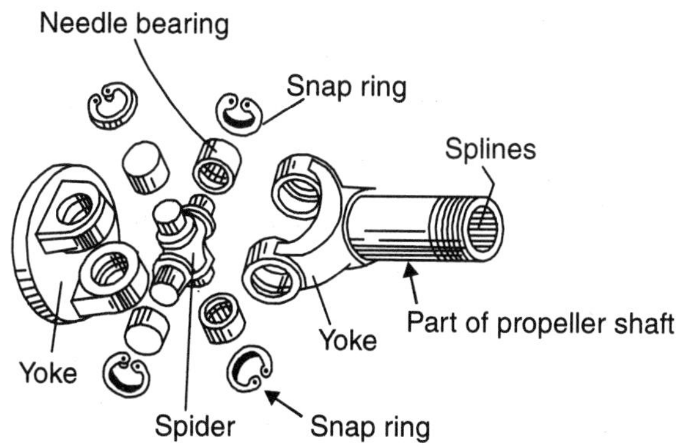

Fig. 13.4 Exploded view of universal joint

As shown in the figure, the shaft 'A' is gear box shaft connected to yoke 'B' of universal joint and shaft 'C' is the propeller shaft connected to yoke D of universal joint. Both ends of yokes B and D are connected to each other by a four arm spider or cross. The shaft 'A' rotates horizontally about its axis but shaft 'C' of propeller shaft can rotate at an angle in up and down movement or both side movement or the combination of the two. The cross uses needle roller bearings in the yokes for support.

It may be noted that this type of single universal joints does not transmit the motion uniformly and speed of propeller shaft does not remain uniform in any full rotation. The driven shaft undergoes cyclic variation as shown in Fig. 13.5.

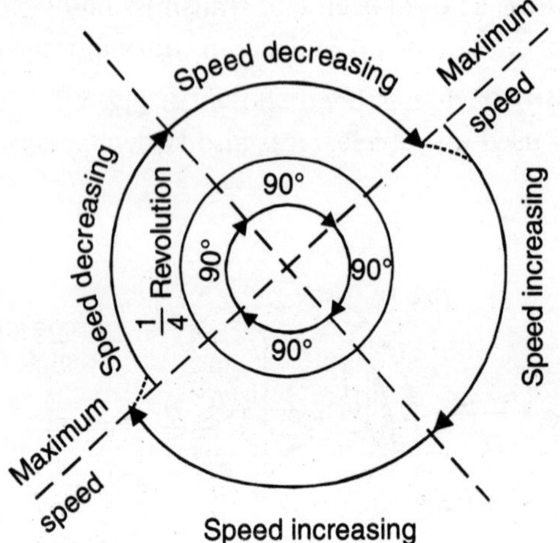

Fig. 13.5 (a) Variation of velocity in Hooke's joint in one revolution

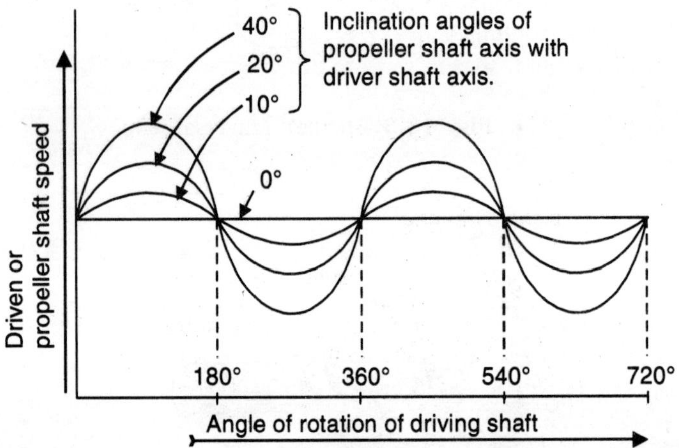

Fig. 13.5 (b) Fluctuation of speed of propeller shaft fitted with Hooke's joint (Graphical representation)

This problem is solved by using another universal joints at the other end of the propeller shaft. In practice two Hooke's joints are employed, one at each end of the propeller shaft.

When two cross-type universal joints are used in the drive line, one joint at each end of the propeller shaft and if the angles of tilt at both the universal cross-joints remain the same, the rear joint provides the direct opposite reaction to the front joint's fluctuation. It nullifies the fluctuation of speed, by having the rear universal joint's half revolution's speed reduction equal to the same amount that the front joint speeds up in the same half revolution. It results in identical speeds of both driver and driven shafts. The universal joints, cross-type, is a double hinged joint consisting of two fork type yokes in whose bearings at ends, rest four arms or trunnions of a cross-shape spider. The bearings should be properly lubricated so that cross-arms may move freely.

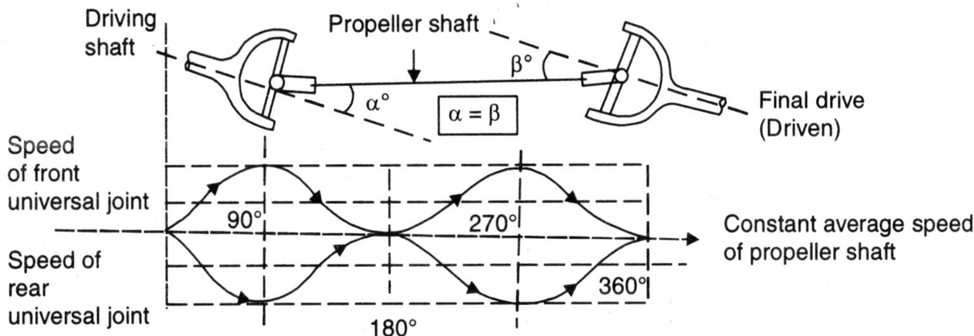

Fig. 13.6 Effect of two universal joints (using two universal joints make velocity uniform)

An improved form of Hooke's Universal Joint is a *Hooke's Rubber Joint.* In this joint, the cross-arms rest in split rubber bushes. The bushes are enclosed by steel sheets. The flexibility of rubber permits the spider to centralise itself as the joint rotates. Thus a constant velocity of the propeller shaft is obtained. It also permits larger angular deviations of propeller shaft.

Flexible Ring Universal Joint: In this joint, a special reinforced fabric flexible rubber ring with six holes is employed between the two shaft fitted with 3–bolt spider at ends. The flexibility of the rubber ring allows the driven shaft to rotate at small angle, but not at an uniform velocity. It does not requires lubrication. For transfer of larger torque, a bigger joint is to be utilised. It is used on small 3-wheelers. It is easy to manufacture them and they are cheaper in cost. The flexible joint is shown in Fig. 13.7.

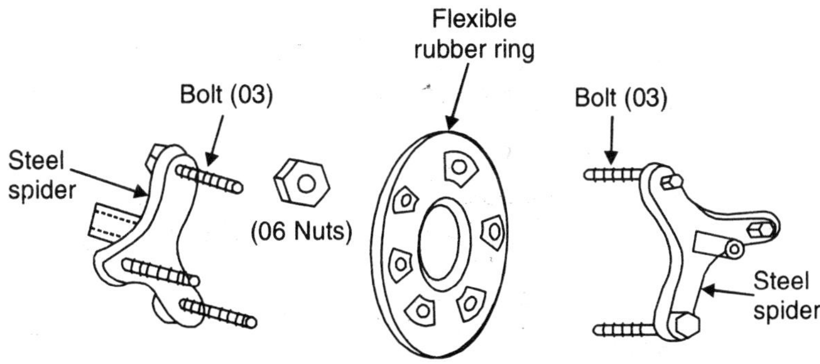

Fig. 13.7 Flexible rubber ring universal joint

CONSTANT VELOCITY UNIVERSAL JOINTS

These joints transmit uniform speeds at sharp angles. Their use is limited to the vehicles having engine located close to the driving wheels and having a short shaft, connecting the drive, and is inclined at a steep angle. Two types of C.V.U. joints are much popular:

(*a*) **Bendix Tracta Joint:** It consists of two centre parts which join to form a sphere. Each piece is so designed and slotted that it can move on the other. The flat yokes are fitted into additional slots in their outer faces. The yokes and centre parts re-align themselves continuously as the joint rotates and such the uniform velocity of the driven shaft is achieved. This type of joint is shown in Fig. 13.8. It is employed on some modern cars. In action, it is similar to a Ball-joint.

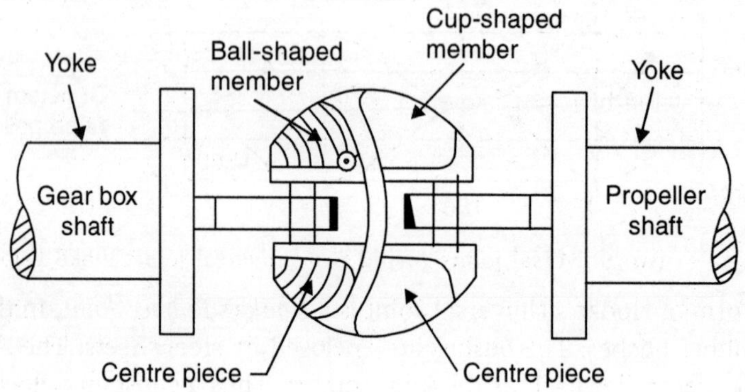

Fig. 13.8 Bendix tracta joint

(*b*) **Rzeppa Universal Joint:** Its yokes are replaced by a cup-shaped and a ball-shaped members. The cup has semicircular grooves in its inner surface, and the ball shaped member has matching grooves in its outer face. Each groove is provided with a steel ball and balls are stayed/ located by a cage. As the joint rotates and the angle between the two shafts changes, the cage keeps the balls in the plane of the bisecting angle and thus the joint forces the driven shaft to rotate uniformly at a constant velocity. The joint is shown in Fig. 13.9. It is used on 4-wheel drive vehicles.

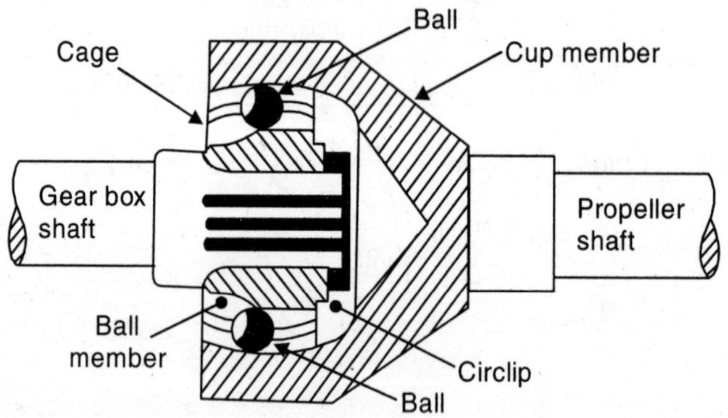

Fig. 13.9 Rzeppa joint (constant velocity)

Some other types of constant velocity joints are also used.

FINAL DRIVE GEAR

The motion and power from gear box reaches a bevel pinion through universal joints and propeller shaft. The shaft of bevel pinion is connected to the rear universal joint which is connected to propeller shaft. The bevel pinion drives a crown wheel which transmits its motion to the differential gears.

The bevel pinion and crown wheel assembly is termed as **final drive** as shown in Fig. 13.10. The function of the final drive may be summarised as below:

1. It turns the power drive by 90° to rear axles.
2. It provides a speed reduction at a pre-fixed value (about 4 : 1 in cars and 10 : 1 in heavy duty vehicles).
3. It raises the value of torque at driving wheels.

Following type of final drives are used on vehicles:

(*i*) Spur bevel gear type (Noisy and wear out easily, teeth inclined at 45°)

(*ii*) Worm and wheel type (Give large speed reduction and more torque in HCV's)

(*iii*) Spiral bevel gear type (Silent and smooth running and stronger curved teeth)

(*iv*) Hypoid gear type: These are widely used. The surface on which the teeth are cut is a hyperboloid. They permit a lower chassis height. This type of final drive gears are employed on almost all the Indian passenger cars. The pinion is of a larger size and therefore, much stronger for the same size of crown wheel on other types. They also run silently because of smooth and gradual meshing of teeth.

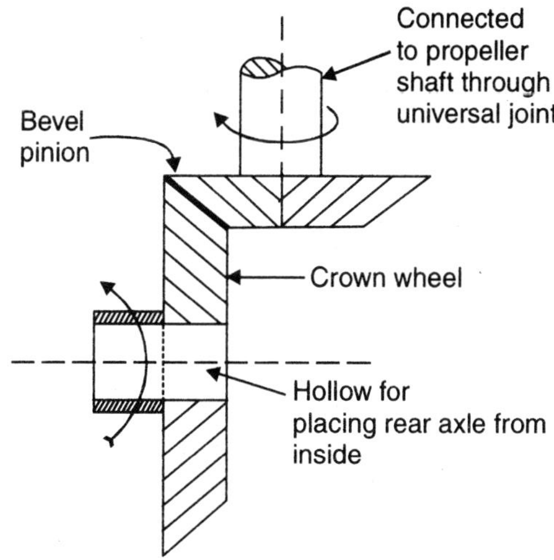

Fig. 13.10 Final drive bevel gears

In the spur and spiral gear type final drives, the axis of crown wheel and axis of bevel pinion have to be kept in the same plane to engage maximum teeth at a time. But in hypoid gears, the axis of bevel pinion may be put below the axis of crown wheel to permit lowering of chassis height. In HCV's, the worm and worm wheel final drives are popular. They provide larger speed reductions and more torque. The height of chassis or ground level clearance may also be varied suitably.

All the above four types of final drive gears are shown in Fig. 13.11.

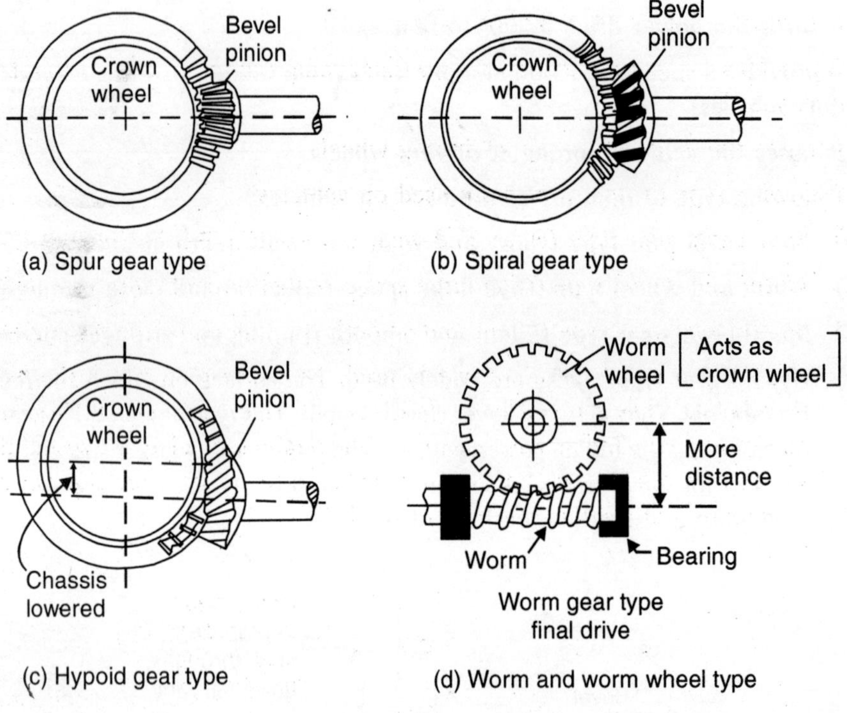

Fig. 13.11 Different types of final drive gears

DIFFERENTIAL GEARS

The differential is located between two half shafts of rear axle. The rear axle used is divided into two parts, each part is called *half shaft*. The differential is a vital part of power transmission to wheels. It performs the following functions:

1. The differential gives a constant reduction in speed to increase torque.
2. It transmits the power to rear axles (two half axles) at right angles to the axis of propeller shaft.
3. When a vehicle negotiates a turn, the outer wheels have to travel greater distance than the distance travelled by inner wheels in the same time slot. It is must to avoid skidding and scrubbing of the tyres. The differential accommodates the two half rear axles, which maintain the straight drive at equal speeds but turn at differents speeds when the vehicle is taking a turn or cornering, see Fig. 13.12. The inner wheels traverse shorter distance, at turns, then the distance travelled by outer wheels to complete the turning smoothly.

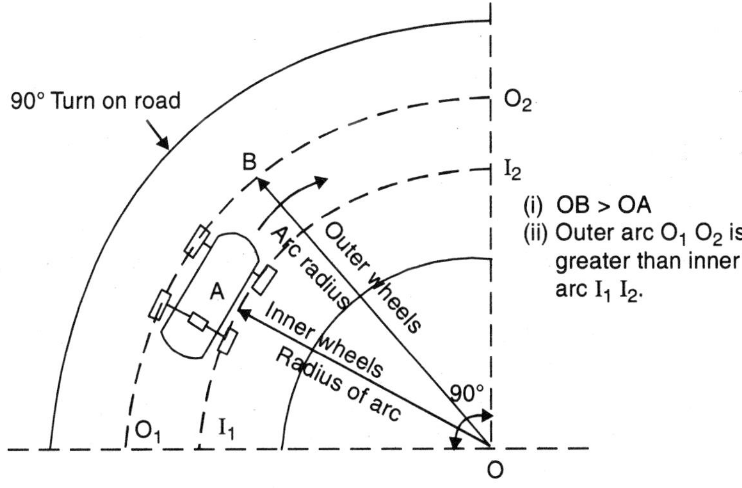

Fig. 13.12 Car taking a turn on road

Construction and Working of Differential

The differential consists of a mechanism having epicyclic gear train which rotates the outer wheel at more speed than the inner wheel while the car is taking a turn or being cornered. In straight running, both outer and inner wheel run at the same speed.

A differential in its simple form is shown in Fig. 13.13. It consists of the following components:

1. *Ring gear or crown wheel* of final drive. It drives cage.
2. *Differential cage* to carry planet pinions. The differential cage is firmly fixed on the crown wheel face. The two planet, pinions, which drive the sun-gears, are mounted on cage with the help of a pin.

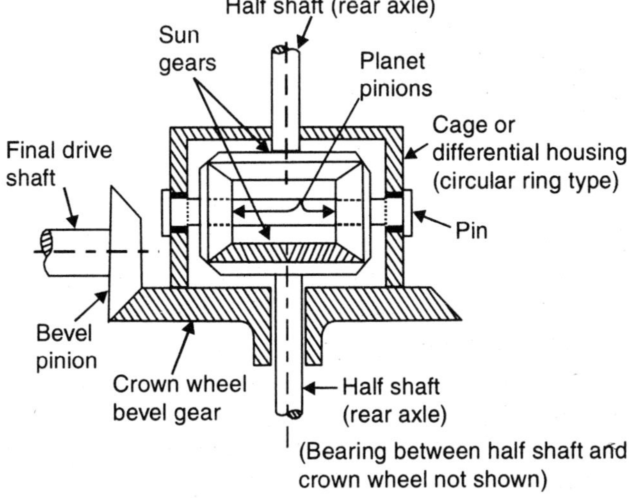

Fig. 13.13 Differential (Gears teeth not shown)

3. *Sun gears*: Both sun gears and planet pinions form the epicyclic gear train. Sun gears are keyed fitted over the inner ends of both half rear axels. The sun gears take drive from planet pinions.
4. *Planet pinions*: The cage consists of a long pin, on which the two planet pinions are mounted with their teeth meshed with sun gears. The planet pinions are free to rotate around the pin and also drive the sun gears.

When the vehicle is running straight, the drive from propeller shaft's bevel pinion is transmitted at 90° to the crown wheel. Along with crown wheel, the differential cage also rotates at the same speed. The cage carries two planet pinions which also rotate with cage at the same speed. The teeth of planet pinions being engaged with teeth of sun gears, transmit power to both rear axle halves through sun gears at uniform speed. *In straight running condition, the planet pinions do not rotate on their own axes or at pin.* The crown wheel, cage and pin with planet pinions rotate as a single unit or as a solid unit with crown wheel. The equal amount of torque and r.p.m. are transferred to the driving wheels.

Suppose that we have only two pinions and two sun gears combination engaged as shown in the differential. Rotate the sun gear by one revolution in clockwise direction, the opposite sun gear, through rotating pinions in between, will also rotate by one revolution in the opposite direction—anti-clockwise as shown in Fig. 13.14. In this case, the pinions have also rotated on their axes or at pin.

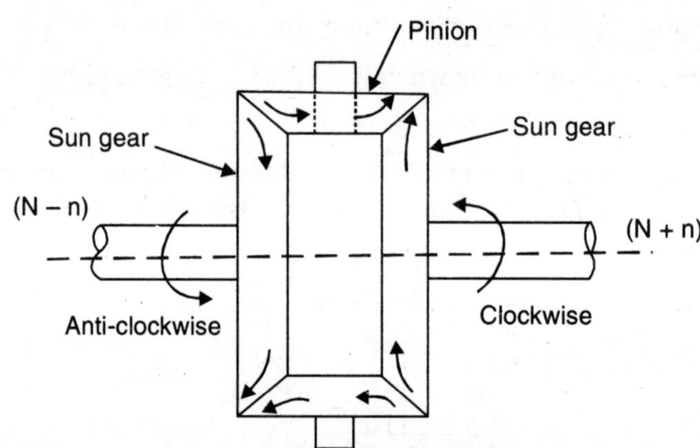

Fig. 13.14 Opposite rotation of sun gears

Now we can easily understand the working of differential on turns:

Suppose a vehicle is going straight and its wheels speed is N r.p.m. Now the vehicle takes a turn—suppose towards left side. When steering system will turn the front wheels in the left direction, there will be a greater road resistance to the motion of the left wheels in comparison to right side wheels. As a result of this differential action, if the left driving wheel slows down by n r.p.m., the planet pinions will start rotating about pin, and the right sun gear will add n r.p.m. to the right wheel's N r.p.m., that is, r.p.m. of right wheel will become (N + n) while r.p.m. of left wheel or the turning side wheel's will be (N – n); this will facilitate the vehicle to follow a curved route on turns. The epicyclic gear train also divides the driving torque equally on the two half-shafts. The planet pinion are free to rotate on pin, therefore, they cannot apply unequal torque through the same number of teeth engaged on both sides with sun gears. The planet pinions act as a balance and divide the driving torque equally to the both half-shafts, even when their speeds are unequal on turns.

REAR AXLES

The rear axle is called *live axle* in case of rear wheel drive and *dead axle* in case of front wheel drive. In four wheel drive, both axles are live axles, as they rotate and transmit power. The live axle is split in two halves for attaching it with differential. Generally a axle housing (most popular type is a **Banjo Type Housing** as shown in Fig. 13.15) in single piece or split type is used to enclose the final drive, differential and half-shafts and their bearings. The axle housing is filled with lubricating oil and properly sealed against leakages. It protects the gears and axles from moisture, dirt and dust. At the two ends of the casing, the rear wheel hubs are also mounted. The differential housing portion has an inspection plate bolted at the casing. Spring pads are provided for spring seats on casing.

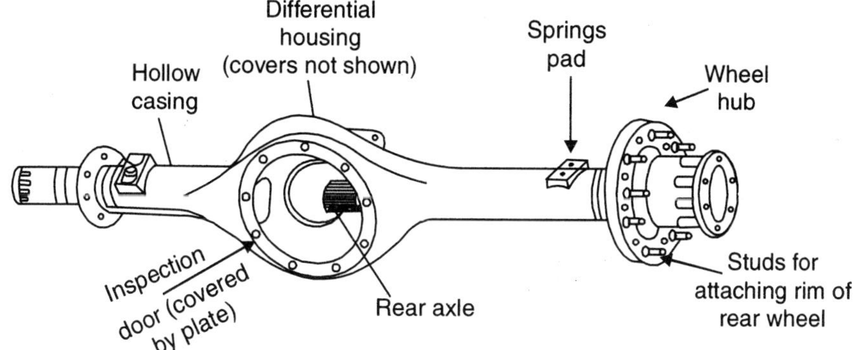

Fig. 13.15 Banjo type rear axle casing

Loads on Rear Axle: The rear axle experiences bears the following loads/stresses:

1. Shear force due to weight of vehicle.
2. Bending loads and bending moment due to side thrust on turns.
3. Driving torque of the engine.
4. Braking forces.
5. End thrust by wheels due to road irregularities and vibrations.

Types of Rear Axles

There are three types of rear axles used on different categories of vehicles:

1. Semi-Floating Axle
2. Three-Quarter Floating Axle
3. Full Floating Axle.

1. Semi-Floating Rear Axle: It is shown in Fig. 13.16. In this axle, the wheel hub is rigidly fixed to the outer end of the axle shaft. The outer race of the wheel bearing fits into the axle casing and inner race fits on the axle shaft. The bearing is held in its position by bearing retainer.

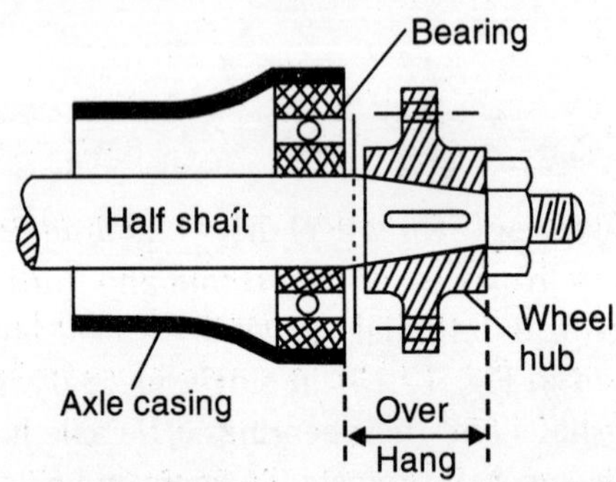

Fig. 13.16 Semi-Floating rear axle

The vehicle load acts on casing and bearings which transmit it to the shaft. The wheel is mounted on the projected portion of the axle which may cause bending of the axle. In case of braking of axle, the wheel is likely to fall out which may cause serious accident. This type of axle was used on old models of Fiats and Jeeps. It is still in use on some light vehicles.

2. Three–Quarter Floating Axle: It is shown in Fig. 13.17. In this type, the bearing is fitted between axle casing and the wheel hub. The wheel hub is firmly attached to the outer end of axle-shaft. This axle does not support any vehicle load and withstands only side/end thrust on turns. Even then some bending stresses are taken by it. Such type of axles are used on Ambassador II cars and many other vehicles.

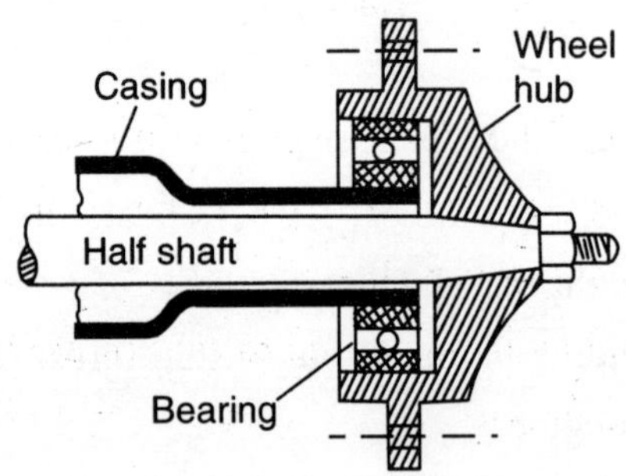

Fig. 13.17 Three-quarter floating axle

3. Full Floating Rear Axle: This type of rear axle transmits driving torque only. The bending load (weight of vehicle and its passengers + luggage) and end thrusts, while cornering, are not carried by this axle. It is shown in Fig. 13.18(a).

Since this type of axle does not have any concern with carrying bending loads and end thrusts and its only function is to transmit the rotating motion and driving torque, it is named as full-floating axle. The wheel hub is supported by two taper roller bearings mounted on the axle casing directly. The axle head is fastened to the wheel hub flange by nut-bolts or the wheel hub is mounted at splines cut at the end of axle as shown in the figure. The wheel hub has internal splines. This type of axle

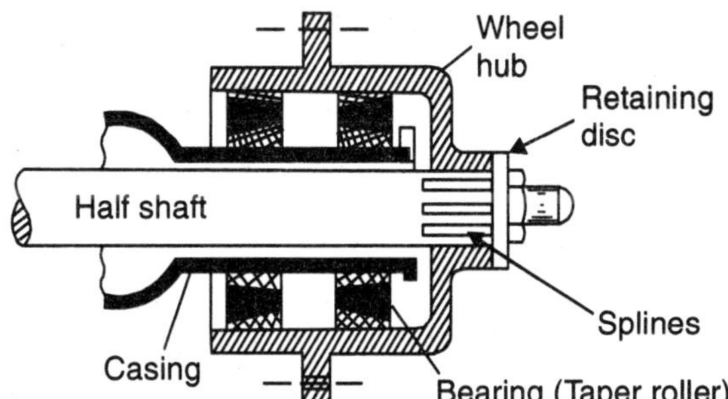

Fig. 13.18 (a) Full floating rear axle

is used on heavy duty vehicles. In case of breakage of axle, the wheel will not come out and the vehicle will not be tilted. The shaft can easily be removed and replaced. It is strong and robust in construction but is costly. Tata LPT 2213–6×2 Trucks, Diesel fired Ambassadors, Mahindra Jeep MM540 and Ford F516 trucks employ these axles. In this axle, the taper roller bearings take up the side thrust. The weight of vehicle is supported by axle casing and wheels.

Double Reduction: In very heavy and large buses, trucks and dumpers, a double speed reduction is used to increase torque. First reduction is obtained as usual by the final drive with a pinion and crown wheel assembly the ends of the half axles (Rear axle) are mounted with pinions which drive larger diameters spur gears, on whose shafts road wheels are mounted. Such arrangement provides the second speed reduction. This arrangement needs road wheels of larger diameters so as to compensate for reduction in ground clearance. Figure 13.18(b) shows a simplified diagram of second speed reduction at rear axle ends.

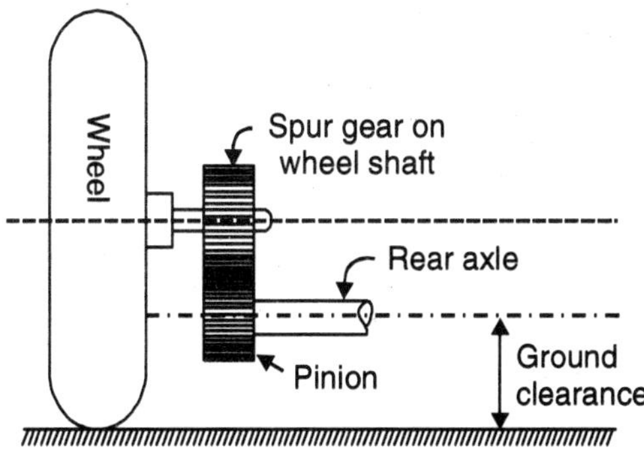

Fig. 13.18 (b) Second speed reduction at wheels

On some very heavy duty commercial vehicles, two speed rear axles are also employed. The driver can shift over to one more speed arrangement of an extra epicyclic gearing besides the conventional differential gearing.

TYPES OF REAR AXLE DRIVES

According to the drive arrangement of propeller shaft, the rear axle drives are of two types:

1. Hotchkiss Drive.
2. Torque-Tube Drive.

1. Hotchkiss Drive: It is a very simple type drive and fairly popular on vehicles with rear wheels drive. In this type, the propeller shaft is conventional telescopic type, uncovered and connected at each end with one universal joint. It is attached with rear axle casing in the middle portion which houses differential. The rear axle casing is supported on semi-elliptical leaf springs with a shackle at rear end. This arrangement enables the leaf spring to sustain torque reaction, driving thrust and side thrust and it also supports the vehicle weights, see Fig. 13.19 (a). In resisting the torque, the leaf spring deflects in its front half and in resisting the braking torque, it deflects in opposite direction as shown in Fig 13.19 (b). Thus, such increase or variation in length of spring is compensated by shackle at rear. The deflection of spring also changes length of propeller shaft which is adjusted by slip joint. The deflection of leaf springs and rear axle movements try to alter the position of final drive shaft, which may cause bending of propeller shaft. The bending of propeller shaft is prevented by the universal joint at rear end. The driving thrust is transferred through springs to the rear end of the frame.

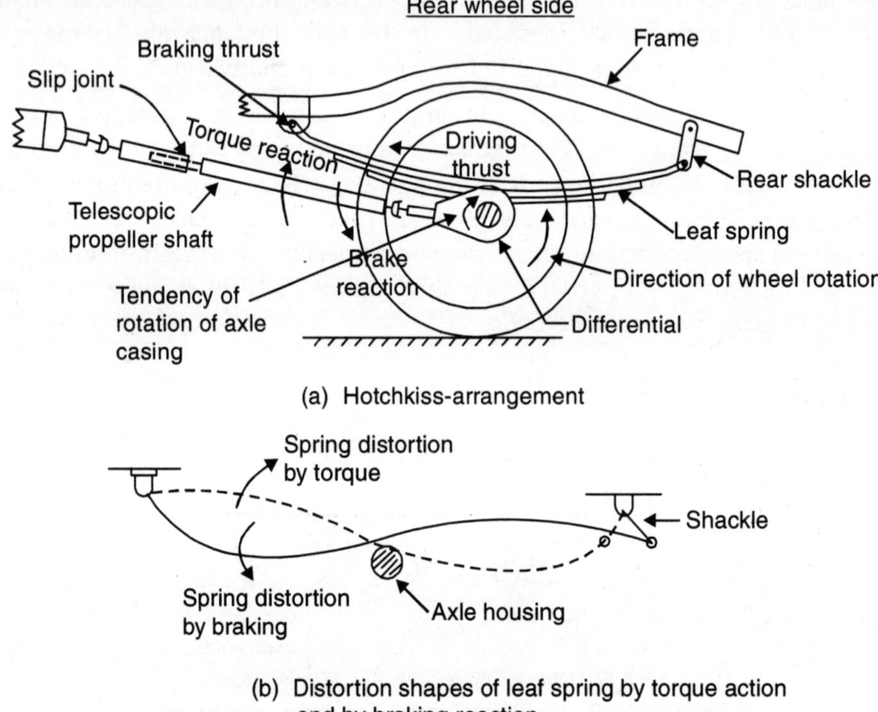

Fig. 13.19 Hotchkiss-arrangement and spring distortion

2. Torque Tube Drive: In this type of drive, the propeller shaft is enclosed inside a hollow tube, called torque-tube, which is rigidly bolted to the rear axle housing at one end. The other end of the tube is held with the help of a ball and socket joint or a spherical joint at the rear of the gear box or at a cross-member of the vehicle frame. Two bracing/radius rods are connected between the axle casing and torque-tube to provide additional rigidity. An universal joint between gear box main shaft and propeller shaft is used to facilitate the angular deflection of the drive. The leaf springs are provided with shackles at both ends. There is no shift in the position of final drive shaft (See Fig. 13.20.)

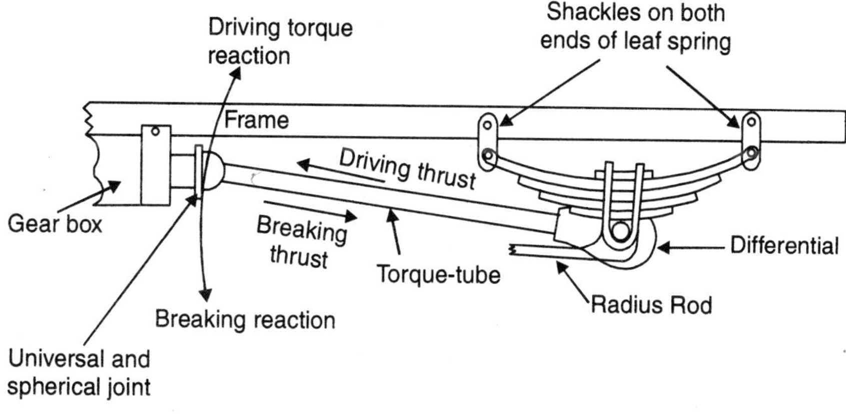

Fig. 13.20 Torque-tube drive

The torque-tube takes up torque reaction, breaking reaction, breaking thrust and driving thrust. The springs take only body weight and side thrust. The driving thrust is transferred to the front side of frame through gear box casing or the cross-member of the frame. The torque-tube drive is generally not used on Indian vehicles. Some foreign car models, such as Electra, Buick, Rambler employ torque-tube drive. The leaf springs with two shackles give good and comfortable ride since they are relieved off torque reaction and driving thrust.

To provide extra rigidity to the torque-tube construction, two radius rods are connected to the rear axle casing and cross-member of the frame. These radius rods can oscillate in small amounts around the pivots at both ends. The radius rods prevent the tendency of rear axle casing to rotate in the clockwise direction due to driving torque reaction (See Fig. 13.21). Torque tube drive is more costly and complicated than Hotchkiss drive.

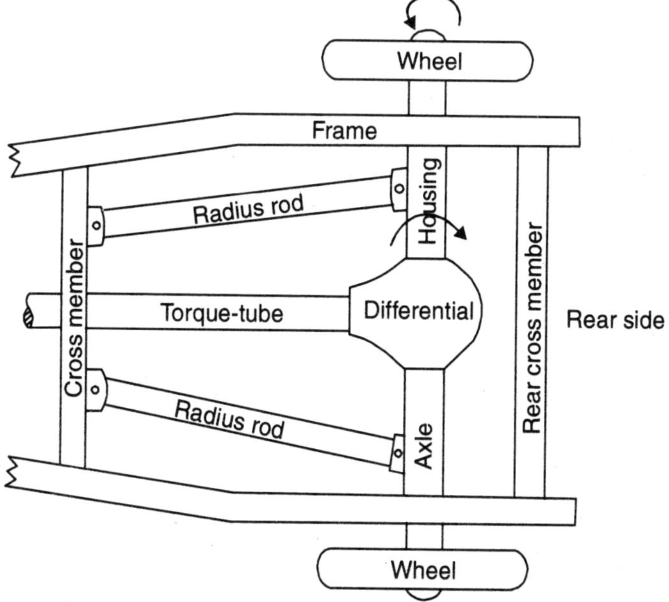

Fig. 13.21 Radius rods in torque-tube drive

QUESTIONNAIRE

1. Describe the construction of a telescopic type propeller shaft. Why is it needed?
2. What is the function of universal joints? Describe with neat sketch a Hooke's universal joints.
3. How variations of velocity in each revolution of propeller shaft is controlled?
4. (*a*) Describe the working of any constant velocity universal joint.

   (*b*) How a flexible rubber ring universal joint works?
5. Describe with a sketch, a Rzeppa joint.
6. What do you understand by Final drive? Describe the working of gears used in final drive.
7. How many types of final gear drives are used?
8. Describe with neat sketch the working of a differential.
9. (*i*) How many types of rear axles are employed on vehicles?

   (*ii*) Describe a full-floating axle.
10. What is the role of axle casing? Describe its functions *w.r.t.*, all the three types of rear axles.
11. Describe arrangements for Hotchkiss and Torque-tube drives.
12. (*i*) On which drive, radius rods and two shackles are used and why?

    (*ii*) What is Double reduction of speed?

CHAPTER 14

Wheels and Tyres

WHEELS

Wheels play a very important role in running the vehicle on road. Without wheels no automobile vehicle may be imagined to work. The wheels along with tyres rotate on their axles and help to run vehicle on road. It also takes the load of vehicle, load of passengers and their luggage. It provides a cushioning effect and help in steering control and directional stability. Wheels also take up braking stresses, side thrusts and road shocks.

They are made of pressed steel, Magnesium-alloy (Elektron) and Iron-Aluminium alloy or Iron-Magnesium alloy.

Characteristics of a Good Wheel

1. It must be structurally strong and rigid.
2. It should be light in weight to minimise unsprung weight.
3. The material of construction should be anti-corrosive and weather-resistant.
4. Wheels should be perfectly balanced–statically as well as dynamically for smooth rolling without vibrations.
5. They should be readily attachable and detachable with much ease.

WHEEL ASSEMBLY

A wheel-tyre assembly consists of the following main parts:

1. Wheel body/disc with rim with holes for tube valve, ventilation and hub.
2. Brake drum.
3. Back or Brake plate fitted with brakes on axle casing.

4. Holes in the wheel disc hub for tightening wheel on axle flange with the help of nuts.
5. Tyre and tube pair or a tubeless tyre.

Wheel is an assembly of disc or body with hub for axle and a rim casted/welded over the disc.

TYPES OF WHEELS

Wheels of various types are in use on automobiles.

1. Disc wheels: They are made of pressed steel and are widely used. Alloy cast wheels are also gaining popularity due to their light-weightness. Iron-Al alloy wheels are also being used due to their lightness and good strength.

The rim surface or seat is designed and made such that the tyre fits over it and forms a 'wedge-fit' on it. The seat is given a small taper towards well or centre to facilitate the tubeless tyres to make proper seal for air. A hole is also provided for projecting out the tube-valve through which air is filled. See Fig. 14.1 (a).

2. Wire spoked wheels: Spoked wheels were used on early cars but now they are almost obsolete for passenger cars but still in use on some sports cars. 2–wheeler bikes are also fitted with wire-spoked wheels. In wire-spokes wheel, wires are connected between rim nipples and holes on hub circle plate. Wire may be lightened to provide proper tension in the rim. Modern bikes these days are having 4–6 ribs casted wheels which efficiently sustain vertical loads, braking loads and road shocks. See Figs. 14.1(b) and (c).

3. 2-piece wheels: 2-pieces of half wheels are fastened together to form a wheel. Such wheels are used on scooters and some tractors (See Fig. 14.1 (d)).

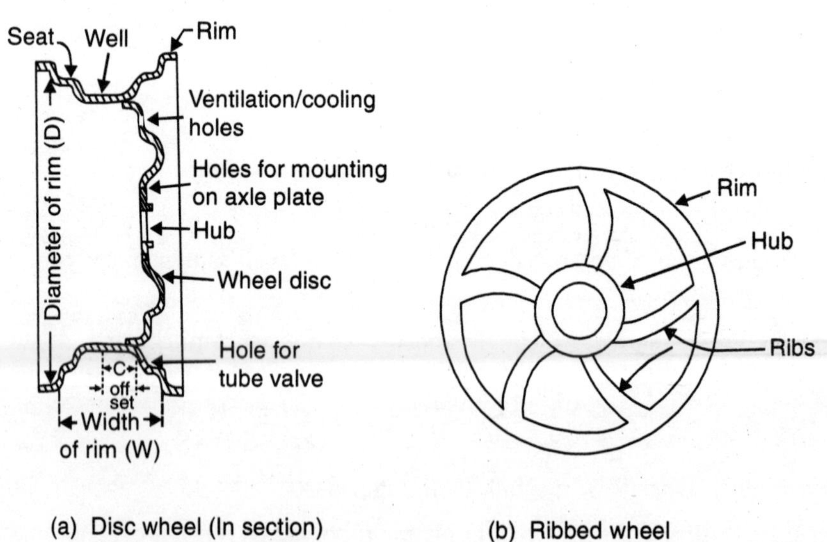

(a) Disc wheel (In section) (Drop centre type rim)

(b) Ribbed wheel

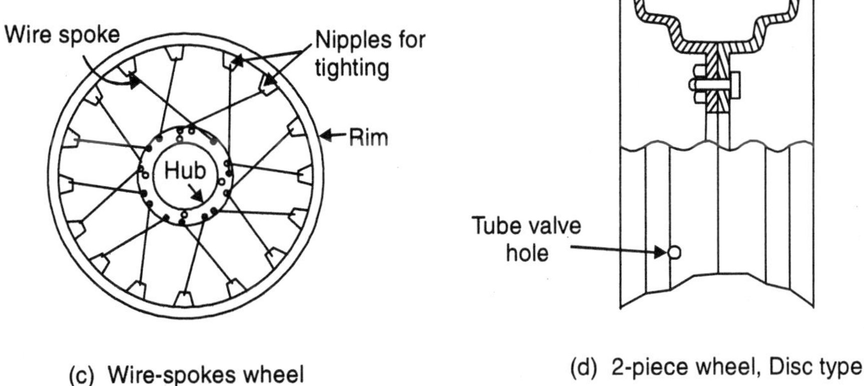

(c) Wire-spokes wheel (d) 2-piece wheel, Disc type

Fig. 14.1 Various types of wheels

WHEEL SIZE/SPECIFICATIONS

Wheel size is generally given in terms of width of rim (W) multiplied by diameter of rim (D) in inches (1″ = 25.4 m.m.). Trucks use 7″ × 20″ size wheels of disc type. Medium segment cars are fitted with 4.5″ × 13″ size wheels.

Referring to Fig. 14.1 (a), the wheel is specified by a code number comprising of the following sequence:

Width (A profile of rim indication) — Diameter of rim (in inches) such as 5.50B – 13 or 4.50J – 14,

where B and J show type of Rim Profile.

Composite wheels: These are wheels used on many vehicles; their hubs and rims are made of steel to give extra strength and rest of the disc is Al-alloy casting. They are light in weight but stronger in load bearing capacity.

Wheel Balancing

To experience a smooth and vibration free steering, all the wheels of the vehicle should be perfectly balanced and circular. An unbalanced wheels makes noise, wobbles and gives a jerky and vibratory motion which affects the suspension and steering systems both.

Wheel balancing is of two types:

(*i*) **Static Balancing:** It ensures uniform distribution of wheel weight around its centre if gravity. Some metal is added by welding process at the proper place on the wheel rim to balance the weight of wheel.

(*ii*) **Dynamic Balancing:** In dynamic balancing, the unbalanced centrifugal forces acting on the wheel, during fast rotations, are balanced. The unbalanced forces create a harmful tendency in the wheel to rotate about two axes which develops unwanted tensions at the king-pin.

A wheel balancing machine is employed in motor garrages for wheel balancing purposes.

TYRES

Automobiles, tyres are designed to carry not only the vehicle loads but also provide a cushioning effect between road and vehicle chassis. The tyres are subjected to so many flexing actions due to road shocks during running. It consists of an outer cover, called tyre and a rubber tube inside. The tyre-tube assembly is mounted on the wheel rim and the air is filled through a tube-valve upto a specified pressure, called *Inflation Pressure.* Due to tightly filled air inside the tube, the tyre carries the vehicle loads and provide cushion. In short a tyre performs the following functions:

1. It smoothly rolls on road to move the vehicle.
2. It takes up vehicle loads.
3. It provides cushion against road shocks.
4. Provides cornering power for easy steering.
5. Transmits driving and braking forces to the road.

Some other desirable properties which a tyre should possess are:

(*i*) It should absorb least power of vehicle in flexing and releasing activities going on continuously on road.

(*ii*) It should not be noisy on road.

(*iii*) It should be perfectly balanced to avoid oscillation's–such as wheel wobble or wheel tramp.

(*iv*) Its tread should be such that it should wear uniformly to minimise skiding.

(*v*) It should be non-skiding. The proper tread pattern design restricts skiding even on wet/ muddy roads.

TYPES OF TYRES

1. Solid tyres—They are now obsolete and not used on automobiles.
2. Pneumatic tyres:

   (*a*) Conventional Tubed Tyres, (*b*) Tubeless Tyres.

(*a*) Conventional Tubed Tyres: There are widely used tyre-tube combinations on most of the vehicles. These tyres employ a thin and flexible rubber tube which is tightly filled with air. The deinflated tube is placed inside the tyre and then air is filled under pressure varying from 26 to 28 p.s.i. (pounds per square inch) or 160 to 190 kPa. The tube acquires the inner shape of the tyre when inflated. An air-valve, fitted on tube, allows the filling of air in and retains it without leakage. The cross-section of a tubed tyre is shown in Fig. 14.2. In order to prevent the tyre from being thrown out of rim, the plies are attached to two rings of high tension steel wires—called BEADS, which keep the tyre firmly attached to rim.

Tube is an endless rubber bag made from natural rubber, Butylene rubber or isoprene rubber etc. It is provided with a one-way tube valve for inflating. It is advised to place a flap of old rubber tube between rim and tube to save it from rust formed on rim of old wheel.

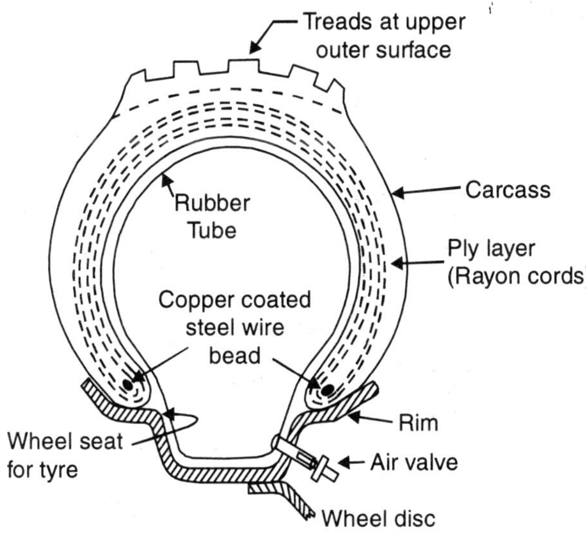

Fig. 14.2 A tubed-tyre in section

Skeleton or/Carcass is the basic structure of the tyre and it bears all types of loads. A number of plies wound in a particular pattern from cords of rayon are placed inside carcass for reinforcement.

(*b*) Tubeless Tyres: It does not employ a separate tube for air-filling. It has a rigid outer cover with special air-retaining inner lining moulded inside the tyre. It is such designed that it holds the air under pressure between itself and the seat on rim. It contains air-tighting inner coating of cord fabric chlorobutyl rubber which retains the high pressure air. The air-valve is fitted in hole of the wheel rim. These types of tyres are lighter in weight and do not heat-up during running. Any nail puncture, with nail still inside the tyre, results in very little or no air-leakage, as long as the nail is not extracted out.

The tubeless tyres have following **advantages** over tubed-tyres:

(*a*) Better cooling due to absence of rubber tube which restricts the flow of heat to atmosphere.

(*b*) Being lighter, unsprung weight is lesser, which gives fuel economy as well as reduces wheel-bounce.

(*c*) Leakage of air is slower or absent as long as nail is retained inside the tyre, it gives improved safety because of no sudden disbalancing on being punctured.

(*d*) It is easy to fit it over the rim or take it out of rim.

The disadvantage of tubeless tyre is that if it is removed from the rim for repairing, the making of an air-tight joint with rim is very difficult. It has to be done with special adhesives and equipment. A tubeless tyre has been shown in cut-section in the Fig. 14.3. In India, tubed tyres are more popular due to ease of puncture repairs. But in advanced foreign countries, tubeless tyres have almost replaced the tubed tyres from all sort of vehicles. In India too, most of the new models of cars are employing tubeless tyres.

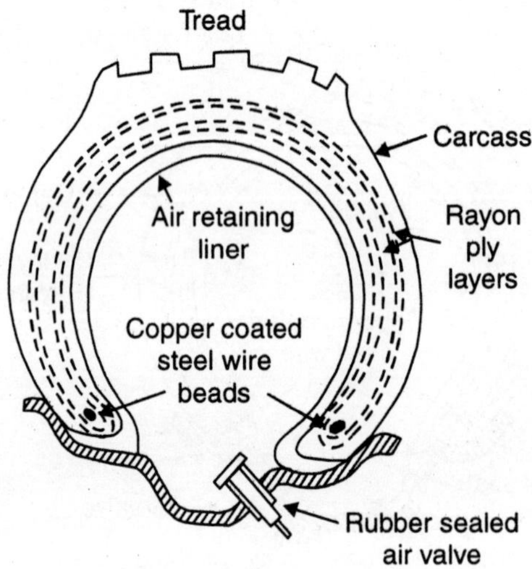

Fig. 14.3 Tubeless tyre

The tyre has three basic parts: Carcass (body), Tread and Beads of metallic wires.

TYPES OF TYRE SKELETON OR CARCASS

Carcass of the tyres may be divided into three categories:

1. Cross ply or Bias ply tyres
2. Radial ply tyres
3. Belted-Bias ply tyres.

1. Cross ply tyres: In this type of tyre, the carcass is made from rubber castings with cotton fibres or staple cotton, or rayon, or nylon or terylene layers. The layers of these fibres are placed unwoven at an angle varying from 30° to 40° from axis of tyre. There are two layers of cords in a ply set which run in opposite directions or cross-wise as shown in Fig. 14.4 (a). Many vehicles are fitted with cross-ply tyres. Number of plies are provided in the tyre according to its load carrying capacity. There are 4 to 30 ply tyres in use on various types of vehicles.

2. Radial ply tyres: In this type, the ply cords run in radial direction, that is, in the tyre-axis direction. It is shown in Fig. 14.4 (b) over these cords, a number of flexible and inextensible material's breaker strips are placed in circumferential direction to provide lateral stability and directional stability. Copper or Bronze coated steel wires casted with tyre rubber, called beads, help in keeping tyre firmly seated inside rim. Radial tyres maintain greater area of contact with road.

3. Belted-bias ply tyres: In this type, a number of stabiliser belts also fitted over the bias or cross plies. Thus a belted bias-ply tyre is a combination of bias and radial tyre. The belted bias ply tyre has following advantages:

(*i*) Tread area is stabilised due to belts which increase the life of the tyre.

(*ii*) Stresses in the carcass are restricted by belts.

(*iii*) The breaker belts hold the tread-flatter against road surface, which gives more traction and safety to vehicle.

(*iv*) Use of belts increases resistance against punctures and tyre cuts. A belted-bias ply tyre is shown in Fig. 14.4 (c).

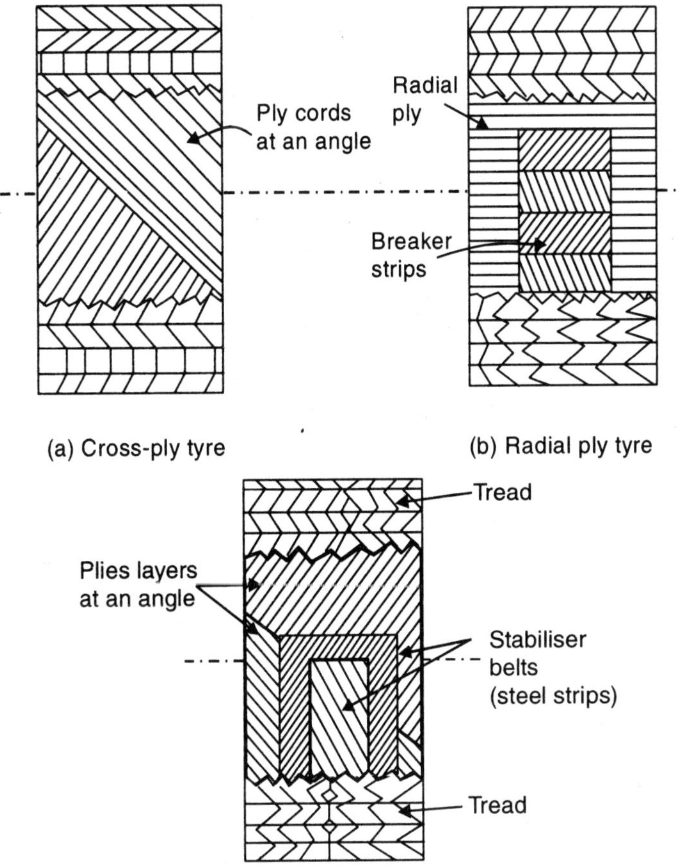

Fig. 14.4 (a,b,c) Various types of carcasses

Aspect Ratio of a Tyre (A_r): It is a geometrical parameter of a tyre section.

$$A_r = \frac{\text{Height of Tyre Section (H)}}{\text{Width of Tyre Section (W)}}.$$

It is expressed in percentage, If H = 0.7 W, the A_r = 70% tyres with A_r as 70% are very popular on passenger cars.

Tyres may also be classified according to *inflation pressure:*

1. **High pressure wide rim tyre:** They are heavy duty tyres. The inflation pressure ranges from 70 p.s.i or 5 kgf/cm^2 to 100 p.s.i. They are employed in heavy duty trucks, buses, load carriers and some military vehicles.

2. **Low pressure tyres:** They are used on light vehicles and cars etc. The pressure ranges from 25 p.s.i. or 2 kgf/cm^2 to 35 p.s.i. They have good frictional grip with road without slip. Steering is also easy.

$$(14.7 \text{ p.s.i.} = 1 \text{kgf/cm}^2 = 98.1 \text{ kPa.})$$

3. **Extra low pressure tyres:** They are more thick and wider. They are also called 'Balloon type tyres'. They are used in sandy road or in fields. The vehicles running in deserts are fitted with extra low pressure tyres. Tractors are also fitted with these tyres. They maintain a good grip on sandy roads and do not slip but steering is tiresome. The inflation pressure is kept at 16 p.s.i. or about 1.1 kgf/cm^2. Racing cars also employ these tyres with powerful hydraulic steering system.

   Bullet proof tyres are used on military as well as V.I.P. vehicles.

Tread Patterns: The grooves of various designs are made on the outer surface of the tyre. They are termed as tread patterns. A proper tread pattern is necessary on tyre for smooth steering, good mileage, low noise during running, increased and perfect road grip without slip, lesser wear, perfect stoppage on braking and increased driving comfort and directional stability. Worn-out treads tyres slip on road while braked; they should be changed by the earliest. Re-treaded tyres are also used but their durability is not satisfactory and guaranteed. Various tread patterns are shown in Fig. 14.5.

The tyre tread also provides good drainage to water and mud lying on roads. The mud or water builds up a wedge in front of the tyre and if the tyre rides this water wedge, it will loose ground contact. This situation is termed as 'Acquaplaning'. The tread channels allow this water and mud to drain out of the tyre surface and such maintain the tyre-grip on road firmly.

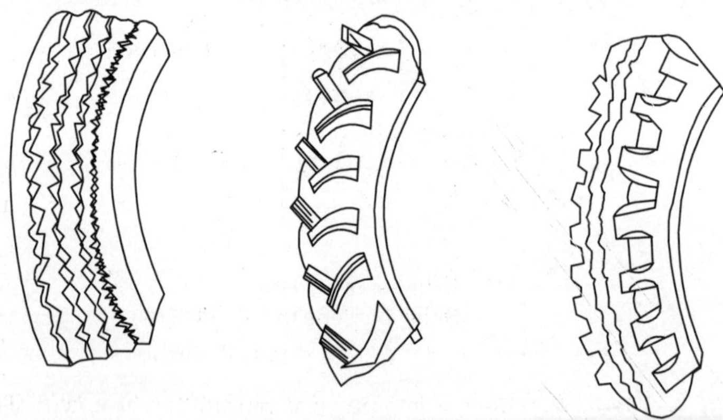

Fig. 14.5 Different tread patterns of tyres

EFFECTS OF INFLATION PRESSURE

The air-pressure inside the tyre must be maintained at the prescribed level. For proper working and longer life of the tyre, correct inflation pressure or air-pressure is must. Both conditions—lower pressure and over-pressure are harmful to the structure of the tyre.

Under pressure tyre or under-inflated tyre has lower pressure than the recommended pressure.

Over-inflated tyre has higher pressure than the recommended limit.

Generally, recommended pressure is kept lower at front wheels and a bit higher at rear wheels due to unequal load-distribution on front and rear axles.

The three states of inflation pressure—(*i*) Under-inflated tyres, (*ii*) Normal inflated tyres and (*iii*) Over-inflated tyres are shown in Fig. 14.6.

But Maruti 800 has 27 p.s.i. in all the four wheels and Maruti Zen also has equal pressure in all four wheels as 30 p.s.i.

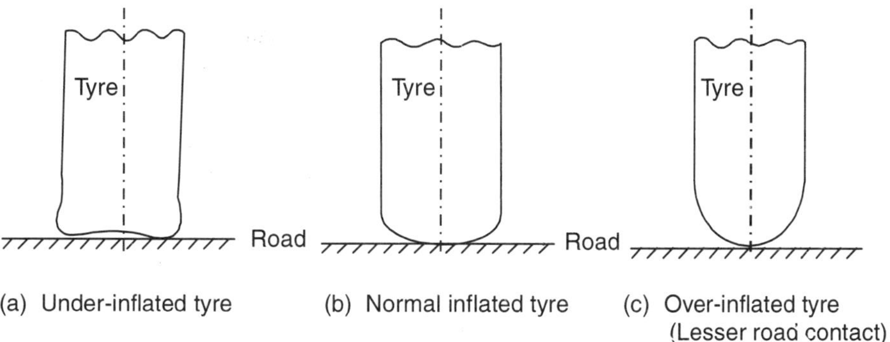

Fig. 14.6 Effect of inflation on tyre

Bad Effects of Under-inflation

(*i*) Excessive flexing develops cracks in the tyre skeleton.

(*ii*) Separation of plies and loosening of cords occurs.

(*iii*) Treads wear more on sides.

(*iv*) Causes impact between rim and side-walls of tyre and gets damaged.

Bad Effects of Over-inflation

(*i*) Rapid wear in the centre.

(*ii*) During fast driving, fabric cords tear and breaking is less effective due to lesser contact with road, also slips on wet road.

(*iii*) Separation of plies and concentration of heat at central layers; undue stresses and strains in tread pattern cause damage to tyre.

(*iv*) Uncomfortable to passengers as cushioning effect is diminished.

(*v*) On over-heating, it may burst and cause accident.

SIZE OF TYRES

(*i*) *Cross-ply tyres:* Specified either by (*i*) Ply-rating (PR) or by (*ii*) Speed rating (SR).

If W = width of tyre and D = diameter (in inches) of the rim of wheel.

Example: '5.20 – 14–8 PR Nylon Tyre'

where W = 5.20 inches, D = 14 inches,

$\therefore$ Diameter of wheel = 5.20 + 14 + 5.20

= 24.40 inches

if H = W (assume)

and it contains 8 plies of Nylon cords (PR rating)

SR ratings specify maximum speeds such as VW → 150 kmph, SR → upto 170 kmph, HR → upto 210 kmph, VR → upto 240 kmph, and ZR → above 240 kmph.

(*ii*) *Radial tyre*. 145/70 SR 13 (tyre specification for Maruti Zen car)

145 mm = Cross-sectional width of tyre.

$$70 = \text{Aspect Ratio}\left(\frac{H}{W} = 0.7 \text{ or } 70\%\right)$$

S → Maximum speed (170 kmph)

R → Radial construction

13 → Diameter of rim in inches.

(*iii*) For 2-wheelers, size is written as 2.75–18 PR4

where W = 2.75 inches, Dia of rim = 18 inches and 4 plies.

Selection of Tyre

The selection of tyre depends on the following three factors:

(*i*) The load to be carried by tyre.

(*ii*) Condition of surface of the road on which vehicle will be moving.

(*iii*) The maximum speed of vehicle.

MAINTENANCE OF TYRE AND TUBES

Tyres and tubes, if maintained properly, may last from 1 lac to 1.5 lac kilometers of run. Few points are being given below for their proper maintenance:

1. Do not overload the vehicle, it causes loosening of side plies.
2. Do not run fast, drive at 50–60 kms per hour.
3. Avoid frequent braking. Control the speed by accelerator in normal driving.
4. Load the vehicle with uniform distribution of load on all over the frame.
5. Tyres should not be under-inflated or over-inflated.
6. Use all the tyres of the same sizes. Never use tyres of different sizes or makes.
7. Periodical checking of Toe-in, Toe-out, Camber, Caster angles is advised for longer life of tyres.

8. Wheel balancing and wheel-alignment at suitable intervals will not only increase the life of tyres but also increase driving comfort and smooth running.
9. To enhance their useful life, tyres must be rotated in proper sequence at long intervals.
10. Keep tyres clean from penetration of stone pieces, nails and other solid particles in the tread pattern.
11. Avoid dropping of oil, grease or battery acid on tyres.
12. Drive slowly on rough roads with least load, if possible.

ROTATION OF TYRES OR WHEELS

The periodic rotation of tyres of a vehicle, increases the useful life of the tyres. A more uniform wear on all over the surfaces of tyres is achieved by rotation. The gross weight of vehicle is more at rear axle (about 60%) and, therefore, rear tyres wear-out rapidly than tyres on front side. Ideally the vehicle should have all the tyres of the same configurations. It is, therefore, essential to change the locations of tyres after certain distance travelled (about 5000 kms). The changing or rotating the tyres is called **Tyre Rotation.** Tyre rotation increases life of tyre by achieving uniform wear over its surface and on all tyres of the vehicle. For rotation of tyres, manufacturers instructions should be followed. An example of tyre rotation is shown in Fig. 14.7.

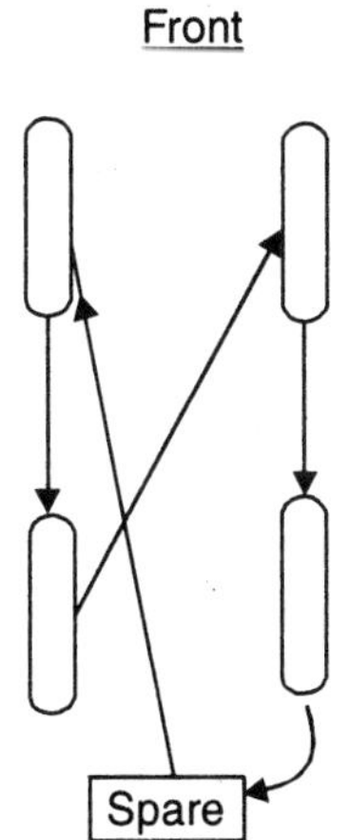

(i) Tyre rotation for front wheel drive vehicles

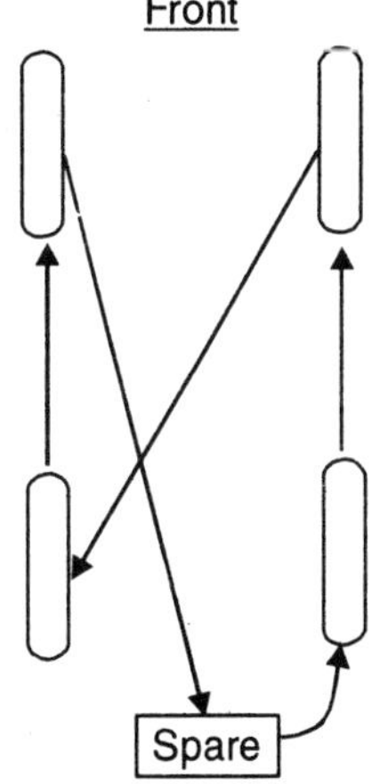

(ii) Tyre rotation for rear wheel drive vehicles

Fig. 14.7 Tyre or wheel rotation in a 4-wheeler

Precautions for Storing Tyres

If tyres are stored in a tyre shop, the following precautions should be observed:

1. The tyres should better be stored standing vertically on their treads. If stacked horizontally, the height of the stacks of tyres should not exceed 6 feet or roughly 2 meters, otherwise their treads may buckle and tyres will get compressed.

2. The tyres should be kept away from sunlight, gas stoves, electric heaters, electric motors or other electric heating appliances. Ultra-voilet rays of sunlight damage the tyres.
3. No fresh air circulation should be permitted in the tyre store because the oxygen speeds up rubber deterioration. The store should be closed for air entry.
4. Petrol, Diesel oil and grease should not come in contact with tyres. Petrol or diesel should not be stored in the same room. Fire-extinguishers also be installed to put-off fire, if lighted by any reason.

Tyre and Tube Repair

(*i*) If damage is not deep and confined to outer tread layer, no repair is required.

(*ii*) In case the damage is deeper up to the upper ply cords, the vulcanised rubber should be used for repair. It will prevent moisture and dirt from reaching the inner cords or casing.

(*iii*) If tyre is showing bubble shape projections over its surface, or the cut is deep and wide, change it with new one.

(*iv*) Holes left by extracting small nails from punctured tyre, need no repairs.

(*v*) Tubes may be repaired by vulcanising or by small round shaped sealing rubber pieces. Tubes having more than 4 sealing patches should be changed. An old tube's rubber strip wrapped round the wheel rim over which the tube rests, increases the life of tube.

TYRE RETREADING

It is a process of reconditioning of an old worn out tyre at tread. The tyre-treading gives a new lease of life to the old tyre. A tyre may be retreaded two to three times. The cost of a retreaded tyre comes out to be just half of the new tyre's cost. A tyre with torn plies or/and casing should not be retreaded.

In retreading process, the tread rubber coating is applied on the buffed surface of tyre and then it is vulcanised in the *curing mould* with the help of heat of steam. While heating the tread rubber, a suitable pressure is applied on it to fix and firmly adhere it over the surface of tyre.

Tube Vulcanising: The tube for tyre is generally made of natural rubber whose monomer (single unit) is isoprene. It has two double bonds—one is used in chain forming and the other in producing an elastomer which is stronger than rubber. It is formed during tube-vulcanising.

Formation of elastomer needs heating of raw natural rubber in the presence of sulphur under a specific pressure. Sulphur forms covalent bonds with carbon and cross links the chain. This process is called *vulcanisation*. The vulcanisation produces a more rigid and stiffer rubber. Different types of vulcanisings are used in various types of rubbers, such as:

(*a*) Sulphur, Magnesium Oxide, Zinc Oxide are used for Neoprene rubber and

(*b*) Di-isocyanates for urethane rubber.

The diagram below shows how the sulphur cross links the chain during vulcanisation.

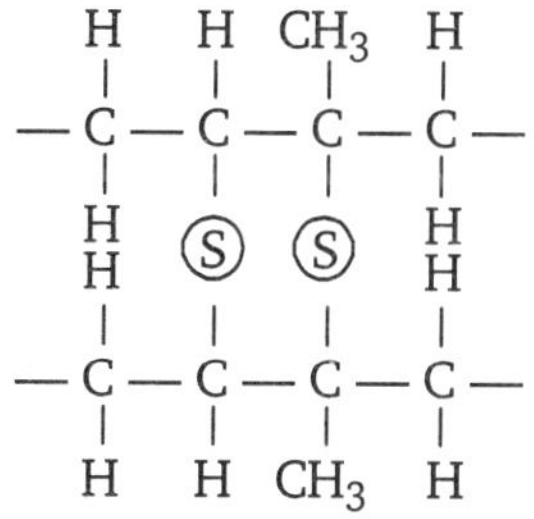

```
    H   H  CH3  H
    |   |   |   |
  — C — C — C — C —
    |   |   |   |
    H  (S) (S)  H
    H   |   |   H
    |   |   |   |
  — C — C — C — C —
    |   |   |   |
    H   H  CH3  H
```

TERMS RELATED WITH WHEELS

1. **Wheel Wobble:** Non-concentric rotation of wheel with jerks and side pulls at low speeds. It may occur due to worn out hub bearings, sagged axle shaft or eccentric wheel and tyre assembly and unequal tyre wear.
2. **Wheel Bounce:** Wheel knocks and makes noise while moving and it does not roll properly. It may be due to unbalanced wheel and improper inflation in any of the wheels.
3. **Wheel Tramp:** It is the vertical movement of front wheels at higher speeds. It may be due to defective front shock absorber, loose track rods and unbalanced wheels.
4. **Wheel Shimmy:** The front wheels oscillate about steering axes or stub axle joints at high speeds. It is caused due to buckled rim, faulty steering adjustments unequal pressures in tyres, and unbalanced wheels. Hydraulic shock absorber may be faulty.
5. **Tyre Wear:** (*a*) In the central tread—due to over inflation pressure.

   (*b*) On sides—due to overloading, wrong toe-in and camber angle, under inflation of tyres.
6. **Wheel Wander:** The vehicle does not run in straight direction but drives to sideways. It may be due to (*a*) Tyre pressure unequal on both sides, (*b*) Steering knuckle bearing may be over tight, (*c*) Incorrect caster angle.
7. **Wheel Alignment:** For the smooth running of the vehicle, all the wheels must roll freely at their axes. The term wheel alignment relates to the relative position of the wheels for achieving a true and free rolling movement over the road. The true wheel alignment gives an easy steering and directional stability. Wheel alignment is the mechanics of adjusting the following inter-related factors:

   (*i*) Toe-in
   (*ii*) Toe-out
   (*iii*) Caster angle
   (*iv*) Camber angle
   (*v*) King-pin inclination
   (*vi*) Combined angle
   (*vii*) Static and dynamic balancing of wheels.

A wheel alignment diagram is given in Fig. 14.8 under straight running conditions. Periodic checking of wheel alignment is advised to enjoy a comfortable ride.

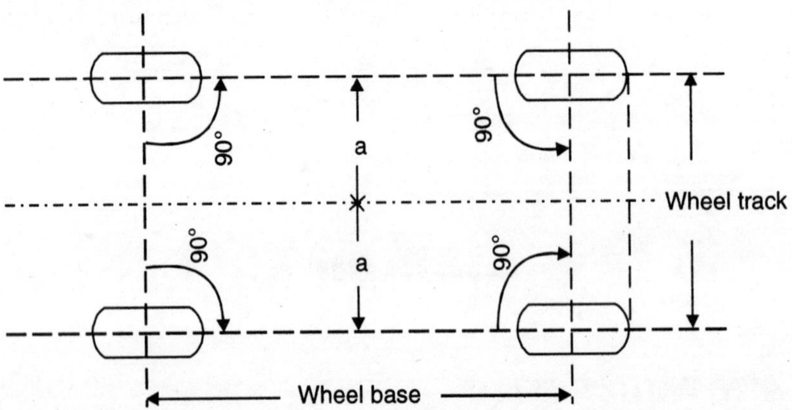

Fig. 14.8 Wheel alignment under straight running condition of a vehicle

QUESTIONNAIRE

1. How many types of wheels are used on automobiles?
2. What are the various requirements of a wheel? From which materials are they made?
3. Describe a wire wheel and its uses.
4. Differentiate between (*a*) Bias-ply tyre and a radial tyre, (*b*) Car wheel rim and a scooter wheel rim.
5. Classify various types of tyres and describe them briefly.
6. Draw a neat sketch of a tubed tyre and describe it briefly.
7. (*i*) How tubeless tyre differs from a tube tyre?
   (*ii*) What are the advantages and disadvantages of tubeless tyres?
8. (*i*) What functions does a tyre serves in a vehicle?
   (*ii*) How tyres are specified?
9. What is rotation of tyres? Is there any advantage of rotating the tyres?
10. Describe the following:
    (*i*) Under inflation. (*ii*) Over inflation.
    What are the effects of the above two?
11. For storing tyres, what precautions should be taken?
12. Write notes on:
    (*a*) High pressure tyres (*b*) Low pressure tyres
    (*c*) Extra low pressure tyres.
13. Describe the constructional details of various types of tyres and write the function of (*a*) wire beads (*b*) breaker strips.
14. Write short notes on:
    (*i*) Wheel wobble, (*ii*) Wheel tramp, (*iii*) Wheel shimmy, (*iv*) Wheel bounce.
15. (*i*) What are the causes of tyre wear? (*ii*) What is wheel balancing?

CHAPTER 15

Air-Conditioning System

INTRODUCTION

Passenger cars with built-in air-conditioning system these days are gaining popularity, even buses are being air-conditioned. The use of air-conditioning systems in cars, passenger buses and railway coaches make the journey more comfortable and less tiresome. The cars, without built in A.C. system, are provided with ready space for fitting air-conditioning system later on.

The air-conditioning system of automobiles is same as domestic air-conditioner. The air-conditioned vehicles has the following plus points:

1. The passenger's compartment is cooled enough to give relief to passengers from bad effects of hot weather.
2. During winter season, the passenger's compartment can also be kept warm by turning the switch knob to heater. The air-conditioning unit adds heat to air by flowing the return air from the cabin over the surface of a heating coil. The coil gets power from battery or alternator. During heating process, the compressor of the cooling system is switched off automatically.
3. Due to keeping window panes closed, inside cleanliness is also achieved.
4. There is proper air circulation inside the closed car during cooling.
5. The humidity inside the vehicle is also controlled.
6. Clean air is continuously provided for breathing in the car cabin.

The air-conditioning capacity needed for a car is larger as compared to domestic air-conditioning because of metallic body of the car and more circulation of the cooled air is required in the hot weather and that too in a moving car. It may vary from 1 Ton to 4 Ton of refrigeration.

AIR-CONDITIONING SYSTEM OF A CAR

A typical automobile air-conditioning system is shown in Fig. 15.1 and its pressure-enthalpy diagram is shown in Fig. 15.2. The complete air-conditioning unit consists of the following parts:

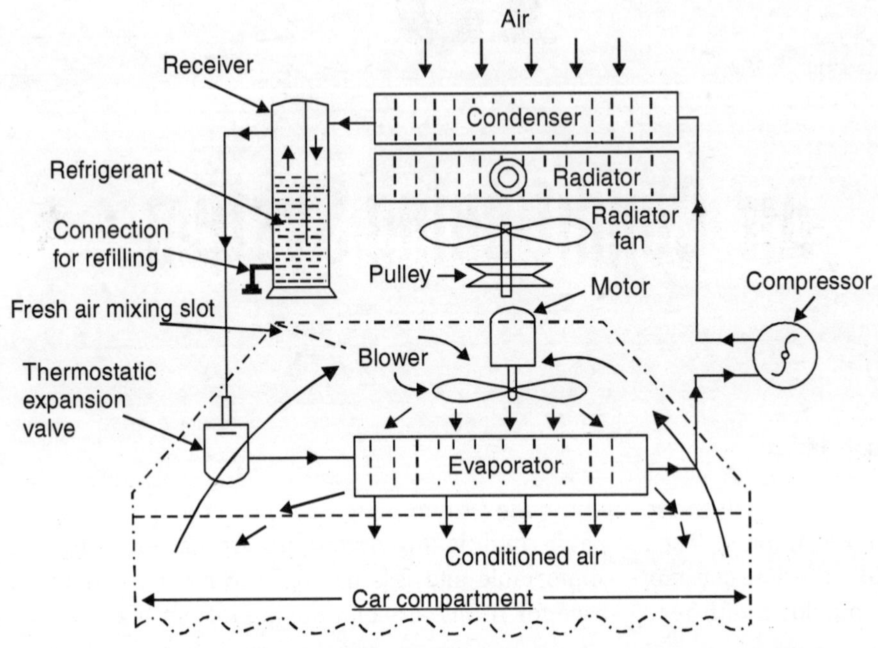

Fig. 15.1 Standard air-conditioning plant (Vapour compression cycle)

1. **Blower:** It circulates the cooled air inside the car compartment. It is fitted behind the evaporator.
2. **Fan:** A separate fan or the fan of the radiator is used for cooling the hot condenser unit.

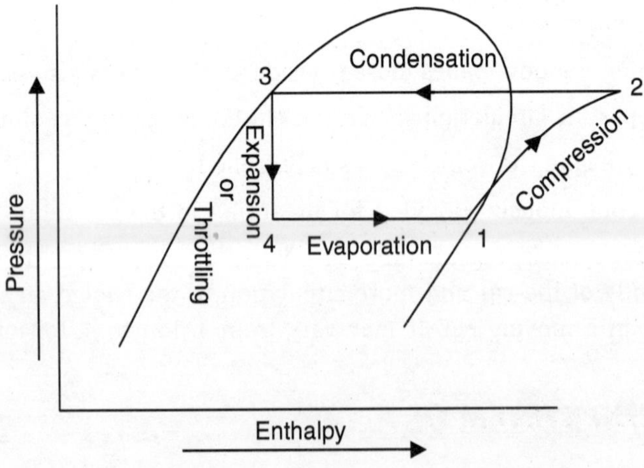

Fig. 15.2 Pressure-enthalpy diagram for vapour compression system

3. **Air-conditioning Unit:** It consists of the following devices:

   (*a*) A reciprocating or rotary/centrifugal type compressor driven by a separate motor or by pulley-belt system from the engine crankshaft.

   (*b*) A condenser—A set of tubes to cool the refrigerant.

   (*c*) A receiver—A container for storage of cooled refrigerant.

   (*d*) An expansion valve—A wire drawing capiliary valve for expansion of refrigerant.

   (*e*) An evaporator—To absorb heat of the car compartment by cooling the air.

4. **Supply Ducts:** The cooled air is directed to enter supply ducts which carry the cooled air to many outlets situated at Dash Board and interior compartment. The cooled air from outlets enters the car compartment and is circulated for its cooling. Some of the fresh air is continuously allowed to mix with the cooling air for easy breathing by passengers.

5. **Return Ducts:** These ducts are having their openings on the dash board which allow the air circulating inside the car to enter into the return ducts which lead to blower for recooling at the evaporator.

6. **Filters:** Simple air-filters of porous paper of cellulose fibre type are fitted in the return ducts to remove dust and dirt particles from air.

   Evaporator, expansion valve, blower and compressor are placed together enclosed behind the dash board. The condenser is placed in front of Radiator and the air from radiator fan cools both parts—condenser and radiator. Receiver is also fitted outside in engine cabin. It is provided a connection for refilling the refrigerant, in case it leaks out.

FUNCTIONING OF THE COOLING OR REFRIGERATION UNIT

(*a*) **The compressor** may be a reciprocating piston type or a centrifugal pump type. It is driven through pulley-belt system by engine crankshaft or may be driven by a separate motor. It draws the low pressure refrigerant vapour from the evaporator and compresses it to a high temperature and pressure and sends it to the condenser. A magnetic clutch is also fitted between compressor and driving shaft to cut it off from driving shaft when cooling is not required.

(*b*) **The condenser** is a heat exchanger. As the hot compressed refrigerant vapour from compressor flows through condenser, it looses its heat to the air passing through the air-passages over the condenser elements. The air to condenser is supplied by radiator fan or by a separate fan in case of larger vehicles. The vapours get condensed and becomes liquid. The liquid refrigerant at high pressure flows into the Receiver-Strainer and Dehydrator, all three functions are in one unit.

(c) **The receiver** acts as storage for the liquid refrigerant as well as a filter and drier. It serves the following purposes:

   (*i*) Receives liquid refrigerant from condenser and keeps it in reserve for supplying to evaporator.

   (*ii*) Absorbs moisture present in the system.

(*iii*) Filters dirt particles/lubricant droplets from the refrigerant. Lubricating oil drops are mixed with refrigerant in the compressor from its lubricant.

(*iv*) To arrest any refrigerant vapour, which could not liquefy in the condenser and hold it till it condenses.

(*d*) Thermostat controlled expansion valve: It controls the flow of refrigerant to evaporator and lowers down the pressure of the high pressure liquid refrigerant by passing it through capillary tube. The thermostatic expansion valve is a variable orifice valve. It has a temperature sensing gas filled bulb placed near the evaporator outlet. The changes in temperature cause the gas inside the bulb to expand or contract. This signal is transformed to the diaphragm to activate the valve. If temperature rises in evaporator, the valve allows more refrigerant to flow for absorbing heat for fast cooling.

(*e*) Evaporator: When the refrigerant at a low pressure enters the evaporator, it starts evaporating by absorbing heat from air passing through the air-passages in the evaporator. This air is sucked from the vehicle compartment and is cooled by passing over evaporator. The such cooled air is circulated in the vehicle compartment and is redrawn back by fan for cooling.

(*f*) Fan: Radiator fan is generally used to supply air to condenser which is mounted in front of the radiator.

(*g*) Blower: It is an air-circulating pump. It is a motor-driven exhaust fan. It draws air for the vehicle compartment through return ducts and passes on to evaporator. The air, when come in contact with surfaces of evaporator, it gets cooled and leads to the vehicle compartment through supply ducts.

REFRIGERANT

It is a special substance used for cooling purpose. It becomes liquid at a very low temperature and also vaporise at fairly low temperature. The refrigerant used in automobiles boils at –30°C at the normal atmospheric pressure. Its chemical name is dichloro-difluoromethane (CCl_2F_2). It is commonly called as FREON–12 or R–12. It is the trade name of the refrigerant.

The car air-conditioning system absorbs about 1 to 10 Horsepower of the engine which affects the fuel economy. The car mileage is dropped by 1.5 km per litre of the fuel (approx.) when the air-conditioning system is being used.

TIPS FOR MAINTENANCE OF A.C. SYSTEM

1. Always clean condenser during normal service of the car. Dirty and clogged condenser results in A.C. tripping and lesser cooling. Do not use A.C. with 'Fresh' mode open. This may lead to mixing of outside air in excess with the cooled air inside the car and reduce the cooling effect.

2. Maintain proper refrigerant level. Less refrigerant results in low cooling. Similarly, excessive refrigerant also result in low cooling. Do not open windows frequently with A.C. 'ON'. This will result in mixing of hot outside air with cool air and reduce the cooling effect.

3. Clean evaporator properly to remove dust and foreign particles. Clogged evaporator results in less air flow and reduced cooling. Also check leakages in evaporator and condenser tubes at a pressure of about 20 kg/cm^2. Do not exceed the pressure limit otherwise A.C. system may be damaged.

QUESTIONNAIRE

1. Describe briefly the various components of an air-conditioning unit of a car.
2. What is the function of an expension valve? If it is removed, the air-conditioning unit will work or not.
3. Describe briefly the functions of the following devices:
   (*i*) Compressor
   (*ii*) Receiver
   (*iii*) Expansion valve
   (*iv*) Temperature sensing bulb.
4. How a car A.C. system differs from the domestic A.C. units?
5. Why car air-conditioning is done? Are there any disadvantages?
6. Describe maintenance procedure of an A.C. system of a car.
7. Draw a neat sketch and describe the working of A.C. system of a car.

CHAPTER 16

Maintenance and Service

Automobiles need day-to-day care for good looking and non-stop smooth working. There is a famous saying, 'a stitch in time, saves from nine'. This is very much true with your vehicle. Besides daily cleaning, you have to keep a close watch on any type of some new unusual noise, which was not present before. As soon as the noise comes to your notice, try to find out the reason and take help of some expert mechanic to rectify it. Proper and timely upkeeps will not allow development of any big trouble in your beloved vehicle.

Maintenance done regularly and properly enhances the good working life of the vehicle. Usefulness of the car or any vehicle depends on two major factors:

(*i*) How the driver has run it?

(*ii*) How the maintenance schedule of vehicle has been looked after?

If the vehicle has been driven with care, by using good driving habits, the wear and tear automatically is reduced and good mileage is maintained and the vehicle seldom asks for any service.

The other important factor is maintenance. In order to preserve full working/useful life of the vehicle, the maintenance must be carried out regularly at the prescribed intervals, as suggested by its manufacturer, and in the prescribed manner.

In order to obtain economical and trouble-free running under all conditions, what is required is the thorough and regular attention to its needs. These days of high and dense vehicular traffic, a well-maintained vehicle is a must or it will create lot of troubles not only to the owner but may also create traffic hazards and danger to fellow vehicles.

'Dealing with the progressive harmful effects of wear and tear and taking required steps to prevent further development of troubles is called *Maintenance.*'

The process of maintenance may be classified as under:

1. Preventive maintenance
2. Breakdown maintenance
3. Overhauling.

PREVENTIVE OR PERIODIC MAINTENANCE

Preventive maintenance aims at checking the slow deterioration in working efficiency which occurs due to running long distances and driving styles. Restoring the lost performance and stopping its development into any major trouble is called preventive maintenance. Preventive maintenance may be carried out in the following sequences:

(*a*) Routine or daily maintenance: It includes the following checks and remedies:

(*i*) Amount of fuel in vehicle, if it is in reserve, refill it.

(*ii*) Check oil-level in crankcase.

(*iii*) Check inflation pressure in tyres.

(*iv*) Check the brakes, if not working properly, immediately get them repaired. Also see Brake-oil level.

(*v*) Check all lights and indicator lights.

(*vi*) See working of horn.

(*vii*) Check level of coolant in the water jacket.

(*viii*) Exterior and interior cleanliness.

(*ix*) Seat Belts, air-conditioning system, music system etc. should also be checked.

(*x*) Check number plates; sometimes the plates are broken/missing/lost. Correct the number plate.

(*xi*) Driving licence, insurance papers and registration papers should be valid and remain in the car.

(*xii*) Movement of wipers of wind shield should be checked on wet glasses only.

(*b*) Weekly maintenance: Check the following:

(*i*) Engine oil level.

(*ii*) Coolant level.

(*iii*) Brake-oil level.

(*iv*) Tyre pressure: Always maintain optimum tyre pressure. It increases life of tyre and improves mileage.

(*v*) Battery condition and cleaning of connector plugs.

(*vi*) Wind shield washer fluid level.

In case of automatic transmission check:

(*i*) Clutch fluid level

(*ii*) Transmission fluid level

(*iii*) Air-bags

(*iv*) ABS (Anti-lock braking system) with EBD (Electronic brake-force distribution)

(*v*) Antitheft devices.

(c) **Monthly/specified distance checks:** Manufacturers of the vehicles supply the maintenance chart for vehicle maintenance. The schedule prescribed should be followed strictly. It will improve the car performance. The services/repairs are carried out according to the distance covered by the vehicle or the time passed between the services. Some of such maintenance schedules are being given below:

(*i*) **During running-in period** for new vehicle (about 1500 kms for 2-wheelers, and 2000 to 5000 kms for 4-wheelers) in which newly assembled engine parts have some degree of surface roughness which slowly smooth down during running of engine, the vehicle must be run at the speed limits provided by the manufacturer and follow the other instructions too. In this period, engine oil and oil filter are to be changed at every free-service to be done by the dealer.

A bundle of instruction for maintenance of car is being given below for periodic maintenance:

Every 500 kms (weekly)

1. Check water level in radiator Add water/coolant if level is low.
2. Check tyre pressures.
3. Check tightness of wheel nuts.
4. If fitted on your car, also check advanced safeguards such as ABS (Anti-lock braking system), anti theft devices, air bags etc.

Every 2000 kms (monthly)

Add to above list, checking of electrolyte level in the battery. These days batteries require no water refilling for more than two years. Wash the car check brakes, brake oil level and coolant level. Also see working of lights and horn and wipers. Interior cleaning must be done.

Every 6000 kms (3 monthly)

Add to above list:

(*i*) Change oil filter and engine oil if it has lost its viscosity, otherwise top-up to the level.

(*ii*) Check fan belt tension. Belt sag should be 1 to 1.5 cm under a pressure of 10 kg.

(*iii*) Check C.B. points gap (0.42 to 0.48 mm). Adjust by sliding the stationary C.B. carrier plate after slackening its screw at distributor unit. Retight screw after adjustment.

(*iv*) Check spark plug points. Clean and adjust gap at 0.7 mm.

(*v*) Check shock absorbers. There should not be any oil leakage.

(*vi*) Check steering column and make the required adjustments and lubricate.

(*vii*) Decarbonise engine head and silencer (ordinary).

(*viii*) Carburetter cleaning or fuel injectors cleaning, checking of F.1-pumps and their calibration should be done.

(*ix*) Clean air cleaner.

(*x*) Lubricate or apply grease at all the lubrication points mentioned by the manufacturer.

Every 12000 kms (or Six monthly)

(*i*) Change engine oil, if required.

(*ii*) Check clutch's free pedal play and adjust.

(*iii*) Steering system should be lubricated at all the moving joints/gears.

(*iv*) Check up wheel bearings play, replace them if needed.

(*v*) Adjust Brakes and their shoes. Bleeding of brakes may be carried out if brakes are not doing well.

(*vi*) Change spark plugs, if required. Also check H.T. cables.

Every 24000 kms (yearly)

Do all the above checks and add the following:

(*i*) Generator's commutator is to be cleaned. Check brushes for wear etc. Change brushes if worn out.

(*ii*) Check working of thermostat in cooling system. It operates between 80°C to 85°C.

(*iii*) Check valve gear mechanism and valve seats, lubricate it.

(*iv*) Check Differential and its bearings. Also check its lubricating oil.

(*v*) Clean self-starter commutator.

(*vi*) Check Back-lash in steering gear and rectify.

(*vii*) Air cleaners cartridge or paper strainer should be changed.

(*viii*) Check Air-conditioning system. Refill refrigerant if it has fully or partially leaked out.

(*ix*) Check seats, seat belts, and dash board instruments.

(*x*) Carry out engine performance testing as:

(*a*) Tune-up testing (general checking of engine parts) and

(*b*) Trouble shooting to repair or remove any type of trouble experienced by the driver.

(*xi*) Change spark-plugs positively. Replace them with plugs of the same specifications.

BREAKDOWN MAINTENANCE

It is a non-periodic maintenance. It is to be done only when some specific trouble is developed in engine or chassis. The vehicle is to be taken to some authorised and recognised garrage for repairs. The trouble-making parts are dismantled, checked for accuracy, repaired or replaced with new one and reassembled. After replacing of parts or repairs, the original performance should be restored.

OVERHAULING OR RECONDITIONING

After long runnings of new vehicles or an old vehicles require frequent breakdown maintenances which become a headache to the owner and he has to spend his lot of time and money both on repairs. In such conditions, it is adviseable to go for complete overhauling of the engine and the chassis and body. Reconditioning of engine is done to bring back it original performance.

(*a*) **Engine Overhauling:** The engine is completely dissembled and the defective parts are brought to original conditions or original specifications by replacing or repairing. The repairs may include reboring of cylinder, grinding of valves and their seats, changing of piston and piston rings, oil seals, bearings, gudgeon pin, crankshaft grinding, clutch repairs, decarbonization of engine cylinder, piston and cylinder head etc., cleaning of fuel supply system, Radiator repair, water jacket washing, repairs in starting motor, Alternator, ignition timing, Battery cleaning or changing. Valve gear repair. Air-conditioning system repairs.

(*b*) **Chassis Overhauling:** It includes gear box servicing, suspension repairs, wheel alignment, wheel balancing, wheel rotation, checking of axles, bearings, wheel mountings, propeller shaft, differential gear, braking system, steering system, lighting system and suspension system.

(c) **Body Repair:** Denting and painting of all body parts. Smooth opening and closing of doors and windows, repair of body from interior side, Dash board equipments. Car interiors such as repair of seats, seats covers, safety belts, glasses, mirrors, music system matings etc.

NUMBER PLATE SPECIFICATIONS

The motor vehicle act specify the size for the numbers and alphabets on number plates as per following norms:

Type of vehicle	**Registration number sizes in m.m.**			
	Side	**Height**	**Thickness**	**Space between the two digits**
Motor Cycle & Scooter	Front	30	5	5
	Rear	Alphabets–35 Digits–40	7	5
Moped	Front & Rear	30	5	5
Auto Rickshaw	Front & Rear	35	7	5
Tempo Taxi	Front & Rear	40	7	5
Car/Jeep/Taxi/ Truck/Bus	Front & Rear	65	10	10

Private vehicles number plates should be written in black numbers at white background.

Commercial vehicles and Taxis should write numbers in black letters at yellow background.

The correct method of writing a number plate is:

UP32 AB0786 or UP32
AB0786

All the digits and alphabets should be upright.

Rate of fine for flouting the above rules is as follows 'as on June, 2008' fixed by R.T.O.

₹ 300/- for 2-wheelers and 3-wheelers.

₹ 600/- for 4-wheelers.

POINTS TO REMEMBER WHEN BUYING A NEW CAR

Think about what car model and options you want and how much you're willing to spend. Do some research. You'll be less likely to feel pressured into making a hasty or expensive decision at the showroom and more likely to get a better deal.

Consider these Suggestions

- Shop around to get the best possible price by comparing models and price in ads and at dealer showroom. You also should contact car-buying services and broker buying services to make comparisons.
- Plan to negotiate on price. Dealers may be willing to bargain on their profit margin, often between 10 to 20%. Usually, this is the difference between the manufacturer's suggested Monroney sticker retail price (M.S.R.P.) and the invoice price.
- Consider ordering your new car if you don't see what you want on the dealer's lot. This may involve delay, but cars on the lot may have option you don't want and that can raise the price. Learning the Terms of Negotiations often have the vocabulary of their own. Here are some terms you may hear when you're talking price.
- Invoice Price is the manufacturer's initial charge to dealer. This is usually higher then the dealer's final cost because dealers receive rebates, allowances, discounts, and incentive awards.
- Base Price is the cost of the car without options, but includes standard equipment and factory warranty. This price is printed on the Monroney sticker.
- Monroney Sticker Price shows the base price, the manufacturer's installed options with the manufacturer's suggested retail price, the manufacturer's transportation charge and the fuel economy (mileage) details should be affixed to the car window, in the showrooms of the dealer. This is required by the federal law, and it may be removed by the purchaser.
- Dealer sticker price, usually on a supplemental sticker, is the Monroney sticker price plus the suggested retail price of the dealers installed options, such as additional dealer markup (A.D.M.) or additional dealer profit (A.D.P.), dealer preparation, and undercoating charges.

Financing Your New Car

If you decide to finance your car, be aware that the financing services obtained by the dealer, even the dealer contacts money lenders on your behalf, may not be the best deal you can get. Contact lenders directly. Compare the financing they offer you with the financing the dealer offers you. Because offers vary, shop around for the best deal, comparing the annual percentage rate (A.P.R.) and the length of the loan. When negotiating to finance a car, be vary of focusing only on the monthly payment. The total amount you will pay depends on the price of the car you negotiate, the APR, and the length of the loan period.

Sometimes, dealers offer very low financing rates for specific cars or models, but may not be willing to negotiate on the price of these cars. To qualify for the special rates you may be required to make a large down payment. Before you sign a contract to purchase or finance the car, consider the terms of the financing and evaluate whether it is affordable. Before you drive off the lot, be sure to have a copy of the contract that both you and the dealer have signed and be sure that all blanks are filled in.

Some dealers and lenders may ask you to buy credit insurance to pay off your loan if you, by the way, die or become disabled. Check your existing policies to avoid duplicating benefits. Credit insurance is not required by federal law. If your dealer requires you to buy credit insurance for car financing, it must be included in the cost of credit. Check with your state Insurance Commissioner or state consumer protection agency.

SOME TIPS FOR CAR MAINTENANCE

Maintenance is something most of us ignore, until our vehicle stops functioning, that is fact. And then we wonder what went wrong and where. Maintenance is one of the most serious aspects of ownership. It determines the longevity, performance and reliability of whichever vehicle you drive. Looking after your car involves more than just taking care of its external coat of paint and keeping it clean and shiny.

Maintenance means taking care of all the parts, even those that are inside the bonnet. These are the ones that are directly concern the performance of your vehicle. Besides taking it to service station at regular periods, it is good idea to go through the owner's manual that will give you a fair idea about its routine maintenance.

Checking the battery, keeping a check on oils, changing the oils, checking the electrical system, are some of the absolutely unavoidable things to keep your vehicle in good shape. Keeping a log book in which you keep all the details regarding repair, maintenance, routine check-ups etc. will not only give you an accurate idea of what needs to be done when, but also help you to keep an eye on the vehicle performance.

Few More Maintenance Tips

A well maintained vehicle speaks volumes about you as an owner. More importantly, it will not desert you when you need it most.

- Get your vehicle serviced only at authorised service stations.
- Check the engine oil level once every two weeks. Also check monthly the levels of coolant and water in the radiator and battery.
- Always use genuine spare parts.

- Give your vehicle anti-rust coating before the monsoons.
- Avoid accelerating and braking abruptly during drive.
- Stop the engine whenever you expect to wait for more than 5 minutes.
- Drive at a moderate speed of 45–55 k.m./hrs. to maximize fuel efficiency.
- Use air-conditioning only when necessary. Don't open doors/window glass when A.C. is 'ON'.
- Avoid riding on clutch pedal, and release the clutch pedal fully while driving.
- Maintain optimum air pressure in your tyres to improve mileage.
- Clean condenser during normal service of the car. Dirty and clogged condenser results in A/C tripping and less cooling. Do not use A/C with fresh mode open. This may lead to mixing of outside air with cool inside thereby reducing cooling effects.
- Maintain proper refrigerant level. Less refrigerant results in less cooling. Excessive refrigerant results in less cooling. Do not open windows frequently with A/C "On". This will result in mixing of hot air with cool air and cooling effect will be reduced.
- Add at least 20 cc of DENSO-6 oil to the compressor during refrigerant charging. Adequate quality of compressor oil prevents poor pumping and premature wearing of internal components.
- Do not operate A/C without refrigerant in the system. This will result in premature failure of compressor.
- Clean evaporator properly to remove dust and foreign particles. Clogged evaporator results in less air flow and reduced cooling. Get it checked for the leakages. Check the A/C system with more than 20 kg./cm^2 pressure. Higher pressure may result in damage to A/C components.
- Adjust proper engine RPM. Less RPM will result in stalling of engine and less cooling.
- Check coolant level and ensure tuning of cooling system. This will prevent overheating of engine.
- Check position of accelerator cut off switch in 1000 cc cars. Improper position of accelerator cut off switch will result in tripping.
- Check proper belt tension. Loose belt will slip and over tight belt will result in noise and premature failure of the magnetic clutch of the compressor.

VARIOUS CAUTIONS DEPICTED ON ROAD SIDES

For safety on roads, carefully observe the road cautions displayed on road sides. The chart shown in Fig. 16.1 is being given for your guidance.

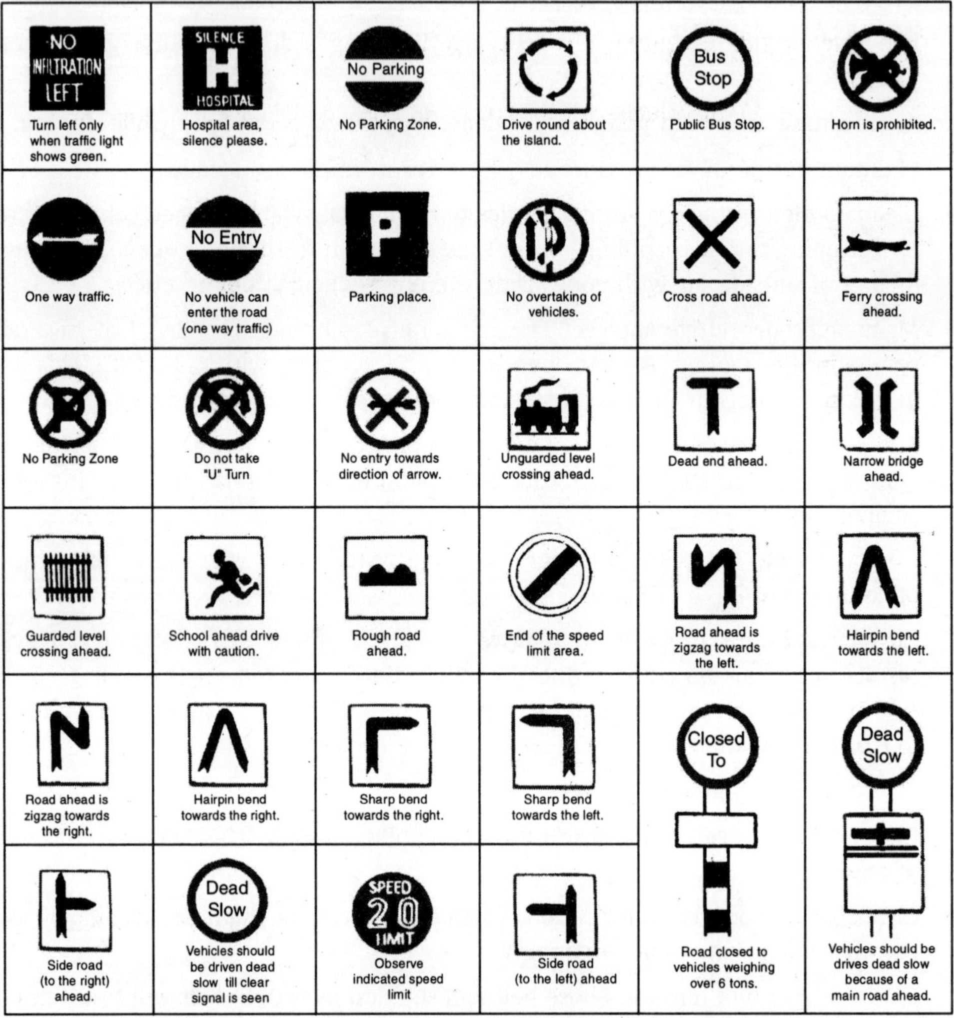

Fig. 16.1 Road cautions

QUESTIONNAIRE

1. Why maintenance of a car is necessary?
2. What activities are carried out daily for the maintenance of vehicle?
3. What is preventive maintenance?
4. Differentiate between breakdown maintenance and overhauling.

5. When Breakdown maintenance is required?
6. What are the operations done in overhauling of an I.C. engine used for driving a car?
7. What repairs and servicings are normally required on chassis and body?
8. After driving upto 6000 kms, what are the main activities to be done for preventive maintenance?
9. How many types of vehicle maintenances do you know? Describe them briefly.
10. When spark plugs should be changed? What is routine servicing of spark plugs?

CHAPTER 17

Exhaust Emissions Control

Industrialisation and faster means of transportation are the two basic parameters to measure the progress and prosperity of any country. Advancements of transportation are directly related with the rise of civilisation in the world. For transportation, we need energy. The general categories of sources of energy employed in the transportation are fossil fuels, nuclear fuels, electrical energy and solar energy. Solar energy is freely available all over the world and it is non-polluting and everlasting but it is very dilute and inconsistent. The nuclear energy is released when there is particle interaction with or within the atomic nucleus. The nuclear power has yet to find its place in the land and air transportation. Fossil fuels were formed thousand many years ago beneath the surface of the earth in deposits of finite extent. These are derived sources and are found in three states:

(*a*) Solid fuels (Coal)

(*b*) Liquid fuels (Petroleum) and

(*c*) Gaseous fuels (Natural gas).

Industrialisation and transportation means are responsible for many types of pollutions, mainly air pollution which may be defined as the presence of one or more contaminants such as dust, fumes, gas-mist, odour, smoke or vapour in the outdoor atmosphere in quantities or characteristics and/or duration such as to be injurious to human, plant or animal life, or to property which unreasonably interferes with the comfortable enjoyment of life. From the point of view of pollution control, measurement of emissions from furnaces and engines is very important. Exhaust emissions from automobile engines are about 50 per cent of the total air pollution. Emission from internal combustion engines may be divided into two groups.

EMISSION FROM I.C. ENGINES

1. *Invisible Emissions*

Such as Carbon dioxide gas, Water Vapours, Oxides of Nitrogen, unburnt Hydrocarbons (UBHC), Carbon monoxide and Aldehydes which are oxygenated hydrocarbons having a pungent smell.

2. Visible Emissions

They are smoke and particulate matter. Both are responsible for breathing troubles and many types of allergies in human beings.

Besides air pollution, sound or noise pollution is also produced by engines. Noise is created by vibrating parts and horns. The vibrations are transmitted to the surrounding air in the form of pressure waves. The high level of noise becomes a source of irritation for the listeners. Good silencing devices fitted on exhaust system are used for sound damping in internal combustion engines.

Nitrogen Oxides (NO_x)

It reacts photo chemically to form a variety of products including powerful irritant smog—a suspended fog like compound. The smog produces eye irritation and affects our respiratory system. It increases chronic fibrosis, emphysema and bronchopneumonia etc.

Sulphur Dioxide (SO_2)

It can further be oxidised to sulphur trioxide (SO_3), which after reacting with water vapour forms sulphuric acid aerosol. The level of sulphate emissions depends on the sulphur content of the fuel used. Average automobiles have safe sulphate emission rate 20 mg/km or less. These emissions have pungent smell and may cause lot of irritation in eyes, nose, throat and lungs. It also develops asthma, coughs, bronchitis and fatigue.

Soot or Suspended Particulate Matter (SPM)

Petrol is often mixed with Tetra-Ethyl Lead (TEL) to increase its anti-knock quality during combustion in spark ignition engines. Compounds of higher molecular weight and lead resulting from use of (TEL) are exhausted by auto engines in the form of very small size particles of the order of 0.2 to 0.6 micron. Few traces of products of partial oxidation are also found in the exhaust gas, of which Formaldehyde and Acetaldehyde are prominent. Other constituents are Acids, Ethers, Ketones and Phenols etc. They are formed due to incomplete combustion of fuel. Lead affects body cells to cause gametotoxicity and carcinogenicity, on embryo it causes embryotoxicity and teratogenicity. It also disturbs digestive system and nervous system and cause miscarriages to expectant mothers. To over come these problems, new unleaded petrol is being supplied to petrol depots along with other higher performance fuels, such as extra mileage fuels etc.

Numerous measures are being adopted to control the air pollution. These are:

1. Enforcement of anti pollution laws.
2. Proper Tuning of engines to control emissions.
3. Effective maintenance of vehicles and their systems.
4. Improvement in design of components such as carburettor.
5. Use of M.P.F.I. system in place of carburettors.
6. Mass plantation of trees.

Every citizen has to be very careful about pollution if he/she wants to live healthy and happy. A panel of road transport and highways ministry, Government of India, has proposed standards that could stop the sale of many of the existing cars. Middle and large segment cars may not qualify the norms and diesel cars will suffer badly. Automobile industry has to take lot of pains to manufacture fuel efficient and almost 'No' pollution vehicles in future. Present pollution levels are shown in Fig. 17.1, the pollution by vehicles is quite alarming.

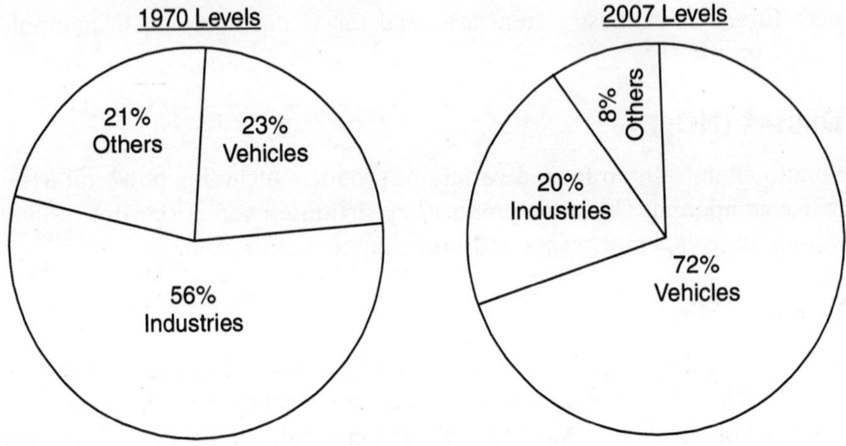

Fig. 17.1 Air-pollution levels in India

EMISSION NORMS

Pure Air contains roughly Oxygen (23% by mass) and Nitrogen (77% by mass) but in actual the atmospheric air is composed of Oxygen, Nitrogen, Carbon monoxide, Carbon dioxide, Water vapour and other gases along with various suspensions of fine solid or liquid particles called Aerosols. For survival of human beings, animals, vegetables and other plants, fresh air is must; which requires a strict pollution control system in the cities. The available air, particularly in cities is always polluted and the amount of pollution above certain levels is highly injurious to health.

Recommendations regarding healthy volume of air which should be available per person around his work place and the fresh air circulation in cubic meter per hour are being given below:

Volume/person in m^3	Fresh air flow/person in m^3	
	Minimum	**Desirable**
5	35	50
10	20	40
15	10	30

In our fast moving society, use of auto vehicles is going on increasing very fast. The thermal efficiency of I.C. engines used in auto vehicles is not more than 30 per cent and the remaining 70 per cent heat is lost to our environment. This emission of energy is responsible for disturbing the ecological balance.

The fuels used in internal combustions engines, petrol and diesel oil, are mainly refined petroleum products which are a complex mixture of various hydrocarbons (HC) with smaller amounts of sulphur. If combustion of fuel inside engine cylinder is complete, the exhaust gases would contain a mixture of Carbon dioxide, Water vapour, Nitrogen and Oxygen and few traces of Sulphur dioxide and some compounds of Lead which are very poisonous. But the combustion process of fuel in the engine cylinders is never complete in actual practice. The incomplete combustion of fuel is responsible for formation of such products which have undesirable properties like toxic substances, smoke, particulate formation etc. These products are called the air pollutants and must be kept within safe limits. For the enforcement of anti-pollution policies, laws have been promulgated by several countries all over the world. European countries follow EURO norms I, II and III and now adopting EURO IV norms. India has enforced emission standards from 01-04-2000. These are known as EURO II or BHARAT 2000 standards.

The following table gives Mass Emission Norms as per EURO II (European Union Research Organization) levels:

Enforced in India with effect from 1-4-2000

Type of fuel	Type of vehicle	Pollution		
		CO gm/km	HC+NO_x gm/km	SPM gm/km
1. Petrol	Small car	2.72–3.16	0.97–1.13	0.14–0.18
	Big car	5.17–6.00	1.40–1.60	0.19–0.22
	L.C.V.	6.9–8.0	1.70–2.00	0.25–0.29
	2 wheeler	2.0–2.4	2.0–2.4	—
	3 wheeler	4.0–4.8	2.0–2.4	—
2. Diesel Oil	All types of Diesel vehicles	4.5–4.9	HC– 1.10–1.23 NO_x–8.00–9.00	0.36–0.680 (For engines above 85 kW)

Carbon Monoxide (CO) is a very toxic and deadly poisonous gas. An increase in the Carbon monoxide content of atmospheric air would result in rise of temperature of the surface of earth. These gases are essentially transparent to short wave length solar radiation but absorb, in pronounced bands, the longer wave length solar radiation, which leads to the Green House Effect. It badly affects lungs and kidneys. It helps to promote heart problems and cancer in human beings and has some marked bad effects such as unconsciousness and sometimes it causes death of occupants of the car. In India, many cases of deaths in car's closed cabins have been reported recently.

United States Standard in the State of California w.e.f. 2000

Emissions gm/mill	NMOG		CO		NO_x	
	Level	Future reduction	Level	Future reduction	Level	Future reduction
(*i*) Current levels	0.250	—	3.40	—	0.40	—
(*ii*) Future levels of						
(*a*) TLEV	0.125	50%	3.40	NIL	0.40	NIL
(*b*) LEV	0.075	70%	3.40	NIL	0.20	50%
(*c*) ULEV	0.040	84%	1.70	50%	0.20	50%

References

NMOG : Non-Methane Organic Gases

TLEV : Transition Low Emission Vehicles

LEV : Low Emission Vehicles

ULEV : Ultra L.E.V.

AUTOMOTIVE EMISSION CONTROL SYSTEMS

There are following possible sources of atmospheric pollution from automobile vehicles:

1. Carburettor and fuel tank emit fuel vapours

They may be collected by a evaporation emission control system (EEC system) or a vehicle vapour recovery (VVR) system. The vapours in carburettor form after stopping the vehicle and in fuel-tank when vehicle is parked in sunshine. These vapours are collected in a canister containing charcoal which absorbs the fuel present in vapours.

2. Crankcase emits blow-by gases and fuel vapour

Positive Crankcase Ventilation (P. C.V.) is used to arrest Blow-By gases. Blow-By Gases, leaking through piston rings into the crankcase, form sludge and acids in the crankcase which clog the lubrication lines. Acids corrode metals. To take away the Blow-By gases, the air is taken from Air cleaner and flown in the crankcase; it takes up fuel and water vapours and then allowed to enter into the Intake manifold through a PCV valve. From Intake manifold it leads to engine cylinder again.

3. Exhaust emission

The exhaust gases coming out of the exhaust pipe or silencer pipe of an auto vehicle, contain:

(*i*) Unburnt fuel (Hydro Carbons)

(*ii*) Partly burnt fuel (CO)

(*iii*) Poisonous Gases (NO_x, SO_2, CO_2)

(*iv*) Some particles of lead (Pb) which are responsible for bad effects on nervous and digestive systems of human beings.

These compounds and lead particles pollute the air. To reduce the exhaust emissions, vehicles are equipped with air-injection system, MPFI systems, exhaust gas recirculation systems, and catalytic (Palladium) convertors which helps in converting NO_x into harmless N_2 and O_2. MPFI system has electronic control over the Air-Fuel ratio which reduces excess supply of fuel and less supply of air to the engine and thereby achieves complete combustion of fuel.

In future Ultra Low Emission Vehicles (ULEV) and Zero Emission Vehicles (ZEV) are likely to be on roads very soon.

Air pollution can effectively be controlled by Nature itself if we pay attention towards mass plantations in the cities and maintain cleanliness. Indoor pollution is also emerging as a major health hazard; it is caused by innumberable nano particles that can not be seen but are harmful. Air conditioned spaces are at a greater risk of nano pollution.

VARIATION OF EXHAUST EMISSIONS

Diesel and petrol are made from mixing various hydrocarbons. If the combustion of fuel is perfect, only carbon dioxide gas, water vapours and some quantity of air will be emitted out. In case the combustion is incomplete then CO, CO_2, N_2 and many unburnt hydrocarbons will be exhausted from the engine cylinder. Nitrogen reacts with oxygen at about 1100°C and forms Nitrogen Oxide (NO_2) which is very toxic and harmful. NO_2 is very high at A/F 15:1, but unburnt hydrocarbon are very less. (A/F $\rightarrow$ Air-Fuel Ratio)

If the air-fuel ratio is 15:1 or 16:1, the combustion is generally complete and polluting gases level is negligible. When rich mixtures 8–10–12 : 1 are used, particularly at low and idling speeds, the percentage of CO is very high and other pollutants percentage also increases. Hydrocarbons are also found more at higher speed decelerations.

Particulate matter and partial oxidation products of organic and inorganic compounds having higher molecular weight and lead compounds, which result from mixing of TEL (A Chemical Containing Lead) are exhausted in the form of very small size (0.02 to 0.06 microns) fine particles of lead compounds. By using catalytic converter in exhaust line, the pollutants are eliminated up to a tolerable level. To lessen pollution, use of lead containing petrol is now prohibited and lead free petrol is on sale.

QUESTIONNAIRE

1. (*i*) What do you understand by Invisible emissions and Visible emissions?

   (*ii*) What are the measures which can be adopted to control air-pollution?
2. What are EURO-II norm?
3. Write short notes on Automotive emission control systems?

(iii) Nitrogen oxides (NO) : NO, NO_2

(iv) Some particles of lead (Pb) which are extremely harmful and are [illegible] to the life of human beings.

These compounds and lead particles can be [illegible] are equipped with an injection system, [illegible] gas recirculation system and catalytic converters which help in converting [illegible] into harmless [illegible] electronic control over the A/F [illegible] supply of fuel and air supply to the engine and thereby achieves complete combustion of fuel.

[illegible] Ultra Low Emission Vehicle (ULEV) and Zero Emission Vehicle (ZEV) are [illegible] reality [illegible].

Air pollution can effectively be controlled by nature itself if we pay attention towards growing plants in the cities and maintain cleanliness. Indoor pollution is also [illegible] health hazard [illegible] caused by [illegible] pollution.

[illegible]

[illegible]

[illegible]

[illegible]

[illegible]

1. (i) What do you understand by [illegible]?

   (ii) What are the measures which can be adopted [illegible]?

2. What are [illegible]?

3. Write short notes [illegible] control [illegible].

Annexure-1

REFERENCES

1. Crouse/Anglin — Automotive Mechanics, TMH, New Delhi.
2. Er. Khaliluddin — Automotive Technology, Nav Bharat Prakashan, Meerut.
3. Dr. Kirpal Singh — Automobile Engineering, Standard Publications, Delhi-6.
4. Harbans Singh Reyat — The Automobile, S. Chand & Co., New Delhi-55.
5. R.P. Sharma — Automobile Engineering, Dhanpat Rai Publications, New Delhi.
6. K.M. Gupta — Automobile Engineering, Umesh Publications, Delhi-6.
7. Workshop Manual of Maruti, Ambassador, Matiz, Tata cars etc.
8. J. Welster — Automobile Engineering, Maintenance & Repairs, California.
9. Periodical, Journals and Magazines on Automobiles (Indian and foreign).
10. Gupta & Kaushisk, Practical Automobile Engineering — Kusum Printers, New Delhi.
11. Srinivasan, Automotive Mechanics, TMH, New Delhi.
12. Agarwal, Bhatia, Singh — Automobile Enginnering, KPH, Ludhiana.
13. KCG Verghese — Motor Vehicle Technology, United Press, Madras.
14. R. Rudramoorthy — Thermal Engineering, TMH, New Delhi.
15. K.K. Ramalingam — Internal Combustion Engines, SCITECH, Chennai.
16. P. L. Kohli—Automotive Chassis and Body, Papyrus Publishing House, New Delhi.

Annexure–2

SPECIFICATIONS OF FOREIGN AND INDIAN VEHICLES (Courtesy: Overdrive, The Popular Auto Magazine)

Some Specifications of Leading Auto Vehicles

Name	Engine				Performance					Efficiency			Space			
Name of Vehicle	Displacement (litres)	Max Power (PS)	Max Torque (Nm)	Gears	Acceleration 0–100 kmph (in sec)	Acceleration 1/4 mile (in sec)	Top Speed (kmph)	Braking 100 kmph-0 (m)	Braking 80 kmph-0 (m)	City (kmpl)	Highway (kmpl)	Overall (kmpl)	Kerb Wt (kg)	Fuel Tank (litres)	Boot (litres)	Remarks
AUDI A6 3.0TDI (Germany)	3.0	236	450	6A	7.4	15.3	241.6	41.1	26.6	7.9	13.6	9.3	1765	80	546	(+) Performance, dynamic handling (–) Stiff ride
325i, BMW (India)	2.5	215	246	6	9.6	16.8	215	39.8	25.1	7	16.1	9.3	1505	67	NA	(+) A driver's car (–) Interior space
Mercedes-Benz S350 (Germany)	3.5	276	350	7A	9.0	16.47	230	44.3	29.0	6.5	10.2	7.4	1925	90	560	(+) Power, luxury, looks, space and status. (–) Price
PALIO Stile 1.1 SLE (Italian)	1.1	57	92	5	20.7	21.4	147	NA	33.2	12.0	16.0	13.0	990	47	260	(+) Refreshed looks (–) Dated interiors
PALIO Stile sport	1.6	101	137	5	12.3	18.4	170	50.1	34.8	7.9	16	10.0	1140	47	260	(+) Performance (–) Fuel efficiency

Ford India Ltd., Via S.P. Koli Post, Chengalpattu-603 204.

Ford's operations in India have maintained a constant pace. The Ikon still manages decent numbers. Now with the launch of the Fiesta and the revamped Fusion, the American major has moved up a few gears.

Name of Vehicle	Displacement (litres)	Max Power (PS)	Max Torque (Nm)	Gears	Acceleration 0–100 kmph (in sec)	Acceleration 1/4 mile (in sec)	Top Speed (kmph)	Braking 100 kmph-0 (m)	Braking 80 kmph-0 (m)	City (kmpl)	Highway (kmpl)	Overall (kmpl)	Kerb Wt (kg)	Fuel Tank (litres)	Boot (litres)	Remarks
Fiesta Duratec ZXi (Ford India)	1.4	82	127	5	14.29	19.25	170	54.37	33.67	12.3	22.3	14.74	1130	45	430	(+) Driver's delight (–) Uncomfortable seats
Fusion (Ford India)	1.6	100	143	5	12.7	18.2	159.7	56.6	35.1	9.1	15.5	11.4	1140	45	337	(+) Exclusively, space (–) Fuel efficiency
Endeavour XLT 4×4 (Ford India)	2.5	143	330	5	15.3	19.6	145.3	56.3	34.1	8.5	13.1	10.1	1995	71	NA	(+) Decent off-roader (–) Size
Chevrolet Aveo 1.4 LS (General Motors, India)	1.4	94	127	5	14.3	19.2	174	42.4	30.1	10.6	21.3	13.2	1095	45	400	(+) Interiors, ride, braking (–) Vague steering

Some Specifications of Leading Auto Vehicles

Name	Engine				Performance					Efficiency			Space			
Name of Vehicle	Displacement (litres)	Max Power (PS)	Max Torque (Nm)	Gears	Acceleration 0–100 kmph (in sec)	Acceleration 1/4 mile (in sec)	Top Speed (kmph)	Braking 100 kmph-0 (m)	Braking 80 kmph-0 (m)	City (kmpl)	Highway (kmpl)	Overall (kmpl)	Kerb Wt (kg)	Fuel Tank (litres)	Boot (litres)	Remarks
Chevrolet Optra 1.6 Elite (G.M.)	1.6	104	148	5	13.1	18.9	177	54.6	34.7	10.4	16.6	12.2	1220	60	405	(+) Styling, ride, handling
Chevrolet Tavera LT L1	2.5	80	186	5	2.9	22	141	54.1	37.1	12.4	20.1	14.3	1660	55	NA	(+) A good option to the Qualis (–) Styling
Chevrolet Aveo U-VA	1.2	76	110	5	15.8	19.9	152	45.5	28.5	11.4	19.6	13.5	1075	45	220	(+) Looks, interior space (–) Lacks 'hot-hatch' punch
Chevrolet Spark	1.0	63	90	5	15.1	19.6	153	58.1	33.4	12.0	18.8	13.7	840	NA	NA	(+) Looks, peppy performance (–) Ride, ergonomics
Ambassador BE MPI AC (Hindustan Motors)	1.8	76	135	5	18.1	21	138	NA	42.3	10.0	19	14.5	1200	45	–	(+) Spacious cabin, spacious boot (–) Brakes
Ambassador SE DLS	1.5	37	83	5	NA	NA	NA	NA	NA	NA	NA	NA	1200	45	–	(+) Low running cost (–) Sluggish
Ambassador 2000 BU	2.0	56	108	5	NA	NA	NA	NA	NA	NA	NA	NA	–	45	–	(+) Nostalgia (–) Dated
Honda City ZX EXI	1.5	78	125	5	13.9	18.2	171	54.2	31.3	14.7	24.9	17.2	1040	42	490	(+) Space, handling, ease of driving, (–) Ride
City ZX GXI CVT	1.5	78	125	CVT	15.6	20.2	164	53	30.4	14.3	26	17.2	1070	42	490	(+) CVT option (–) Price
Honda Accord VTIL A/T	2.4	144	196	5A	11.4	18.1	205	48.3	30.2	9.8	16.1	11.4	1450	65	420	(+) Possibly the best luxury sedan (–) Quality issues
Hyundai Accent GLE	1.5	94	123	5	13.8	19.4	175	50.2	38.4	9.9	18.5	12.0	1023	45	328	(+) Price, equipment (–) Looks a little dated

Some Specifications of Leading Auto Vehicles

Name	Engine				Performance					Efficiency			Space			
Name of Vehicle	Displacement (litres)	Max Power (PS)	Max Torque (Nm)	Gears	Acceleration 0–100 kmph (in sec)	Acceleration 1/4 mile (in sec)	Top Speed (kmph)	Braking 100 kmph-0 (m)	Braking 80 kmph-0 (m)	City (kmpl)	Highway (kmpl)	Overall (kmpl)	Kerb Wt (kg)	Fuel Tank (litres)	Boot (litres)	Remarks
Accent GLS	1.6	102	141	5	13.2	19	172	48.3	29.8	8.0	17.6	9.5	1038	45	328	(+) Space, ergonomics, VFM (–) Ride
Verna 1.6 XI (Hyundai)	1.6	103	145	5	12.4	18.4	178	51.3	34.9	9.1	17.4	11.1	1152	45	352	(+) Performance, ride (–) Gear shift
Getz Prime 1.3 GLS (Hyundai)	1.3	82	116	5	14.5	19.6	166	60	35.9	11.9	16.8	13.1	1095	45	290	(+) Comfort, space (–) Pricey
Getz Prime 1.3 GLX ABS	1.3	82	116	5	14.5	19.6	166	60	35.9	11.9	16.8	13.1	1095	45	290	(+) Safety, features (–) Pricey
Santro Xing XK (Hyundai)	1.1	63	89	5	16.7	20.3	142	NA	38.4	16.1	24.3	20.2	778	35	NA	(+) Great city car, looks, space, ergonomics
Santro Xing XK (AC)	1.1	63	89	5	16.7	20.3	142	NA	38.4	16.1	24.3	20.2	778	35	NA	(+) Space at the rear, headroom (–) Ride
Santro Xing XL	1.1	63	89	5	16.7	20.3	142	NA	38.4	16.1	24.3	20.2	778	35	NA	(+) Seating (–) Handling, performance
Santro Xing X0	1.1	63	89	5	16.7	20.3	142	NA	38.4	16.1	24.3	20.2	778	35	NA	(+) Equipment, optional ABS
Elantra GT	1.8	127	168	5	11.9	18.1	190	50.6	32.7	9.5	15.9	11.1	1276	55	327	(+) Space, comfort, ride, boot space
Mahindra & Mahindra Ltd. Mahindra Towers, Dr. G.M. Bhosale Marg, Worli, Mumbai-400 018.																www.mahindra.ocm
Scorpio STD (M & M)	2.6	115	255	5	21.5	21.4	142	71.2	41.1	9.0	14.6	10.4	1980	55	NA	(+) Performance, mile munching abilities

Some Specifications of Leading Auto Vehicles

Name	Engine				Performance					Efficiency			Space			
Name of Vehicle	Displacement (litres)	Max Power (PS)	Max Torque (Nm)	Gears	Acceleration 0–100 kmph (in sec)	Acceleration 1/4 mile (in sec)	Top Speed (kmph)	Braking 100 kmph-0 (m)	Braking 80 kmph-0 (m)	City (kmpl)	Highway (kmpl)	Overall (kmpl)	Kerb Wt (kg)	Fuel Tank (litres)	Boot (litres)	Remarks
Scorpio DX	2.6	115	255	5	21.5	21.4	142	71.2	41.1	9.0	14.6	10.4	1980	55	NA	(+) Features, luggage space (–) Low speed ride
Scorpio SLX	2.6	115	255	5	21.5	21.4	142	71.2	41.1	9.0	14.6	10.4	1980	55	NA	(+) Road presence, (–) Handling, braking
Scorpio Getaway	2.6	110	270	5	21.7	21.8	157	56.4	33.1	8.2	12.5	9.3	2050	80	NA	(+) Adventurous looks (–) Woeful braking
Bolero SLX DI	2.5	63	180	5	30.3	23.5	114	62.9	38.9	13.6	20.3	15.0	1670	60	NA	(+) Practicality, aircon, ride (–) Performance
Bolero XLS	2.5	72	134	5	43.4	25.0	107	NA	53.2	7.53	14.3	9.22	1760	60	NA	(+) Fuel efficiency (–) Image, ergonomics, brakes
Mahindra Renault Ltd., Ashok Nagar, Chakravati Ashok Road, Kandivali (E), Mumbai-400 101.																www.mahindraenault.ocm
French auto maker Renault begins its innings in India. And it has M&M for company–one of India's fastest growing home grown companies. It started off with a budget sedan but expect better things soon.																
Logan 1.6 GLS	1.6	85	128	5	15.2	19.7	151	64.4	41.8	NA	NA	NA	1080	50	510	(+) Value for money (–) Underpowered
Logan 1.5 DCI DLE	1.5	65	160	5	17.2	20.4	149	65.1	41.8	14	27.3	17.3	1140	50	510	(+) Great space (–) Built to a price
Logan 1.5 DCI DLS	1.5	65	160	5	17.2	20.4	149	65.1	41.8	14	27.3	17.3	1140	50	510	(+) VFM, fuel efficient (–) Ergonomics
Logan 1.4 GL	1.4	75	110	5	NA	NA	NA	NA	NA	NA	NA	NA	1760	60	NA	(+) Cheap to buy (–) Style
Logan 1.4 GLE	1.4	75	110	5	NA	NA	NA	NA	NA	NA	NA	NA	1760	60	NA	(+) Cheap to run (–) Handling

Some Specifications of Leading Auto Vehicles

Name	Engine				Performance					Efficiency			Space			
Name of Vehicle	Displacement (litres)	Max Power (PS)	Max Torque (Nm)	Gears	Acceleration 0–100 kmph (in sec)	Acceleration 1/4 mile (in sec)	Top Speed (kmph)	Braking 100 kmph-0 (m)	Braking 80 kmph-0 (m)	City (kmpl)	Highway (kmpl)	Overall (kmpl)	Kerb Wt (kg)	Fuel Tank (litres)	Boot (litres)	Remarks
Maruti Udyog Ltd., 11th Floor, Jeevan Prakash Building, 25, Kasturba Gandhi Marg, New Delhi - 110 001.																www.marutiudyog.com
The company that started it all, MUL is the biggest car manufacturer in India and also serves as an export hub for Suzuki. The company rides high on its unmatched dealer and service networks.																
Esteem LX	1.3	86	106	5	12.8	18.5	164	NA	35.1	8.2	15.4	11.8	860	40	376	(+) Fuel efficiency, performance, VFM (–) Dated
Esteem LXI	1.3	86	106	5	12.8	18.5	164	NA	35.1	8.2	15.4	11.8	860	40	376	(+) Price (–) Interiors
Esteem VXI	1.3	86	106	5	12.8	18.5	164	NA	35.1	8.2	15.4	11.8	860	40	376	(+) Good resale, low maintenance
800 STD MPFI (BS3)	0.8	38	59	4	20.8	39.5	144.2	NA	42.8	18.9	26.8	22.9	665	30	–	(+) Cheap to buy and run
800 STD MPFI (BS3)	0.8	38	59	4	20.8	39.5	144.2	NA	42.8	18.9	26.8	22.9	665	30	–	(–) Tinny, dated, Alto a better option
Omni 5-seater BS3	0.8	38	62	4	NA	NA	NA	NA	NA	NA	NA	NA	740	36	–	(+) Price, commercially useful, space
Omni 8-seater BS3	0.8	38	62	4	NA	NA	NA	NA	NA	NA	NA	NA	740	36	–	(–) Safety issues, image
Omni LPG BS3	0.8	38	62	4	NA	NA	NA	NA	NA	NA	NA	NA	740	36	–	(+) Price, commercially useful (–) Safety issues
Swift LXI	1.3	88	113	5	12.9	18.8	156	50.4	32.3	12.2	20.7	14.3	1010	43	232	(+) Style, handling, seats (–) Space, ride quality
Swift VXI	1.3	88	113	5	12.9	18.8	156	50.4	32.3	12.2	20.7	14.3	1010	43	232	(+) Peppy performance (–) Build
Swift ZXI	1.3	88	113	5	12.9	18.8	156	50.4	32.3	12.2	20.7	14.3	1010	43	232	(+) Fully loaded, great VFM (–) Build quality

Some Specifications of Leading Auto Vehicles

Name	Engine				Performance					Efficiency			Space			
Name of Vehicle	Displacement (litres)	Max Power (PS)	Max Torque (Nm)	Gears	Acceleration 0–100 kmph (in sec)	Acceleration 1/4 mile (in sec)	Top Speed (kmph)	Braking 100 kmph-0 (m)	Braking 80 kmph-0 (m)	City (kmpl)	Highway (kmpl)	Overall (kmpl)	Kerb Wt (kg)	Fuel Tank (litres)	Boot (litres)	Remarks
Swift LDI (Diesel)	1.3	76	190	5	13.2	18.7	152	56.6	34.6	16.9	19.3	21.6	1075	43	232	(+) Torquey and refined engine. Very fuel efficient.
Swift VDI (Diesel)	1.3	76	190	5	13.2	18.7	152	56.6	34.6	16.9	19.3	21.6	1075	43	232	(+) Better ride than petrol (–) No airbags.
Alto STD (BS3)	0.8	46	62	5	21.5	21.7	137	NA	35.8	15.9	23.2	19.6	740	35	–	(+) Modern, good for city use
Alto LX (BS3)	0.8	46	62	5	21.5	21.7	137	NA	35.8	15.9	23.2	19.6	740	35	–	(+) Low maintenance (–) Bland, cramped inside
Alto LXI (BS3)	0.8	46	62	5	21.5	21.7	137	NA	35.8	15.9	23.2	19.6	740	35	–	(+) Economical (–) Better options sprouting up
Wagon R LX (BS3)	1.1	65	84	5	16.2	20	137	NA	33	17.5	23	18.8	885	35	–	(+) Roomy interiors, good ride, visibility
Wagon R LXI (BS3)	1.1	65	84	5	16.2	20	137	NA	33	17.5	23	18.8	885	35	–	(+) Compact (–) Nervous handling, vague steering
Wagon R VXI (BS3)	1.1	65	84	5	16.2	20	137	NA	33	17.5	23	18.8	885	35	–	(+) Ease on ingress, egress, features, boot space
Wagon R DUO (BS3)	1.1	58	77	5	19.5	21.3	124	NA	39	11.2*	21.7*	13.7*	850	35/22	–	(+) Low running cost (LPG) (–) Performance
Zen Estilo LX	1.1	67.8	84	5	16.3	20	154	52.3	32	12.7	22.4	15.2	875	35	–	(+) Refreshing looks, price (–) Interior space
Zen Estilo LXI	1.1	67.8	84	5	16.3	20	154	52.3	32	12.7	22.4	15.2	875	35	–	(+) Refreshing looks, price (–) Interior space
Zen Estilo VXI	1.1	67.8	84	5	16.3	20	154	52.3	32	12.7	22.4	15.2	875	35	–	(+) Refreshing looks, price (–) Interior space

Some Specifications of Leading Auto Vehicles

Name	Engine				Performance					Efficiency			Space			
Name of Vehicle	Displacement (litres)	Max Power (PS)	Max Torque (Nm)	Gears	Acceleration 0–100 kmph (in sec)	Acceleration 1/4 mile (in sec)	Top Speed (kmph)	Braking 100 kmph-0 (m)	Braking 80 kmph-0 (m)	City (kmpl)	Highway (kmpl)	Overall (kmpl)	Kerb Wt (kg)	Fuel Tank (litres)	Boot (litres)	Remarks
SX4 VXI	1.6	103	145	5	11.8	18.1	180	47.1	27	10.1	18.1	12.1	1170	50	NA	(+) Beefy looks, price (–) Interior space
SX4 ZXI	1.6	103	145	5	11.8	18.1	180	47.1	27	10.1	18.1	12.1	1200	50	NA	(+) Beefy looks, price (–) Interior space
Grand Vitara	2.0	120	170	5	–	–	–	–	–	9.4	12.7	10.1	1580	66	NA	(+) Ride, low ratio 4×4 (–) Underpowered

Mitsubishi Motors India Ltd., HM Chennai carplant, Adhigathur, Kadambathur P.O., Tiruvallur District, Tamil Nadu www.mitsubishi.motors.co.in

Mitsubishi Motors is getting to grips and a slew of new launches is being worked out for the near future. The launch of the revamped Montero displays the company'y serious intentions.

Name of Vehicle	Displacement (litres)	Max Power (PS)	Max Torque (Nm)	Gears	Acceleration 0–100 kmph (in sec)	Acceleration 1/4 mile (in sec)	Top Speed (kmph)	Braking 100 kmph-0 (m)	Braking 80 kmph-0 (m)	City (kmpl)	Highway (kmpl)	Overall (kmpl)	Kerb Wt (kg)	Fuel Tank (litres)	Boot (litres)	Remarks
Mitsubishi Lancer LX	1.5	88	129	5	14.9	19.8	155	NA	37.4	10.1	12.4	11.3	1095	50	500	(+) Customisation possibilities (–) Performance
Mitsubishi Lancer LXD	2.0	66	118	5	21.3	NA	140	NA	35.7	11.5	13.8	12.7	1010	50	500	(+) Low running cost, ride (–) Handling
Mitsubishi Lancer Invex LE	1.8	118	162	4A	13	18.9	173	49	35.2	9.4	13.8	10.5	–	50	420	(+) Automatic option (–) Fuel efficiency
Mitsubishi Lancer Cedia Select	2.0	115	175	5	10.45	17.3	190.3	50.1	28.3	8.8	18.9	11.3	1210	–	430	(+) Rally heritage, space, (–) This is no EVO
Mitsubishi Lancer Cedia Sports	2.0	115	175	5	10.45	17.3	190.3	50.1	28.3	8.8	18.9	11.3	1210	–	430	(+) Great car to be driven in (–) Not a driver's car
Mitsubishi Lancer Cedia Spirit	2.0	115	175	5	10.45	17.3	190.3	50.1	28.3	8.8	18.9	11.3	1210	–	430	(+) Ride, engine, gearbox

Some Specifications of Leading Auto Vehicles

Name	Engine				Performance					Efficiency			Space			
Name of Vehicle	Displacement (litres)	Max Power (PS)	Max Torque (Nm)	Gears	Acceleration 0–100 kmph (in sec)	Acceleration 1/4 mile (in sec)	Top Speed (kmph)	Braking 100 kmph-0 (m)	Braking 80 kmph-0 (m)	City (kmpl)	Highway (kmpl)	Overall (kmpl)	Kerb Wt (kg)	Fuel Tank (litres)	Boot (litres)	Remarks
Mitsubishi Montero	3.2	167	405	5	14.78	19.58	172	48.2	28	8.1	13.6	9.4	2380	88	1050	(+) Butch, go-anywhere ability (–) Price
Mitsubishi Pajero	2.8	119	292	5	19.9	21.4	152	62.3	36.8	5.5	10.5	8.0	2060	92	NA	(+) Great off-roader (–) Too long in the tooth
Nissan Motor India Pvt. Ltd., 36, Maker Chambers III, Nariman Point, Mumbai–21. www.nissan.in																
Since the launch of X-Trail in 2005, Nissan has been conservative about its operations here. No concrete plans for new model launches and absence from the headlines have raised to a lot of speculations about its next offering.																
X-Trail Comfort	2.2	136	314	6	13.14	19.05	173	51.03	35.26	13.6	20	15.2	1629	60	NA	(+) Efficiency, handling (–) Lacks macho image
X-Trail X-special	2.2	136	314	6	13.14	19.05	173	51.03	35.26	13.6	20	15.2	1629	60	NA	(+) Interiors, features (–) Price
X-Trail Elegance	2.2	136	314	6	13.14	19.05	173	51.03	35.26	13.6	20	15.2	1629	60	NA	(+) Car-like proportions (–) Road presence
Teana	2.3	170	224	4A	NA	NA	NA	NA	NA	NA	NA	NA	1509	70	476	(+) Interiors, comfort (–) Price
Rolls-Royce Motor Cars, The Drive, Westhampnett, Chichester, Sussex PO18 OSH, UK www.rolls.roycemotorcars.com																
'The spirit of Ecstasy' has landed on Indian shores to lure the super-rich and has been successful in its own way. Incomparable in luxury, value and build, a Rolls Royce immediately transforms you into royalty.																
Phantom	6.8	460	720	6A	7.00	–	240	–	–	–	–	–	2550	100	–	(+) Everything (–) A scratch on it would hurt a lot

Some Specifications of Leading Auto Vehicles

Name	Engine				Performance					Efficiency			Space			
Name of Vehicle	Displacement (litres)	Max Power (PS)	Max Torque (Nm)	Gears	Acceleration 0–100 kmph (in sec)	Acceleration 1/4 mile (in sec)	Top Speed (kmph)	Braking 100 kmph-0 (m)	Braking 80 kmph-0 (m)	City (kmpl)	Highway (kmpl)	Overall (kmpl)	Kerb Wt (kg)	Fuel Tank (litres)	Boot (litres)	Remarks
Skoda Auto India Pvt. Ltd., Plot No. A-1/1, Shendra Five Star Industrial Area, MIDC, Aurangabad - 431 201. www.skoda.auto.co.in																
The Czech manufacturer currently boasts a range including variants of Octavia and Laura as well as its flagship model, the Superb. Plush and loaded, these cars spell VFM and are good alternatives to more expensive ones in their respective segments.																
Superb 2.5 TDI	2.5	163	350	5A	13.22	20.07	NA	53.06	39.04	9.15	13.2	10.7	1350	62	528	(+) Diesel option (–) Lacks snob value
Octavia L&K	1.9	89	210	5	13.6	19.2	175	58.3	21	13.3	18.6	14.4	1330	55	530	(+) Good diesel engine, L&K fully loaded
Octavia Elegance TDI AT	1.9	89	210	4A	14	NA	196	NA	26	10.5	18	14.3	1270	55	530	(+) The only diesel automatic in the segment
Laura L&K DSG	1.9	105	250	6A	12.13	20.68	177	43.43	29.28	12.5	17	16.2	1350	65	560	(+) Superb gearbox (DSG), driver appeal (–) Ride
Tata Motors, 27th Floor, World Trade Center 1, Cuffe Parade, Mumbai.																
From being a truck maker of repute to one of India'a leading car makers, there is no looking back for Tata Motors now. The company is all set to launch India's most cost effective car soon.																
Indica DLE	1.4	54	85	5	25.7	22.6	135	NA	36.6	14.5	22.3	16.5	980	37	217	(+) Economical to buy and run
Indica DLS	1.4	54	85	5	25.7	22.6	135	NA	36.6	14.5	22.3	16.5	980	37	217	(+) Ride, visibility, looks (–) Performance
Indica DLG	1.4	54	85	5	25.7	22.6	135	NA	36.6	14.5	22.3	16.5	980	37	217	(+) Handling, (–) Image, too many around

Some Specifications of Leading Auto Vehicles

Name	Engine				Performance					Efficiency			Space			
Name of Vehicle	Displacement (litres)	Max Power (PS)	Max Torque (Nm)	Gears	Acceleration 0–100 kmph (in sec)	Acceleration 1/4 mile (in sec)	Top Speed (kmph)	Braking 100 kmph-0 (m)	Braking 80 kmph-0 (m)	City (kmpl)	Highway (kmpl)	Overall (kmpl)	Kerb Wt (kg)	Fuel Tank (litres)	Boot (litres)	Remarks
Indica DLX	1.4	54	85	5	25.7	22.6	135	NA	35.6	14.5	22.3	16.5	980	37	217	(+) Space (–) Ergonomics, quality
Indica Turbo DLS	1.4	68	130	5	17.7	20.68	141	62.1	44.0	15.1	20.8	16.5	1050	35	217	(+) VFM, power (–) Seats, steering
Indica Turbo DLG	1.4	68	130	5	17.7	20.68	141	62.1	44.0	15.1	20.8	16.5	1050	35	217	(+) Features, power (–) Ownership
Indica Turbo DLX	1.4	68	130	5	17.7	20.68	141	62.1	44.0	15.1	20.8	16.5	1050	35	217	(+) Features, power (–) Ownership
Indica Turbo DLX ABS	1.4	68	130	5	17.7	20.68	141	54.8	31.3	15.1	20.8	16.5	1050	35	217	(+) ABS (–) Cab image
Indica Xeta GLE(P)	1.2	70	122	5	17.2	20.4	153	59.5	35.5	12.1	18.4	13.6	995	37	217	(+) Cheap to buy, decent fuel efficiency, space
Indica Xeta GLS(P)	1.2	70	122	5	17.2	20.4	153	59.5	35.5	12.1	18.4	13.6	995	37	217	(+) Ride, handling, planted feel
Indica Xeta GLG(P)	1.2	70	122	5	17.2	20.4	153	59.5	35.5	12.1	18.4	13.6	995	37	217	(+) VFM, big car feel (–) Seats, steering
Indica Xeta GLX(P)	1.2	70	122	5	17.2	20.4	153	59.5	35.5	12.1	18.4	13.6	995	37	217	(+) Features (–) Ownership
Indigo GLE (P)	1.4	86	110	5	15.8	20	158	57.5	33.5	10.8	18.7	12.8	1065	42	450	(+) Good handling, interior room, boot space
Indigo GLS (P)	1.4	86	110	5	15.8	20	158	57.5	33.5	10.8	18.7	12.8	1070	42	450	(+) Easy ingress, egress (–) Unrefined engine, ride
Indigo GLX (P)	1.4	86	110	5	15.8	20	158	57.5	33.5	10.8	18.7	12.8	1070	42	450	(+) VFM (–) Interior design, fit and finish, seats

Some Specifications of Leading Auto Vehicles

Name	Engine				Performance					Efficiency			Space			
Name of Vehicle	Displacement (litres)	Max Power (PS)	Max Torque (Nm)	Gears	Acceleration 0–100 kmph (in sec)	Acceleration 1/4 mile (in sec)	Top Speed (kmph)	Braking 100 kmph-0 (m)	Braking 80 kmph-0 (m)	City (kmpl)	Highway (kmpl)	Overall (kmpl)	Kerb Wt (kg)	Fuel Tank (litres)	Boot (litres)	Remarks
Indigo LS (DICOR)	1.4	70	140	5	15.9	21	157	63	37	12.7	19.5	14.5	1070	42	450	(+) Diesel engine makes for economical running
Indigo LX (DICOR)	1.4	70	140	5	15.9	21	157	63	37	12.7	19.5	14.5	1070	42	450	(+) Space (–) Unrefined, top end performance
Indigo XL (P)	1.4	101	124	5	13.3	18.8	NA	NA	33.1	–	–	–	1150	42	450	(+) Rear legroom (–) Uninspiring top end
Indigo XL (DICOR)	1.4	70	140	5	16.9	20.5	NA	NA	33.1	–	–	–	1185	42	450	(+) Rear legroom (–) Throttle lag
Safari 2.2 VTT DICOR VX 4WD	2.2	140	320	5	16.6	NA	153	74.1	36.3	12.2	17.9	13.6	2115	65	NA	(+) Comfort, SUV appeal, space (–) Mushy gearbox
Safari 2.2 VTT DICOR LX 4WD	2.2	140	320	5	16.6	NA	153	74.1	36.3	12.2	17.9	13.6	2115	65	NA	(+) Off roading and mile munching abilities
Safari 2.2 VTT DICOR EX 4WD	2.2	140	320	5	16.6	NA	153	74.1	36.3	12.2	17.9	13.6	2115	65	NA	(+) SUV looks (–) Narrow powerband driveability
Safari TCIC	2.0	90.3	186	5	25.5	NA	136	NA	NA	8.6	13.3	10.6	1935	65	–	(+) looks, ride (–) Vague steering, sluggish
Safari 2.1 EXI (Petrol)	2.1	135	195	5	16.8	20.6	154	49.5	31.9	6.5	12	8	1700	65	–	(+) Performance (–) Fuel efficiency
Sumo VICTA LX	2.0	68	117	5	NA	NA	NA	NA	NA	NA	NA	NA	1730	65	–	(+) Space, price, versatility (–) NVH, ride
Sumo VICTA EX	2.0	90	190	5	22.8	22.0	142	NA	43.2	8.6	13.3	10.9	1700	65	–	(+) Practicality (–) Performance, image, brakes
Sumo VICTA GS	2.0	90	190	5	22.9	22.1	134	NA	43.2	8.6	13.3	10.9	1105	65	–	(+) Features for a price (–) Handling

Some Specifications of Leading Auto Vehicles

Name	Engine				Performance					Efficiency			Space			
Name of Vehicle	Displacement (litres)	Max Power (PS)	Max Torque (Nm)	Gears	Acceleration 0–100 kmph (in sec)	Acceleration 1/4 mile (in sec)	Top Speed (kmph)	Braking 100 kmph-0 (m)	Braking 80 kmph-0 (m)	City (kmpl)	Highway (kmpl)	Overall (kmpl)	Kerb Wt (kg)	Fuel Tank (litres)	Boot (litres)	Remarks
Indigo Marina GLE (Petrol)	1.4	85	110	5	14.2	19.5	161	66.3	42.1	9.2	14.8	12	1105	42	410	(+) Larger storage space, flexibility, exclusively
Indigo Marina GLS (Petrol)	1.4	85	110	5	14.2	19.5	161	66.3	42.1	9.2	14.8	12	1135	42	410	(+) VFM, good handling (–) Interior design
Indigo Marina GLX (Petrol)	1.4	85	110	5	14.2	19.5	161	66.3	42.1	9.2	14.8	12	1135	42	410	(+) Interior room (–) Fit and finish, seats
Indigo Marina LS (DICOR)	1.4	70	140	5	18.5	21	144	63.9	42.2	12.1	20.4	16.3	NA	42	410	(+) Diesel engine makes for economical running.
Indigo Marina LX (DICOR)	1.4	70	140	5	18.5	21	144	63.9	42.2	12.1	20.4	16.3	NA	42	410	(+) Space (–) Unrefined, top end performance
Toyota Kirloskar Motor Ltd., Plot No.1, Bidadi Industrial Estate, Ram Nagar Taluk, Bangalore-09. www.toyotabharat.com																
Toyota entered India with the Qualis which turned out to be a huge success. Riding high on the success of Innova, it offers luxury cars like Corolla and Camry, while Land Cruiser Prado takes the CBU route.																
Innova E (Diesel)	2.5	102	200	5	17.6	20.5	151	43.2	29.7	10.1	16.2	11.8	1585	55	NA	(+) Price (–) Driver appeal (no power steering)
Innova G1 (Petrol)	2.0	136	182	5	13	18.9	181	44.4	27.3	8.8	16.9	10.8	1510	55	NA	(+) Interior room, luggage space, practicality
Innova G1 (Diesel)	2.5	102	200	5	17.6	20.5	151	43.2	29.7	10.1	16.2	11.8	1585	55	NA	(+) Versatility, mile munching abilities, diesel
Innova G4 (Petrol)	2.0	136	182	5	13	18.9	181	44.4	27.3	8.8	16.9	10.8	1510	55	NA	(+) Handling, creature comfort

Some Specifications of Leading Auto Vehicles

Name	Engine				Performance					Efficiency			Space			
Name of Vehicle	Displacement (litres)	Max Power (PS)	Max Torque (Nm)	Gears	Acceleration 0–100 kmph (in sec)	Acceleration 1/4 mile (in sec)	Top Speed (kmph)	Braking 100 kmph-0 (m)	Braking 80 kmph-0 (m)	City (kmpl)	Highway (kmpl)	Overall (kmpl)	Kerb Wt (kg)	Fuel Tank (litres)	Boot (litres)	Remarks
Innova G4 (Diesel)	2.5	102	200	5	17.6	20.5	151	43.2	29.7	10.1	16.2	11.8	1585	55	NA	(+) Ride (–) Size in the city
Innova V (Petrol)	2.0	136	182	5	13	18.9	181	44.4	27.3	8.8	16.9	10.8	1510	55	NA	(+) Safety (–) Higher running cost
Innova V (Diesel)	2.5	102	200	5	17.6	20.5	151	43.2	29.7	10.1	16.2	11.8	1585	55	NA	(+) Equipment
Camry V6	2.4	167	224	5A/5	11.59	18.29	202	41.2	25.08	–	–	–	1515	70	535	(+) Space, comfort, looks, ride (–) Fuel efficiency
Corolla H1	1.8	125	158	5	9.8	17.1	192	66.2	35.5	9.6	16.5	11.4	1160	50	430	(+) Ride, space, ergonomics (–) High speed stability
Corolla H4	1.8	125	158	4A	12.1	18.7	176	66.2	35.5	7.4	14.3	9.5	1185	50	430	(+) Features (–) Large steering
Corolla H5	1.8	125	158	5	9.8	17.1	192	66.2	35.5	9.6	16.5	11.4	1160	50	430	(+) Engine, performance (–) NVH
Land Cruiser Prado VX	4.0	235	362	5A	10.1	17.3	188	55.1	32.5	5.0	9.5	6.1	2070	87	NA	(+) Off-road ability, presence (–) Price, efficiency

Volkswagen, Volkswagen AG, Brieffach 1972, D-38436, Wolfsburg, Germany. www.volkswagen.com

Volkswagen is finally in India and spearheading the marque's entry is the Passat. Currently operational through Skoda's manufacturing facility, VW plans to start full fledged operations soon in the country.

Name of Vehicle	Displacement (litres)	Max Power (PS)	Max Torque (Nm)	Gears	Acceleration 0–100 kmph (in sec)	Acceleration 1/4 mile (in sec)	Top Speed (kmph)	Braking 100 kmph-0 (m)	Braking 80 kmph-0 (m)	City (kmpl)	Highway (kmpl)	Overall (kmpl)	Kerb Wt (kg)	Fuel Tank (litres)	Boot (litres)	Remarks
Passat	2.0	140	320	6A	12.2	18.8	204	42.6	27.1	13.3	18.6	14.6	1476	70	565	(+) Sporty engine, DSG box (–) stiff ride

Specifications of 2-Wheelers

Name	Engine				MISC			Performance			Efficiency				
Name of Vehicle	Displacement (CC)	Max Power (PS)	Max Torque (Nm)	Injected/Carburetted	Gears	Kerb Wt (kg)	Fuel tank (litres)	Acceleration 0-60kmph (sec)	Top Speed (kmph)	Acceleration 0-400 m (sec)	Braking 60-0 (metres)	City (kmpl)	Highway (kmpl)	Overall (kmpl)	Remarks
Bajaj Auto Ltd., Mumbai-Pune Road, Akurdi, Pune 411035. wwww.bajajauto.com															
The manufacturer causing the maximum stir in Indian biking scene has shown commitment by constantly upgrading its products and knows what the Indian biker wants. Outstanding sales figures signal where the company is headed to. It has been playing the technology card for some time too. Expect good things in the future from the Pune based manufacturer.															
Avenger	179	16.0	14.7	C	5	152	14	6.04	107.5	20.6	18.4	42	56	46	(+) Only true-cruiser, comfort, price (–) Gearbox, handling
Pulsar 220 DTS-FI	220	20	19.1	I	5	150	15	4.4	127	18.15	17.1	36.9	43.1	38.3	(+) Smooth power delivery, ride VFM (–) unsatisfying performance
Pulsar 200 DTS-I	198.8	18	17.1	C	5	145	15	4.5	118.7	18.45	20.1	38	47	40	(+) Styling, performance, VFM (–) Marginally better than the 180
Pulsar 180 DTS-I	179	16.5	14.7	C	5	139	15	4.8	115.9	18.70	19.93	36	58	47	(+) Features, styling, handling, performance (–) Gearbox
Pulsar 150 DTS-I	149	14.2	11.7	C	5	134	15	5.6	111	20.03	16.5	56.5	65	58.6	(+) Performance, handling, efficiency, features (–) Gearbox
Discover 135 DTS-I	134.2	13.1	11.8	C	4	133	10	5.7	103.4	20.7	18.4	NA	NA	NA	(+) Performance, smooth engine (–) Resemblance to Discover 125
Discover 125 DTS-I	124	11.5	10.8	C	4	125	10	6.4	104.8	20.8	22.9	63	72	65	(+) Styling, ride, engine (–) None
XCD 125	124.5	9.6	10.7	C	4	112	NA	6.9	94	NA	NA	73.6	65.3	71.5	(+) Fuel efficiency, bottom and grunt, VFM (–) Performance

Specifications of 2-Wheelers

Name	Engine				MISC			Performance				Efficiency			
Name of Vehicle	Displacement (CC)	Max Power (PS)	Max Torque (Nm)	Injected/Carburetted	Gears	Kerb Wt (kg)	Fuel tank (litres)	Acceleration 0-60kmph (sec)	Top Speed (kmph)	Acceleration 0-400 m (sec)	Braking 60-0 (metres)	City (kmpl)	Highway (kmpl)	Overall (kmpl)	Remarks
Platina	99	8.3	8.0	C	4	113	13	9.68	88	24.36	20.3	62	73	65	(+) Looks, price (–) Riding position
CT 100	99	8.3	8.0	C	4	109	10.5	9.0	88.7	24.8	25.7	70	83	73	(+) Fuel efficiency, price (–) Quality
Kristal	94.8	7.3	7.6	C	V	99	4.5	NA	NA	NA	NA	NA	NA	NA	(+) Styling, innovative features, price (–) Quality, footboard space

Hero Honda Motors Ltd., 34, Basant Lok, Vasant Vihar, New Delhi - 110 057. www.herohonda.com

India's largest manufacturer of motorcycles continues its peaking performance but is threatened by Bajaj Auto. Though Hero Honda is still banking on the success of the Splendor and its clones, new models and technologies keep making appearances. The X-treme we expect is the first step towards greater things to come from the number one two-wheeler maker.

Name of Vehicle	Displacement (CC)	Max Power (PS)	Max Torque (Nm)	Injected/Carburetted	Gears	Kerb Wt (kg)	Fuel tank (litres)	Acceleration 0-60kmph (sec)	Top Speed (kmph)	Acceleration 0-400 m (sec)	Braking 60-0 (metres)	City (kmpl)	Highway (kmpl)	Overall (kmpl)	Remarks
Karizma	223	16.9	18.3	C	5	150	15	4.7	125.6	18.79	19.48	40	54	–	(+) Performance, handling (–) Nothing to complain about
Achiever ES	149.1	13.6	12.8	C	5	138	12.5	6.2	108.1	20.58	18.16	60.3	71	63	(+) Smooth engine, great finish (–) Looks
Pleasure	102	7.0	7.8	C	V	104	5	10.1	76.8	24.51	21.41	46.6	58.4	49.6	(+) Peepy scooter, easy to ride (–) A bit pricey
CBZ Xtreme	149.1	14.4	12.8	C	5	143	12.3	5.7	111.9	20.2	23.6	56	61.2	57.3	(+) Looks, fit and finish, performance (–) Vibrations
Hunk	149.1	14.4	12.8	C	5	146	14	5.6	108	20.2	23.6	56.4	62.3	57.8	(+) Extremely good looking, fit and finish (–) Standard equipments

Specifications of 2-Wheelers

Name	Engine				MISC			Performance			Efficiency				
Name of Vehicle	Displacement (CC)	Max Power (PS)	Max Torque (Nm)	Injected/Carburetted	Gears	Kerb Wt (kg)	Fuel tank (litres)	Acceleration 0-60kmph (sec)	Top Speed (kmph)	Acceleration 0-400 m (sec)	Braking 60-0 (metres)	City (kmpl)	Highway (kmpl)	Overall (kmpl)	Remarks
Splendor NXG	97.2	7.7	7.5	C	4	107	10.3	8.6	90.1	22.9	23.5	73	68	72	(+) Proven engine, refreshed looks (–) Dated underpinings
Splendor Plus	97.2	7.51	7.2	C	4	116	12.8	12.2	82.1	26.1	29.3	56	70	63	(+) Reliability, Fuel efficiency (–) Dated looks
Passion Plus	97.2	7.3	7.2	C	4	100	10.1	12.2	82.1	26.1	29.3	56	70	63	(+) Reliability, Fuel efficiency (–) Ageing looks
S Splendor	125	9.13	10.35	C	4	117	12	7.4	98.7	22.4	20.1	73	80	75	(+) Fuel efficiency, refinement (–) Uninspired styling
S Splendor DS ES	125	9.13	10.35	C	4	121	12	7.4	98.7	22.4	19.1	73	80	75	(+) Fuel efficiency, refinement (–) Uninspired styling
Glamour	125	9.13	10.35	C	4	121	12	7.4	98.7	22.4	19.1	73	80	75	(+) Fuel efficiency, refinement (–) Slightly over-styled
Glamour FI	125	9.13	10.35	FI	4	121	12	7.4	98.7	22.4	19.1	73	80	75	(+) Fuel efficient, smooth engine (–) expensive
CD-Deluxe	97.2	7.7	7.5	C	4	107	10.5	7.8	89.3	22.9	23.0	76	78	77	(+) Best in-class performance, efficiency (–) Nothing much actually
CD-Dawn	97.2	7.7	7.5	C	4	107	10.5	7.8	89.3	22.9	23.0	76	78	77	(+) Best in-class performance, efficiency (–) Nothing much actually

Honda Motorcycle & Scooter India Pvt. Ltd., Plot No.1, Sector 3, IMT Manesar, Dist. Gurgaon-122 050. www.hondawheelersindia.com

The most promising two-wheeler manufacturer has still kept us waiting for its higher capacity bikes. The current range on offer adheres to Honda standards of refinement and reliability though.

Specifications of 2-Wheelers

Name	Engine				MISC			Performance				Efficiency			
Name of Vehicle	Displacement (CC)	Max Power (PS)	Max Torque (Nm)	Injected/Carburetted	Gears	Kerb Wt (kg)	Fuel tank (litres)	Acceleration 0-60kmph (sec)	Top Speed (kmph)	Acceleration 0-400 m (sec)	Braking 60-0 (metres)	City (kmpl)	Highway (kmpl)	Overall (kmpl)	Remarks
Unicorn	149	13.5	12.7	C	5	139	13	5.9	108	20.6	16.2	57	68	60	(+) Refinement, gearbox, dynamics (–) Not actually desirable
Unicorn ES	149	13.5	12.7	C	5	139	13	5.9	108	20.6	16.2	57	68	60	(+) Refinement, gearbox, dynamics (–) Not actually desirable
Shine	124.7	10.2	10.4	C	4	123.3	10.5	5.84	97.5	22.2	24.7	61	68	63	(+) Handling, build quality (–) Looks, basic engine
Shine ES	124.7	10.2	10.4	C	4	123.3	10.5	5.84	97.5	22.2	24.7	61	68	63	(+) Handling, build quality (–) Looks, basic engine
Eterno	148	8.3	10.6	C	4	125	4.5	11.9	85	25.6	22.9	56	70	59	(+) Steel body, engine, gearbox (–) Styling, ride
Activa	102	7.0	7.8	C	V	111	6	15.8	77	27.3	24.3	49	58	53.5	(+) Smooth engine, reliable (–) Ageing image
DIO	102	7.0	7.8	C	V	99	6	15.6	78	27.0	24.1	47	56	51.5	(+) Looks, quality (–) Stiff ride, No warranty on fibre parts
Indus Elec-Trans, 72, Palodia (via Thaltej), Ahmedabad-382 115.															www.induselectrans.com
In the age of increasing fuel prices, electric seems to be the buzz word. The current offerings by Indus are high on economy, but low on performance. The future looks promising with new models in the pipeline.															
Yo smart	Elec	NA	NA	–	NA	79	–	6.64*	26.7	65.3	5.22**	NA	NA	NA	(+) Ease of riding, inexpensive to run (–) Battery range cx3e, performance
Yo spin	Elec	NA	NA	–	NA	76	–	NA	NA	NA	NA	NA	NA	NA	(+) Ease of riding, inexpensive to run (–) Battery range, performance *time for 0-20 kmph **time for 20-0 kmph

Index

D

E

F

N

O

P

R

S

T

U

V

W